Role of Microbes in Industrial Products and Processes

Scrivener Publishing
100 Cummings Center, Suite 541J
Beverly, MA 01915-6106

Publishers at Scrivener
Martin Scrivener (martin@scrivenerpublishing.com)
Phillip Carmical (pcarmical@scrivenerpublishing.com)

Role of Microbes in Industrial Products and Processes

Edited by
Sanjay Kumar
Narendra Kumar
and
Shahid-ul-Islam

Scrivener Publishing

WILEY

This edition first published 2023 by John Wiley & Sons, Inc., 111 River Street, Hoboken, NJ 07030, USA and Scrivener Publishing LLC, 100 Cummings Center, Suite 541J, Beverly, MA 01915, USA

For more information about Scrivener publications please visit www.scrivenerpublishing.com.

Wiley Global Headquarters
111 River Street, Hoboken, NJ 07030, USA

For details of our global editorial offices, customer services, and more information about Wiley products visit us at www.wiley.com.

Library of Congress Cataloging-in-Publication Data

ISBN 978-1-119-90101-3

Cover image: Pixabay.Com
Cover design by Russell Richardson

Set in size of 11pt and Minion Pro by Manila Typesetting Company, Makati, Philippines

Printed in the USA

10 9 8 7 6 5 4 3 2 1

Contents

Preface

Microbes are everywhere imaginable in our environment. For example, just one gram of soil contains billions of microorganisms. Therefore, as an alternative to the limited conventional natural resources being overburdened by the increasing world population, these microbes are seen as future crops for the production of various products of industrial importance. Due to advances in industrial biotechnology, they have become popular as cell factories; and with the recent discoveries in recombinant DNA technology, the application of microorganisms in various sectors has exponentially increased for the development of various processes and products.

In addition to highlighting the role that microbes play in industrial products and processes, this book discusses the recent breakthroughs engendered by their use. Chapter 1 highlights the tremendous potential of microbes in the production of diverse pharmaceutical and medicinal products like antibiotics, antivirals, antifungals, and vaccines. Chapter 2 focuses on the health benefits of nutraceuticals and their utilization for treating diseases occurring in humans. The challenges in promoting and regulating their production are presented, along with the opportunities they provide. Chapter 3 summarizes the applications of microbes in different food industries, and Chapter 4 reviews the vital functions and corresponding action mechanisms of natural preservatives obtained from various microbiological sources. Chapter 5 highlights marine-derived microbic enzymes for industrial applications, and Chapter 6 mainly focuses on novel bioactive compounds obtained from cyanobacteria. Chapter 7 covers the occurrence, habitat, and diversity of actinobacteria in the natural environment as well as novel bioactive compounds and major enzymes synthesized by these actinomycetes. Chapter 8 includes thorough information on recent breakthroughs in the pretreatment of lingo-cellulosic materials, sustainable bioethanol production from parali, bioethanol yields and properties, and the possible use of bioethanol as an alternative to fossil fuels. Chapter 9 provides an overview of various modeling approaches and will familiarize the reader with different optimization techniques. Chapter 10 concentrates on vegetable by-products with bioactive properties and their potential applications in the health sector (nutraceuticals) and food production (functional foods). Chapter 11 provides a comprehensive insight into the role of microorganisms in the modern-day production of bioelectricity and biohydrogen, including its future prospects and associated challenges. Chapter 12 discusses the biotechnological strategies involved in the recovery and pretreatment of microalgal biomass to maximize the production of biogas. Future trends in the context of the biorefinery are also addressed for the sustainable production of biogas from microalgae. Chapter 13 discusses the feasibility of algal wastewater treatment from different industrial sources. Finally, Chapter 14 highlights the possibility of removing pharmaceutical pollutants with the help of a low-cost, environmentally friendly biological approach using microorganisms or the enzymes derived from them. The scalability of these treatments is also discussed, along with the challenges involved and mitigation methods.

The information presented in this book is targeted towards research scholars, scientists, and faculty working in the domain of microbial fermentation, organic waste treatment, and value-added products.

We are grateful to all the authors for their commitment and for sharing their knowledge and professional experience to make this project a reality. Special thanks to Mr. Martin Scrivener, President of Scrivener Publishing, USA, who accepted and supported this project.

The Editors
August 2022

1

Role of Microbes in the Pharmaceutical Industry

Alisha Chunduri[1], Niveditha Donthula[1], M. Jahanavi[1], Sowmya Golla[1], Pooja Aich[2], Sahithya K.[3], Bishwambhar Mishra[1], Sanjeeb Kumar Mandal[1]* and Anuranjeeta[4]

[1]*Department of Biotechnology, Chaitanya Bharathi Institute of Technology, Hyderabad, India*
[2]*Department of Medicinal and Applied Chemistry, Kaohsiung Medical University, Kaohsiung City, Taiwan*
[3]*Department of Microbiology, Indian Academy Degree College, Bengaluru, Karnataka, India*
[4]*Ambala College of Engineering and Applied Research, Devsathali, Ambala, Haryana, India*

Abstract
Microorganisms are widespread, living pioneer species of the earth. They are indispensable for the ceaseless existence of various other organisms, including humans. Their interaction ranges from cycling nutrients in the environment to manufactured applications that include food production, medicinal, energy development, wastewater treatment, and warfare. Although various microbes are employed in different fields, they are proven to be promising in the medicinal field with their discoveries and applications for the well-being of people globally. The feasibility in culturing, selecting, and manipulating microbes makes them comparatively more advantageous in large-scale bioprocess industries. Mostly, the application of microbes in this industry is for antibiotics. Amoxicillin and penicillin are commonly known antibiotics produced from the fungus *Penicillium*. This chapter is an effort to highlight the tremendous potential of microbes in the pharmaceutical industry and their principal role in producing various pharmaceutical and medicinal products like antibiotics, antivirals, antifungals, and vaccines. Along with these, the role of microbes, like microalgae, in medicinally important food products and nutraceuticals are also elaborated.

**Corresponding author*: sanjeeb.vit@gmail.com; ORCID: 0000-0002-3920-9456

Sanjay Kumar, Narendra Kumar and Shahid-ul-Islam (eds.) Role of Microbes in Industrial Products and Processes, (1–34) © 2023 Scrivener Publishing LLC

Keywords: Antibiotics, bioprocess, microbes, nutraceuticals, pharmaceutical industry

1.1 Introduction

The pharmaceutical industry has been revolutionized ever since the association of pharmaceutical products and microbes was discovered. These microorganisms comprise a large genetic diversity, and their tremendous potential has proven to have a critical impact on the pharmaceutical sector. Microorganisms have been used for decades in the large-scale production of various products like antibiotics, drugs, alcohol, food processing, etc. [1]. Microbes are considered an eminent source of pharmaceuticals as they have more significant potential as drugs from natural sources to treat various kinds of diseases such as cancer, anemia, diarrhea, atopic dermatitis, etc. They are also promising sources of antioxidants, immunosuppressants, enzymes, vitamins, antibiotics, etc. [1, 2].

The microbes are the largest biomass on earth, enlisting more than 10^{20} microbes, including bacteria, fungi, algae, viruses, etc. These microbes have provided an excellent means for synthesizing a wide range of molecules, including antibiotics, antivirals, antifungals, vaccines, recombinant products, etc. [2]. The enormous discovery, development, and introduction of microbes into our healthcare system have undoubtedly immensely aided our fight against various diseases caused by infectious microorganisms and their contribution to individuals and societal well-being. These microorganisms are used to produce various commercial products, including riboflavin and vitamin-K, produced by *Escherichia coli* (*E. coli*). They are used as nutritional supplements to treat diseases due to vitamins and minerals or regulate the body's processes. *E. coli* is also used to produce D-p-hydroxyphenylglycine (a D-amino acid), is an essential intermediate for the synthesis of amoxicillin, a semisynthetic antibiotic. Cholera and plague are common bacterial infections that affect one-third of the world's population. Treatment with Doxycycline tetracycline was the best choice for cholera and chlorhexidine for plaque [3].

Sir Alexander Flemming discovered penicillin, the first antibiotic from the fungus species *Penicillium notatum*, which kick-started the golden era of antibiotic discovery. Fungi have recently been involved in producing an innovative and vital drug called Cyclosporin. Cyclosporin is an anti-rejection substance that aided in the development of organ transplant surgery [4]. Streptomycin is a substance obtained from *Streptomyces griseus* and is of great value in medicine. It can destroy many organisms which cannot be killed by penicillin, particularly gram-negative organisms. *Aspergillus nidulans* species are used for the isolation of secondary metabolites, and these metabolites are used as drugs [5]. Ergot alkaloids and lovastatin are the drugs isolated. Ergot is used as experimental psychiatry, and lovastatin is a popular cholesterol-lowering drug. Secondary fungal metabolites have antibacterial, antiviral, antitumoral, and immunosuppressant activities [6].

Algae are proving to be a source of many potential new drugs and bioactive compounds. Therapeutic use of algae has a long history dating back to Indian Ayurveda medicine, Chinese medicine. Chondria, digenia, and alsidium are used as antihelminthics against ascariasis. The chemicals extracted from algae are agar, carrageenan, and alginates. These are widely used in the food industry. Commercial compounds extracted from microalgae such as carotenoids (beta-carotene), astaxanthin, and docosahexaenoic acid (a long-chain polyunsaturated fatty acid) are being used as a nutritional supplement and nutraceutical [7]. The extracts from marine and freshwater algae often show antibiotic activities such as antibacterial, antifungal, anti-algal, etc.

The first antibiotically active compound to be found was Acrylic acid from the green algae *Chlorella vulgaris*, used in various industries such as chemical, pharmaceuticals, water treatment, etc. Different algae tend to produce specific classes of drugs, chemicals, and antibiotics. The compounds like bromoform and iodoform from red algae *Asparagopsis* and *A. taxiformis* are disinfecting solid agents. The red algae, *Delisea fimbriata*, and fimbrolides have solid antibiotic activity. In brown algae, a wide range of compounds like phenols and tannins from terrestrial plants have antibacterial activity. The Dictyotales produce several aliphatic and show antimicrobial activity. The bioactive compounds from green algae are sesquiterpenoids flexilin from *Caulerpa flexilis* and diterpenoids halimedatrial from *Halimeda spp.* have antibiotic activities. The blue-green algae also produce a wide range of antibiotic compounds, especially antifungal compounds. Spirulina is blue-green microalgae widely used as a natural source in the food, pharmaceutical, medicine, cosmetics, and perfumes. *Spirulina platensis* is used to develop novel drugs for treating neurodegenerative disorders such as Parkinson/Alzheimer's disorders [8, 9].

A virus is an infectious agent of simple composition that can multiply in a host cell. Viruses use cellular parts of the host cell for replication, reproduction, nutrition, etc. The name virus is derived from a Latin word called "*Poison or slimy liquid*". Beijerinck first surmised a new kind of infectious agent which he designated as *Contagium vivumfluidum* in tobacco plants [10]. Viral agents are substances that inhibit the growth of viruses by inhibiting DNA or RNA synthesis. The virus infections classify these diverse groups of antiviral agents and some of the antiviral agents are tenofovir, abacavir, and stavudine are used for treating HIV and emtricitabine, telbivudine, tenofovir is used for treating Hepatitis-B virus [11].

The development of effective medicine is achieved since there is progress in the scientific world which acted as a significant driving force for vaccines. Vaccines are the biological material/compound that enhances immunity against diseases or treats the disease [12]. Germ theory and live attenuated vaccines/inactivated pathogens were established by Pasteur, Koch, Ramon, and Merieux, often referred to as the golden age of vaccines. The vaccine is coined from a Latin word called Variolae vaccine, which Edward Jenner demonstrated to prevent smallpox using cowpox [13]. Vaccines have eliminated many infectious diseases, which caused millions of deaths. The critical contributions made by vaccines include the eradication of poliomyelitis, smallpox, and a decrease in 95% incidence of diseases such as tetanus, measles, rubella, hepatitis-A, B vaccines, diphtheria, mumps, etc. [14].

1.1.1 Microbes with Medical Significance

Microbes supplement various domains like agriculture, biotechnology, and the medical industry by catalyzing the product development necessary for humankind. Given that the focus of this chapter is the pharmaceutical industry, the following section emphasizes products developed from the microbe *E. coli*, which is closely related to human beings both internally and externally [15].

1.1.1.1 Production of Insulin

Bacterial cells have been playing a significant role in the production of human insulin using genetic engineering techniques. Genetic engineering is a technique where commercially important products can be produced by transforming bacterial cells. It is also termed recombinant DNA technology. In this process, a small piece of the desired gene is taken and introduced into the host cells, which acts as a vector. This desired/foreign DNA gets incorporated as a permanent feature of the host cell. These cells transform and replicate into daughter cells and are passed on to further generations [16].

The insulin gene's mRNA from cells of Langerhans' islets is isolated from the diabetic patient. The plasmid DNA/vector DNA is taken out of bacterial cells, *E. coli* is widely used for insulin production. The enzymes used

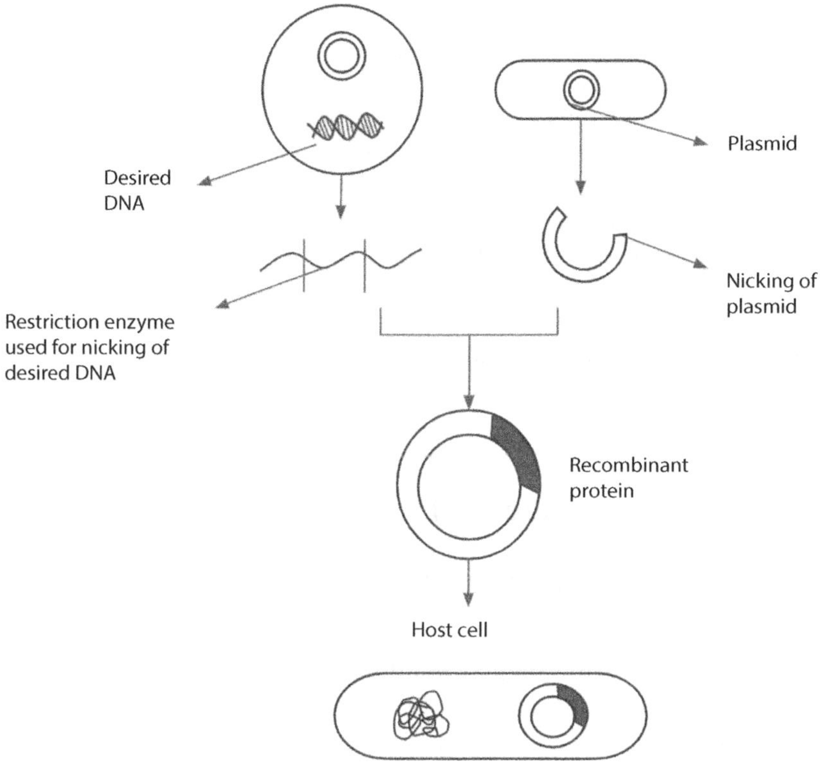

Figure 1.1 Microbes in recombinant DNA technology.

for cutting the DNA at a specific recognition site are restriction enzymes. The vector DNA is cut using restriction enzymes to produce sticky ends at both ends. With the vector ring open, the gene of interest/gene responsible for insulin production is incorporated into the vector ring and sealed. After combining human insulin genes and bacterial vectors, the transformed cells are inserted into the host cells back. Cells are provided with nutrients for their growth and the production of human insulin [17].

The enzymes used for cutting the DNA at a specific recognition site are restriction enzymes (Figure 1.1).

1.1.1.2 Production of Somatotropin

Pharmaceutically extracted somatotropin hormone from the pituitary gland could not meet the annual demand; thus, somatotropin biosynthesis was achieved using genetic engineering technique [15, 18].

Double-stranded cDNA was produced from mRNA precursors of human Growth Hormone (hGH) cells which were then incorporated into bacterial cells. At this step, the bacterial cells are unable to attain a biologically active form/unable to convert peptides to the active form. Thus, a fragment that has a complete hGH sequence after 24 codons is cleaved from a recombinant plasmid that has full-length hGH. The cleaved fragment of DNA and cDNA molecules are ligated together to create a complete hGH coding fragment. Then, this complete hGH fragment is ligated into a restriction site just below the lac promoter on a plasmid. The recombinant plasmid is allowed to transform into *E. coli* cells. The newly synthesized hGH fragments are subsequently purified and produced commercially by this method [18].

1.2 Role of Microbes in Drug Discovery

1.2.1 Microbes in Antibiotic Production

Antibiotics are antibacterial substances produced by microbes as part of their natural defense mechanism to inhibit or kill bacteria growing in their vicinity. Due to the extensive range of antibiotics produced by microorganisms, these antibiotics are commercially exploited for treating infectious diseases caused by morbific microorganisms. Hence, microorganisms are a prominent source for numerous antibiotics produced either as primary metabolites or secondary metabolites during the growth phase and stationary phase, respectively.

The revolution in the microbial drug era began with penicillin being the first natural antibiotic discovered by Alexander Fleming, and it was obtained by a fungus, *Penicillium notatum*, that antagonizes the growth of *Staphylococcus aureus* [19]. This has paved the way for developing antibiotic substances to be valuable against infectious bacteria. After that, a remarkable quantity of antibiotics was explored. Another such antibiotic discovered was streptomycin from the species *Streptomyces griseus*, discovered by Selman Waksman, which was found to induce immunization against tuberculosis [20, 21]. This has revolutionized the pharmaceutical industries to exploit microorganisms that synthesize antibiotics. Later, many other antibiotics like cephalosporin, chloramphenicol, and kanamycin were explored [22].

The industrial production of these antibiotics is produced by different techniques like fermentation, semisynthetic, and synthetic techniques [23]. Production of antibiotics using fermentation technique involves various stages like preparation of inoculum, fermentation, scaling up of fermentation process, recovery, and purification of the product, and formulation of the product. Parameters like oxygen concentration, temperature, pH, and nutrient levels should be constantly monitored for the maximum yield of the product. Penicillin is the first antibiotic to be commercially extracted using this technique [24]. Cephamycin C produced by *Streptomyces clavuligerus*, tetracycline produced by *Streptomyces aureofaciens*, Griseofulvin produced by *Penicillium griseofulvum*, surfactin produced by *Bacillus subtilis*, and mycophenolic acid produced by *Penicillium brevicompactum* are a few antibiotics that are commercially produced using fermentation techniques [25]. In a semisynthetic approach, the naturally produced antibiotics are chemically modified to enhance their activity. This approach is proven to produce valuable antibiotics but at a meager success rate. A few such valuable antibiotics that are semi-synthetically derived are clindamycin from parent compound lincomycin, by chloro substitution of hydroxyl group, was found to be effective in treating osteomyelitis and skin infections and toxoplasmosis [26]. Other semisynthetic antibiotics developed from penicillin are amoxicillin developed by replacing one H atom in penicillin with $CH(NH_2)C_6H_4OH$ group, ampicillin (a beta-lactam antibiotic), developed by the addition of an amino group to the benzylpenicillin molecule, benzylpenicillin (penicillin G), and phenoxymethylpenicillin (penicillin V) [27]. These new penicillins give a broader spectrum of use and are more potent than penicillin. Another such approach that does not involve the use of microorganisms in synthesizing antibiotics is the Synthetic technique. Antibiotics are chemically synthesized based on the structure of natural antibiotics. Chloromycetin is the first synthetic antibiotic [28]. The list of common antibiotics and their sources are mentioned in Table 1.1.

Table 1.1 Some of the antibiotics produced from the microorganisms and their antibiotic activity against the mentioned species of bacteria.

Antibiotic compound	Source	Effective against	Reference
Penicillin	*Penicillium notatum*	Gram-positive bacteria	[19]
Streptomycin	*Streptomyces griseus*	Gram-positive and Gram-negative bacteria	[20]

(*Continued*)

Table 1.1 Some of the antibiotics produced from the microorganisms and their antibiotic activity against the mentioned species of bacteria. (*Continued*)

Antibiotic compound	Source	Effective against	Reference
Cephalosporin	*Acremonium chrysogenum*	Gram-positive and Gram-negative bacteria	[36]
Bacitracin	*Bacillus subtilis*	Gram-positive bacteria (*staphylococci, streptococci, corynebacteria*, and *clostridia*)	[37]
Polymyxin B	*Bacillus polymyxa*	Gram-negative bacteria (*Pseudomonas aeruginosa, Acinetobacter baumannii*, and *Klebsiella pneumoniae*)	[38]
Erythromycin	*Streptomyces erythraeus*	Gram-positive bacteria (*Streptococci, Staphylococci, Clostridia, Corynebacteria, Listeria, Haemophilus* sp., *Bacillus anthracis*, and *Neisseria meningitides*)	[39]
Neomycin	*Streptomyces fradiae*	Gram-positive and Gram-negative bacteria	[40]
Tetracycline	*Treptomycesaureofaciens*	Gram-positive and Gram-negative bacteria	[41]

(*Continued*)

Table 1.1 Some of the antibiotics produced from the microorganisms and their antibiotic activity against the mentioned species of bacteria. (*Continued*)

Antibiotic compound	Source	Effective against	Reference
Vancomycin	*Amycolatopsisorientalis*	Gram-positive cocci and bacilli, such as *Staphylococcus aureus*, *Staphylococcus epidermidis*	[42]
Gentamicin	*Micromonospora purpurea*	Gram-negative bacteria (*Pseudomonas*, *Proteus*, *Escherichia coli*, *Klebsiella pneumoniae*, *Enterobacter aerogenes*, *Serratia*), and Gram-positive (Staphylococcus)	[43]
Rifamycin	*Amycolatopsisrifamycinica*	*Mycobacteria*	[44]

Antibiotics are categorized into a few classes based upon chemical or molecular structures, including beta-lactams, macrolides, tetracyclines, quinolones, aminoglycosides, sulphonamides glycopeptides, and oxazolidinones [29]. However, microbes are playing a key role in synthesizing antibiotics of only a few classes, including beta-lactams, macrolides, tetracyclines, aminoglycosides, and glycopeptides.

1.2.1.1 Beta-Lactams

These are the most prevalent class of antibiotics. These beta-lactams antimicrobials have a beta-lactam ring, 3-carbon, and 1-nitrogen ring, as a common feature. This class of antibiotics includes:

1. Penicillins and their members include penicillin G, penicillin V, ampicillin, amoxicillin, carbenicillin, piperacillin, mezlocillin, and ticarcillin [30]. These penicillins contain a four-membered beta-lactam ring, with an acyl side chain

attached to it, along with a five-membered thiazolidine ring. These are most prominently used in medicine. Due to their wide range of activity, these are effective against many bacterial infections [29].

2. Cephalosporin: These are isolated from the fungus *Cephalosporium acremonium* possess a dihydrothiazine ring and a beta-lactam ring with an acyl side chain [29].
3. Monobactams: Unlike other members of beta-lactams antibiotics this class possesses a monocyclic structure, that is, the beta-lactam ring is not fused with any other ring. This class of antibiotics includes aztreonam produced by *Chromobacteriumviolaceum*. This class of antibiotics is effective against a wide range of gram-negative bacteria [29].
4. Carbapenems: These members have a five-membered ring along with a β-lactam ring. Thienamycin is the only carbapenem with medicinal importance, yet it is the most effective, non-toxic antibiotic discovered. It is produced by strain *Streptomyces cattleya* [29].

1.2.1.2 Macrolides

This class of natural antibiotics possesses a large macrocyclic lactone ring to which one or more deoxy sugars may be attached. Erythromycin is the first macrolide discovered from the species of *Saccharopolyspora erythraeus* [31]. Later, clarithromycin and azithromycin were semi-synthetically derived from erythromycin. These bacteriostatic antibiotics effectively treat infections caused by gram-positive by inhibiting protein synthesis of growing bacteria. It is also used to substitute for β-lactam antibiotics in allergic patients [32].

1.2.1.3 Tetracyclines

These broad-spectrum antibiotics contain a linear fused tetracyclic nucleus. The first tetracycline discovered was a natural antibiotic named aureomycin (chlortetracycline) from *Streptomyces aureofaciens*. The semi-synthetic antibiotic developed from aureomycin was tetracycline. This tetracycline had a better solubility than aureomycin, making it a more favorable pharmacological product. These tetracyclines were also proven to have anti-inflammatory, anti-neurodegenerative, and mammalian activity [33].

1.2.1.4 Aminoglycosides

These antibiotics are generally made up of 3-amino sugars linked by glycosidic bonds. Streptomycin produced by a soil-borne microorganism, *Streptomyces griseus*, has been greatly used in treating bubonic plague, tularemia, and tuberculosis. The highly toxic nature of this antibiotic has led to the discovery of other antibiotics of less toxicity, including gentamicin from *Micromonospora purpurea*, neomycin from *Streptomyces fradiae*, tobramycin from *Streptomyces tenebrarius*. Actinomycetes have been a source for this class of antibiotics [34].

1.2.1.5 Glycopeptides

Glycopeptides are characterized by glycosylated tricyclic or tetracyclic heptapeptides. Actinomycetes mainly produce these, hence called actinomycete-derived antibiotics. *Amycolatopsis orientalis* produces vancomycin as a secondary metabolite, is one such glycopeptide antibiotic, that is used to treat severe bacterial infections by inhibiting the cell wall synthesis of growing gram-positive bacteria and these infections include colitis, endocarditis, and hemodialysis shunt prophylaxis. Teicoplanin, produced by *Actinoplanesteichomyceticus,* is the best choice for infections in which the disease-causing gram-positive microorganisms are resistant to other drugs [35].

1.2.2 Microbes in Antivirals and Antifungals

1.2.2.1 Microbes in Antiviral Production

A viral cell consists of nucleic acid, a protein coat, and viral enzymes. A virus is an obligate intracellular pathogen that utilizes the cellular machinery of the host to replicate, which is usually considered as a complex character to develop drugs with selective toxicity against viruses [45]. Antiviral drugs inhibit the replication of viruses in the replication cycle at various stages. These are the analogs of nucleosides and function by inhibiting the nucleic acid biosynthesis, i.e., they act as chain terminators [46]. Viral infections like cytomegalovirus, hepatitis B virus (also known as serum hepatitis), and hepatitis C virus (liver inflammation) are present threats to human health due to the lack of effective therapeutic agents. Some bacteria like myxobacteria and δ-proteobacteria which belong to the Myxococcales order have unique secondary metabolites. The major part of which are biogenetically derived from polyketide synthase (*Streptomyces spp.* produces PKS) and non-ribosomal peptides synthetases (NRPS which is produced by cyanobacteria) or a hybrid of PKSs and NRPSs (i.e., haliamide which is obtained from *Haliangium ochraceum* SMP-2) which are mentioned in detail in Table 1.2. These secondary metabolites were found to have antiviral activity and antagonize the growth of cytomegalovirus, including glandular fever and pneumonia. Siderophores are the secondary metabolites isolated from strains of myxobacteria under limiting conditions of iron. Some of the siderophores include nannochelins and hylachelins which should be screened for various antiviral activities, especially for anti-HCMV, and should study the structure-activity relationship for the possible discovery of more potential antivirals. Some antivirals like soraphen A are used to see the possibility to treat the co-infection [47].

Another copious source of pharmacologically active natural products is Marine microbes, which constantly provide promising antiviral agents. They are known to be producers of antiviral agents as they provide an unlimited biological resource for the production of therapeutic drugs from various resources like marine bacteria, algae, marine sponges, fungi, and other miscellaneous marine organisms. These microbes produce compounds like Exopolysaccharides Sulphated exopolysaccharides (EPS) obtained from *Bacillus sp.*, Naviculan from *Diatom*, vidarabine from *Tethyacripta, Halovir A–E* from *Scytidium sp., Eudistomin* from *Eudistoma olivaceum* respectively shows a biological activity against Herpes simplex virus, and many

Table 1.2 Some of the antiviral compounds from secondary metabolites.

Type	Compound name	Source	Biological activity against	References
Polyketides	Balticolid	*Streptomyces spp.*	Influenza A virus and HSV-I	[59, 60]
Non-ribosomal peptides	Cyanovirin-N	*Cyanobacteria*	Fluconazole-resistant *C. albicans*	[61]
NRPS-PKS hybrids	Haliamide	*Haliangiumochraceum SMP-2*	Cytotoxicity towards the HeLa cell line	[62]
Terpenoids	Indole diterpenoids	*Penicillium camemberti*	H1N1	[63]
Indole alkaloids	Aspernigrin C	*Aspergillus niger*	HIV	[63]

others which are used for the treatment and control of many viral diseases like HIV, HCMV in humans [48] which have a strong potential market value with more novelty and there are 89 antiviral compounds till 2019 of which eight structural classes have been isolated from marine microorganisms, and 35 of them exhibit anti-H1N1 activity [49]. In 1998, an antiviral compound stachyflin was obtained from marine fungi, i.e., *Stachybotrys* sp. against the influenza A virus (H1N1). After 2006, there were many antiviral compounds obtained from marine fungi against Enterovirus-71, Herpes Simplex Viruses (HSVs), Human Immunodeficiency Virus (HIV), Influenza Virus, Porcine Reproductive and Respiratory Syndrome Virus, Molluscum Contagiosum Virus, Respiratory Syncytial Virus, Tobacco Mosaic Virus [50]. The antiviral compounds derived from marine microorganisms can be classified eight biosynthetic categories: alkaloids like trichobotrysins, quinones include Anthraquinones, peptides include rubrumlines, polyketones include asteltoxins, pyrones, sterols, terpenoids

Table 1.3 Antiviral compounds from marine microbes [49].

Type	Antiviral compound	Source
Alkaloids	Trichobotrysins A, B, D	*Trichobotrys effuse*
	Trypilepyrazinol	*Penicillium sp*
	Acridanone	*Streptomyces fradiae strain VITMK2*
Quinones	Anthraquinones, aspergilols	*Aspergillus versicolor*
	Anthracene-9,10-dione	*Nocardia alba*
Peptides	Eutypellazines A–L	*Eutypella sp.*
	Rubrumlines A–O	*Eurotium rubrum*
Polyketones	Asteltoxins	*Aspergillus sp.*
	Wailupemycin J	*Streptomyces sp.*
Sterols	3α-hydroxy-7-ene-6,20-dione	*Cladosporium sp.*
Terpenoids	Talaromyolide D	*Talaromyces sp.*
	Chrodrimanins	*Penicillium sp.*
Others	Isoprenylated cyclohexanols, Truncateols	*Truncatellaangustata*

like chrodrimanins, and others like isoprenylated cyclohexanols, truncateols C, cytosporin L, and vaccinol J which are mentioned with few more examples in Table 1.3. Among this classification, peptides are said to be the most common group of secondary metabolites with potentially useful activities. Marine natural products are also used as an important source of discovery of lead compounds for antiviral drugs [49].

1.2.2.2 Microbes in Antifungal Production

Antifungal compounds are used to treat fungal infections and these compounds mainly focus on targeting either the formation or the function of ergosterol [51]. Ergosterol is an important component of the fungal cell membrane. Antifungal compounds can be divided into four main classes; polyenes, azoles, allylamines, and echinocandins which are mentioned in detail in Table 1.4. Polyenes antifungal agents involve amphotericin B, the standard of therapy for systemic fungal infections. Here, the interaction of antifungal with the sterol presents in the membrane results in the production of aqueous pores which consists of a ring of eight molecules of amphotericin B which are linked hydrophobically to membrane sterols that lead to altered permeability and leakage of cytoplasmic components

Table 1.4 Antifungal compounds from microbes.

Type	Compound name	Source	Biological activity against	References
Echinocandins	Caspofungin B	*Aspergillus rugulovalvus*	*C. albicans spp., Cryptococcus neoformans*	[55]
	Pneumocandin	*Zalerionarboricola*		
	Papulacandin	*Papulariasphaerosperma*		
Allylamine	Terbinafine	*Streptomyces sp. KH-F12*	*C. albicans spp., Cryptococcus neoformans, Dermatophytes*	[56]
Polyene	Amphotericin B	*Streptomyces nodosus*	*Aspergillus spp., C. albicans spp.*	[57]
Azoles	Fluconazole	-	*C. albicans spp., Cryptococcus neoformans, Aspergillus spp.,*	[51]
Others	Griseofulvin	*Penicillium griseofulvum*	*Pneumocystis jirovecii*	[58]

which leads to death of the organism. Azoles are 14α-demethylase inhibitors. They have two subclasses, triazole and imidazole. Triazole derivatives, for example, fluconazole and itraconazole that has antifungal activity at least in part of inhibiting the cytochrome P-450-dependent 14 α sterol demethylase. Ergosterol is a sterol that is present in the cell membrane of fungi and acts as a bioregulator of membrane fluidity and integrity in the cells. So, inhibition of 14 α demethylase leads to reduction of ergosterol and accumulation of sterol precursors, which results in the formation of a plasma membrane with altered structure and function. Allylamines include terbinafine and naftifine which are newly classified in ergosterol biosynthetic inhibitors that are highly potent against dermatophytes *in vivo* and *in vitro*. Terbinafine has good activity against some of the azole-resistant *Candida albicans* strains and *Cryptococcus neoformans*. Here, the inhibition takes place due to the accumulation of sterol precursor squalene epoxidase (target for allylamine activity) and the absence of sterol intermediate. As the level of squalene increases, the membrane permeability also increases, which leads to the disruption of cells [51]. There are a few antifungal drugs that target other fungal structures; Echinocandins including caspofungin is used for treating mucosal and candidiasis infections. It is available only as an intravenous formulation. As they are not metabolized through cytochrome p450 enzymes, there will be minimum drug–drug interactions except for cyclosporine as it has a risk of raised liver transaminase levels [52]. Depending on the usage and availability of different classes of antifungal agents, it can be predicted that an increase in the number and variety of fungal species resistant to these agents. An antifungal drug that is naturally produced is used to treat skin infections, i.e., Griseofulvin, which specifically disrupts cell wall by interfering with microtubules involved in spindle formation during mitosis. There are a few drugs like atovaquone, sulfamethoxazole-trimethoprim that act as antimetabolites against fungal and bacterial processes, respectively. Atovaquone is used to treat pneumocystis pneumonia caused by *Pneumocystis jirovecii*.

Some of the antifungal agents can be used against plant pathogens fungi such as *Lactobacillus plantarum*, a Gram-positive bacterium that has a broad characteristic in producing antifungal compounds. From this species, two compounds were extracted, Benzene acetic acid and 2-propenyl ester which are pH-dependent and thermostable. Additionally, *L. plantarum* has a role in inhibiting the growth of many microorganisms like *Alternaria solani*, *Phytophthora drechsleri Tucker*, *Fusarium oxysporum*, *Glomerellacingulata*, and *Botrytis cinerea* [53]. A large number of microorganisms are being tested for their antifungal activity against food spoilers as bioprotective tools in food products. Lactic acid bacteria are tested in many applications such as dairy and bakery production against fungi. Many of the antifungal species have characteristics of microbial weeds which dominate in a community in open microbial habitat. *L. plantarum* strains are widely used in foods representing labs such as milk, malted barley, and water for malt production. This strain is mostly present in fermented foods. Every strain of *L. plantarum* has its role such as TK9 active against *Penicillium* which is present in citrus, apple, and yogurt spoilage (fermented food products) and UFG 121 retards the growth of *Fusarium culmorum* in an oat-based beverage. However, these strains with antifungal activity can be isolated from a wide variety of fermented foods like kimchi (fermented vegetables), koumiss (beverage), and tempeh [54].

1.3 Microbes in Vaccines

Vaccines are a major domain in the pharmaceutical and health industry, and their importance is prevalent in various areas at the global level as either a preventive or a prophylactic measure for diseases. Starting from vaccine research to manufacturing, the key player involved here is a microbe. Vaccines are microbial preparations that act against the microbes themselves, providing active acquired immunity to the vaccinated human [14]. There are various classes of vaccines depending on the type of preparation and the nature or part of the microbe being considered for production. The

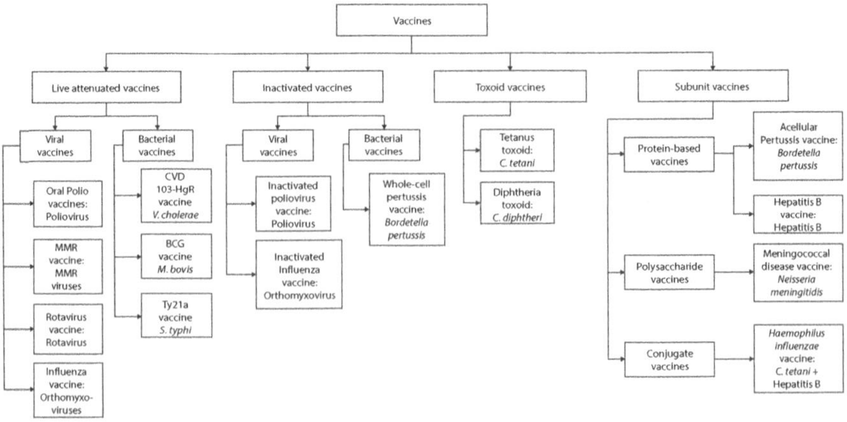

Figure 1.2 Flowchart showing the types of vaccines and the microbe involved in the production of the vaccine.

microbe took, for example, a bacterium or a virus, if present in a live and weak condition is a live attenuated vaccine [64], if killed then is an inactivated vaccine. In some cases, instead of a full microbe only a part or unit is taken for production which includes subunit and conjugate vaccine [65], and when the toxin secretions of bacteria are used then they are inactivated toxins, and here the microbe itself is not present in the vaccine [14, 66]. This section provides an overview of vaccines and their types as displayed in Figure 1.2, emphasizing a few of the contributions of microbes to this industry.

1.3.1 Live Attenuated Vaccines

One of the most successful treatment interventions includes live attenuated vaccines. These vaccines contain a live microbe; hence, they are altered to milden the virulence or disease-causing ability of the microbe chosen [67, 68]. The first vaccination attempt by Edward Jenner is an example of a live attenuated vaccine where he used mild cowpox viral material to elicit an immune reaction against smallpox, and the successful implementation has opened the door for vaccination [13]. The immunity that is developed as a result of the administration of such a vaccine is more effective and enduring than inactivated vaccine, and therefore is widely used. Similar mechanisms are utilized for the attenuation of viruses and bacteria. Hence, the host's immune system develops a memory of the pathogen [69] while causing unchallenging elimination or destruction of the attenuated and weak pathogen, with any unfavorable side-effects being extremely rare or negligible. Administration of this type of vaccine can be done through intradermal, subcutaneous injections and through oral and nasal mucosal pathways [69–71].

1.3.2 Bacterial Attenuated Vaccines

As the name suggests, bacterial attenuated vaccines contain a whole weakened bacterium which causes an immune response leading to the development of memory which helps in the prevention of an actual bacterial infection later. The most important live attenuated bacterial vaccines which are also approved for human use are CVD 103-HgR, a live vaccine against *Vibrio cholerae* O1 administered orally [72], Bacillus of Calmette and Guerin (BCG) which is a *Mycobacterium bovis* strain against human tuberculosis [73], Ty21a strain of Salmonella Typhi [74], an orally administered vaccine against typhoid fever.

1.3.3 Viral Attenuated Vaccines

Live viruses are treated and altered to reduce the virulence and then administered through a route depending on the vaccine. Most administered live attenuated viral vaccines are Measles, mumps, rubella (MMR) vaccine which is a combined vaccine for three infectious conditions measles, mumps and rubella [75], Varicella vaccine or the chickenpox vaccine, Influenza vaccine given as either flu shots through a nasal spray, and Rotavirus vaccine to protect against rotavirus infections which are a commonly known cause of diarrhea among children. MMR vaccine is a combination of live attenuated viruses of measles, mumps, and rubella, namely Measles morbillivirus, Mumps orthorubula virus, and Rubivirusrubellae virus, respectively [76]. Preparation of measles and mumps virus is carried out in chicken embryo cell cultures and, for rubella virus, it is done in human diploid lung fibroblasts [76]. Varicella vaccine contains live attenuated varicella virus [77]. Another example of using live attenuated microbes is the yellow fever virus for the vaccine against yellow fever [78].

Some vaccines like the Influenza vaccine and Polio vaccine are present in both live attenuated and inactivated forms. In the case of the Influenza vaccine, the administration is different in the case of live attenuated and inactivated forms with live attenuated influenza vaccine (LAIV) administered as a nasal spray (intranasally) and inactivated form being injected intramuscularly [79]. Oral poliovirus vaccine (OPV) is a live attenuated version of polio vaccine and contains three poliovirus serotypes of the Sabin strains. Similarly, inactivated poliovirus vaccines are also produced from three poliovirus strains [80]. The most recent example is the Covishield (ChAdOx1 version produced by Serum Institute of India) is a vector-based viral vaccine developed from attenuated adenovirus which carries SARS-CoV-2 virus spike protein's genetic material, with the vector not having the ability of replication [81].

1.3.4 Inactivated Vaccines

Microorganisms are killed through physical or chemical techniques and further processed for vaccine production. The process of killing inactivates the virus or bacteria (e.g., through a chemical like formaldehyde), due to which the pathogenic microbe cannot return to virulent, disease-causing form [82].

An example of a bacterial inactivated vaccine is the Whole-cell pertussis vaccine. Pertussis is also known as whooping cough and is an acute respiratory tract disease [83]. This is also termed the whole-cell DTP vaccine and is comprised of inactivated *Bordetella pertussis* cells along with toxoids of tetanus (obtained from *Clostridium tetani*) and diphtheria (obtained from *Clostridium diphtheri*) [84]. Inactivated poliovirus vaccine (IPV) as mentioned earlier is a commonly known inactivated viral vaccine containing inactivated poliovirus strains [80]. Inactivated type of Influenza vaccine has also been previously mentioned is available other than the live attenuated form and contains killed influenza viruses [79]. Hepatitis A vaccine is made of inactivated viruses of Hepatitis A and is a preventive vaccine against this contagious liver infection [85]. Vaccines based on inactivated toxins have already been described along with the whole-cell DTP vaccine where toxoids of diphtheria and tetanus are obtained from the respective microorganisms *C. diphtheri* and *C. tetani* [84]. The Covaxin vaccine against COVID-19 is a recent example of inactivated vaccine and consists of the whole-virion antigen of SARS-CoV-2 in inactivated state [86].

1.3.5 Subunit Vaccines

Unlike the vaccines containing the whole pathogen, some vaccines only contain a part of the pathogen which is necessary for recognition by the host to generate an immune response and immunological memory for the future infection. Subunit vaccines can be protein-based, polysaccharide, or conjugate vaccines. Protein-based subunit vaccines contain only the antigen(s) (protein) that is specifically necessary to generate the immune response. Examples of this type include Acellular pertussis which is a bacterial vaccine and Hepatitis B vaccine which is a viral vaccine. Acellular pertussis vaccine is made from the endotoxin released by *Bordetella pertussis* bacteria, which is proteinaceous in nature [87]. Hepatitis B vaccine is made from the surface antigen of Hepatitis B virus [88]. Polysaccharide vaccines are made up of a surface polysaccharide or sugar produced by a microbe or pathogen. An example of this case is the polysaccharide from the surface of *Neisseria meningitidis* taken for producing a vaccine against Meningococcal disease [89]. Conjugate vaccines have a polysaccharide similar to polysaccharide vaccine but along with it, there is a carrier protein attached. This leads to a better immune response when compared to the polysaccharide vaccine. *Haemophilus influenzae* type b (Hib) vaccine is a conjugate vaccine that is made by the combination of tetanus toxoid from *C. tetani* with a surface polysaccharide of hepatitis B virus [90].

Viral vector vaccines: To enable the elicitation of an immune response, specific constituents of the virus (protein particles that are altered to not cause disease) can be delivered by another virus that is safer and acts as a carrier. These are viral vector vaccines [91]. The Ebola vaccine is an example of this type with the vector Chimpanzee Adenovirus type 3 (which is incapable of replication) [91]. Apart from conventional technology, recombinant technology is used for the production of vaccines.

1.4 Microbes in Medically Important Food Products or Nutraceuticals

Food and dietary supplements are becoming more popular as a means of preserving one's health. Microbes are found everywhere as a source of both good and hazardous substances for humans, as they are employed in a variety of businesses and to create other valuable items. On the other hand, they are triggering a variety of diseases in humans at varying rates. As a result, they are employed in the production of yogurt, cheese, wine, and bread [92, 93]. Microbes are utilized in a variety of sectors to make vinegar and various forms of alcohol due to their high rate of activity. They are employed in the pharmaceutical industry to make chemical compounds and in pharmaceutical research [94–96]. Nowadays, microorganisms are used in the chemical industry to synthesize industrial goods also including. There are several gaps in the literature about the involvement of different strains of bacterial species in medical sectors. Some key aspects of microorganisms in the food industry are still unknown. This study emphasized the importance of bacteria in industries for synthesis in both medicinal and industrial settings, although some species have a significant influence on human health, producing major illnesses in humans [97].

Bio-preservation is the process of extending the shelf life of food by utilizing microorganisms along with antagonistic abilities and/or their antimicrobial metabolic activity. Fermentation is the oldest bio-preservation technology, and it is used to preserve milk products, meat, fish, and vegetables. Fermentation is the process of using microorganisms like yeasts, molds, and bacteria to transform raw components into finished items. The morphological, metabolic, physiological, and taxonomic characteristics of these microorganisms vary. The most prevalent bacteria developed in popular fermented foods are LAB. LAB, which generates lactic acid as an end product of sugar fermentation, can represent large variations of Gram-positive and catalase-negative bacteria.

1.4.1 Acidophilus

Milk is a product of Europe that has been infected with *Lactobacillus acidophilus* strains. The acidity of the final goods is defined by the presence of roughly 1% lactic acid.

1.4.2 Buttermilk

Is a drink that originated in the Middle East (Balkans), and Central Asia. It is a fresh dairy beverage created by inoculating *Streptococcus, Leuconostoc,* and *Lactococcus lactis* subsp. *lactis/cremoris* into a butter churning liquid and fermenting it at 20–25 degrees Celsius until the pH hits 4.5.

1.4.3 Dahi

Is an Indian dairy product that has been fermented. It's a yogurt-like product prepared with *Lactobacillus acidophilus, Streptococcus lactis, Lactococcus lactis,* and *Lactococcus lactis* subsp. *diacetylactis,* as well as *Lactobacillus acidophilus, Streptococcus lactis, Lactococcus lactis,* and *Lactococcus lactis* subsp. *diacetylactis.* These strains are mixed into boiled and refrigerated milk before being fermented at 30 degrees Celsius for 7 hours. A sharp acid flavor and a yellowish cream white appearance characterize the completed product.

1.4.4 Kefir

Is prepared by inoculating fresh milk with the "kefir grain," which comes from the Caucasus area of Asia. It's a fermented dairy beverage that's sour and moderately alcoholic. Kefir's beneficial properties and composition are mostly determined by the type of milk that was fermented, and it may include a high level of vitamin B12, *Lactobacillus kefiranofaciens* and *Lactobacillus kefir* and *Lactococcus* spp., *Streptococcus* spp., *Leuconostoc* spp., *Kluyveromycesmarxianus, Candida inconspicua, Candida maris, Torulopsis kefir,* and *Saccharomyces cerevisiae.* In kefir grains, there is a 1000:1 ratio of LAB to yeast. These strains are mixed into boiled and refrigerated milk before being fermented at 30 degrees Celsius for 7 hours. A sharp acid flavor and a yellowish cream white appearance characterize the completed product. Fifteen percent total solids, 1.8% protein, 5% fat, and 12% sugar are typical Dahi compositions.

1.4.5 Yogurt

Which originated in the Balkans and the Middle East, is one of the most famous fermented dairy products in the world. *Lactobacillus delbrueckii* subsp. *bulgaricus* and *Streptococcus thermophilus,* both naturally present in milk or added as starter culture by the "back-slopping method," ferment cow, goat, sheep, or buffalo milk to form yogurt. *Streptococcus thermophilus* develops quicker than *Lactobacillus delbrueckii* subsp. *bulgaricus*, fermenting lactose in the existence of dissolved oxygen as well as fabricate more acid lactic and formic acid.

Oceans encompass more than 70% of the planet's surface, as it is already known. According to experts, the biological richness of the maritime environment is greater than that of tropical rainforests. Marine water is rich in biodiversity, making it a great source of unique bioactive chemicals in large quantities and types. Because the bulk of the marine microbiota is soft-bodied as well as lives an inactive lifestyle, it must rely on alternative defensive mechanisms, namely the production of biochemical substances that are harmful to other species. The variety of marine microflora is staggering; there is a pressing need to investigate and harness their biotherapeutic potentiality [98, 99].

1.4.6 Algae

A kind of marine microflora, is commonly exploited in the creation of nutraceuticals as well as a food source or component [100]. Algae come in both micro and macro forms, and both are employed as nutraceuticals. They are high in beta-carotene (vitamin A), vitamin C, E, H, B1, B2, B6, and B12, astaxanthin, polysaccharides, and polyunsaturated fatty acids, among other minerals and vitamins [101]. Microalgae-derived bioactive compounds are used as food additives, baby milk fortifiers, and other nutritional supplements [102]. Those chemicals have also demonstrated promising pharmacological capabilities, indicating that they might be exploited to generate novel cancer treatments. Other chemicals, such as ziconotide, are utilized as pain relievers and incendiary agents in mollusks (*Conus magus*). Natural products found in marine invertebrates' symbiotic bacteria or cyanobacteria have strong structural similarities to familiar microbial metabolites, suggesting that bacteria and microalgae both are involved in the production and are the genuine suppliers of these metabolites. Molecular approaches are now being utilized to investigate the microbial heterogenicity in marine sponges and the part of bacteria in bryostatin manufacture in the bryozoan *Bugula neritina* [103].

1.4.7 Marine-Derived Fungi

Bioactive metabolites produced by marine fungus have been shown to have anticancer, antiplasmodial, anti-inflammatory, antibacterial, and antiviral properties [104, 105]. This is because they include certain rare and unusual carbon structures. These unique compounds are employed in medicine and plant protection as new lead structures. Kjer *et al.* [106] outlined the technique for isolating and cultivating them from different types of marine species (sponges, algae, and mangrove plants). Aspergiolide A [107] is a new anthraquinone derivative with a naphthol [1,2,3-de] chromene-2,7-dione skeleton derived from *Aspergillus glaucus*, a marine filamentous fungus found in Fujian, China. It was discovered to be cytotoxic to K562 and P388 cell lines.

In the presence of meleagrin D and E and roquefortine H and I, *Penicillium* sp., isolated from deep ocean residue, which was discovered to have anticancer activity in their alkaloid extracts [108]. Other new chemicals and metabolites from the marine fungus *Ampelomyces* sp., which has antibacterial and antifouling properties, were also isolated and described. Presence of the chemical 3-chloro 2,5-dihydroxy benzoyl alcohol, which successfully prevented the larval settling of the tubeworm. *Hydroides elegans* and cyprids of the barnacle *Balanus amphitrite* were responsible for the anti-larvicidal activity. The substance is non-toxic and acts as an antifoulant and/or antibiotics [109].

Cladosporium sp., another marine-derived fungus, also has antibacterial, antifouling properties. Presence of xylose or glucose, the fungus generated beneficial chemicals in a nutrient-rich growing medium [110]. Another study found that ethanol extracts from the marine-derived fungus *Fusarium* sp. have unique antimitotic and antifungal properties. Novel antibacterial dihydroxy-bis-dethio-bis-methylthio gliotoxin, dioxopiperazine, was found in the broth of another marine-derived fungus, *Pseudallescheria*. Gliotoxin demonstrated considerable radical scavenging action against 1,1-diphenyl-2-picrylhydrazyl (DPPH) [111]. Antibacterial activity against methicillin-resistant and multidrug-resistant *Staphylococcus aureus* is strong in all three substances.

1.4.8 Marine-Derived Actinomycetes

Secondary metabolite production is well-known among actinomycetes. They generate many well-known antibiotics with significant biological activity, including streptomycin, erythromycin, and tetracycline [112]. Actinomycetes from the sea have been identified to create new anticancer [113], antimalarial [114], and antibacterial chemicals [115, 116]. *Nocardiopsis lucentensis*, a marine-derived actinomycete, developed four new 3-methyl-4-ethylidene proline-containing peptides known as Lucentamycins A-D from their fermentation broth [26]. Lucentamycins A and B were the only compounds that showed high *in vitro* cytotoxicity against HCT-116 human colon cancer [117]. From their ethyl acetate extract, marine-derived isolation of *Streptomyces* sp. produced four novel compounds, Mansouramycin A-D, and the known 3-methyl-7-(methylamino)-5,8-isoquinoline dione. These bioactive compounds showed effective cytotoxicity with high selectivity for non-small breast cancer, cell lung cancer, prostate cancer, and melanoma cells [118], indicating that they might be used as anticancer medicines. Perez and associates [119] discovered the macrodiolide tartrolon D from *Streptomyces* sp. MDG-04-17-069. Colon (HT29), lung (A549), and breast (MDA-MB-231) tumor cell lines were revealed to have high cytotoxic action against the isolated tartrolon [119].

1.4.9 Other Marine Sources of Nutraceuticals

Sponge, coelenterates, algae, echinoderms, tunicates, mollusks, and bryozoans, among other marine flora and fauna, have proven the ability to manufacture a variety of medicinal chemicals, including several unique anticancer agents. In addition, these compounds have been demonstrated to be useful in the treatment of infectious diseases and inflammation [120].

The last three decades have witnessed remarkable breakthroughs in nutraceuticals production by metabolic engineering of microbial-based platforms in response to the rising demand for value-added nutraceuticals to fortify people against ailments. In addition to laboratory-scale preparation, metabolic engineering in microbes allows the large-scale manufacture of complex natural chemicals such as carotenoids from carbon sources. Novel or more complex value-added chemicals might be produced spontaneously in microorganisms thanks to the discovery of biosynthetic pathways and the development of synthetic biology technology.

1.5 Major Challenges and Future Prospects

Microorganisms are known to be ubiquitous, occupying every space on earth from aquatic and terrestrial ecosystems to high atmospheres, including geothermal vents or hot springs. Microorganisms have been involved in various processes from ancient times before mankind. They are also prolific sources of bioactive compounds, including phenolic compounds, amines, organic acids, pigments, growth factors, antibiotics, and alkaloids. The idea that the microorganisms could be employed as elementary chemical reactors to produce various bioactive compounds became widely accepted and the capability to manipulate these microorganisms has led to the invention of several technologies [121]. The application of technical aspects to microbial mediated processes comprises Microbial Technology or Micro-biotechnology or Industrial Microbiology or Industrial Fermentation Technology. The acceleration of industrial fermentation technology resulted in the production of various beneficial products that have dramatic effects on society and the world economy [122].

The primary challenge of microbial technology is the discovery of potential microbial strains from diverse habitats that can produce industrially important products. Taxonomic knowledge is essential to differentiate the microorganism to avoid the rediscovery of known microbes. Interestingly, the approaches available to isolate and select novel microbial strains have been empirical and limited to cultural methods. In essence, it is necessary to develop innovative and specific procedures for the selection of valuable and uncommon microorganisms that aim to improve the quality of the material screened. According to recent evaluations, more than 98% of the microorganisms remain unculturable using conventional methods, which makes the selection process more complicated. Advanced molecular biological approaches are being used to investigate the functionalities of unculturable microorganisms [123]. A continuous detection, identification, and evaluation of microorganisms is an integral part of microbial technology that allows the discovery of industrially valuable microbes.

The capability to regulate microbial metabolism through manipulation of cellular molecules such as DNA and mRNA also determines the optimum yield of the product. A key barrier in achieving this goal is the lack of efficient molecular approaches that provide rigorous information on the microbial cellular process. Advanced computational technologies are required that enable the needed changes in multiple molecular targets simultaneously within the cells. Recent developments in metagenomics, metabolic engineering, and quantitative proteomics, provided the tools such as CRISPR-based methods, have eliminated some of the drawbacks of conventional approaches [124].

The beginning of an industrial operation of microbial product production also meant that optimization of variable parameters has been at most priority ever since their influence on the rate of microbial product formation was understood. Parameter's optimization focuses on achieving the maximum rate of product produced per unit time per substance volume which is crucial to make the process economically feasible since it lowers the cost of downstream processing. Introducing an improved or new microbial strain that is adapted to functional analytics of processes could solve the problem. Even though microorganisms have been employed in industrial processes for several years, an adaptation of microbial strains is still achieved by performing a series of experiments at batch and pilot scales. Guidance from Process Analytical Technology (PAT) composed of spectroscopic studies is essential to reveal the heterogeneity of the process for a better understanding [125]. An increasing number of microbial metabolites have been used over the years. While the industrial significance of these metabolites is well known, new insights and advances continue to be discovered.

1.6 Conclusions

The pharmaceutical industry is a well-established and applied field of biology with a wide range of opportunities and uses. It makes use of microbe's capacities to manufacture a great variety of bioproducts in large quantities for medical, industrial, and agricultural purposes. This chapter gives a brief overview of the various microorganisms that are used in the pharmaceutical industry and the roles they play. In today's world, pharmacies are used to produce large or high-quality quantities of important medical materials, such as recombinant proteins, amino acids, new drugs, antibiotics production or viral vaccines, antivirals and antifungals, food products, or nutraceuticals to improve the screening and development to understanding the biological system. Many possible new medicinal applications are developed in eukaryotic living cells which express higher value products acquired by recombinant-DNA rather than classical methods. Highly developed cell culture technologies and advanced product recovery methods must be devised to fully realize the potential of biotech-based pharmaceutical manufacture. Furthermore, recent advances in high-throughput analysis of DNA and protein expression, protein function, and signal transduction will aid in the elucidation of cellular behavior methods and provide tremendous opportunities for microbe optimization, as well as a new opportunity in the pharmacy and medicine fields. The pharmaceutical industry encompasses many diverse aspects of biotechnology and genetic engineering, including fundamental study in biomolecular interactions and a wide range of applications, such as pharmaceutical manufacture, antibiotics, vaccines, food preservation, and food processing. In the pharmaceutical industry, a variety of microorganisms is used, including naturally occurring organisms, laboratory-selected mutants, and even genetically modified organisms (GMOs). The demand for microorganisms in medicine is increasing. They serve as a link between the research lab and the pharmaceutical industries and food production systems' economic large-scale implementations. However, the industrial usage of microorganisms is strongly ingrained in today's society and allows for the transfer of products that benefit society.

References

1. Kalsoom, M., Fazal, U.R., Talha, S., Sanwal, J., Khalid, N., Adnan, M., Irfan, Z., Abdullah, T.M., Adnan, R.M., Zahra, A., Ali, H., A review: Biological importance of microbes in agriculture, food, and pharmaceutical industry. *Innovare J. Life Sci.*, 8, 1, 2020.
2. Surendar, A., Short communication: Role of microbiology in the pharmaceutical and medical device. *Int. J. Pharm. Res.*, 10, 433, 2018.
3. Arslan, A., Irfan, M., Sanaullah, K., Arif, H., Sarfraz, K., Javed, R., Ali, S., Microbes: Role in industries, medical field, and impact on health. *Saudi J. Med. Pharm. Sci.*, 7, 278, 2021.
4. DiDomenico, B., Novel antifungal drugs. *Curr. Opin. Microbiol.*, 2, 509, 1999.
5. Solomon, L., Tomii, V.P., Dick, A.A.A., Importance of fungi in the petroleum, agro-allied, agriculture and pharmaceutical industries. *N. Y. Sci. J.*, 12, 8, 2019.
6. Steinberg, F.M. and Raso, J., An overview: Biotech pharmaceuticals and biotherapy. *J. Pharm. Pharm. Sci.*, 1, 48, 1998.
7. Borowitzka, M.A., Pharmaceuticals from algae, in: *Biotechnology, Vol. VII*, p. 94, 2009.
8. Cardozo, K.H., Guaratini, T., Barros, M.P., Falcão, V.R., Tonon, A.P., Lopes, N.P., Campos, S., Torres, M.A., Souza, A.O., Colepicolo, P., Pinto, E., Metabolites from algae with economical impact. *Comp. Biochem. Physiol. C Pharmacol. Toxicol. Endocrinol.*, 146, 60, 2007.
9. Meenakshi, B., Pharmaceutically valuable bioactive compounds of algae. *Asian J. Pharm. Clin. Res.*, 9, 43, 2016.
10. Sokullu, E., Soleymani, A.H., Gauthier, M.A., Plant/bacterial virus-based drug discovery, drug delivery, and therapeutics. *Pharmaceuticals*, 11, 211, 2019.
11. Bamogo, P.K.A., Brugidou, C., Sérémé, D., Tiendrébéogo, F., Djigma, F.W., Simpore, J., Lacombe, S., Virus-based pharmaceutical production in plants: An opportunity to reduce health problems in Africa. *Virol. J.*, 16, 167, 2019.
12. Ada, G., Vaccines and vaccination. *N. Engl. J. Med.*, 345, 1042, 2001.
13. Wacker, M. and Casimiro, D.R., Synthesizing vaccines with microbes, in: *Development of Novel Vaccines*, p. 125, 2012.
14. Plotkin, S.A., Vaccines: Past, present and future. *Nat. Med.*, 11, 5, 2005.

15. Huang, C.J., Lin, H., Yang, X., Industrial production of recombinant therapeutics in *Escherichia coli* and its recent advancements. *J. Ind. Microbiol. Biotechnol.*, 39, 383, 2012.
16. Sanchez-Garcia, L., Martín, L., Mangues, R., Ferrer-Miralles, N., Vázquez, E., Villaverde, A., A 2015 update-Recombinant pharmaceuticals from microbial cells. *Microb. Cell Fact.*, 15, 1, 2016.
17. Baeshen, N.A., Baeshen, M.N., Sheikh, A., Bora, R.S., Ahmed, M.M.M., Ramadan, H.A., Saini, K.S., Redwan, E.M., Cell factories for insulin production. *Microb. Cell Fact.*, 13, 1, 2014.
18. Flodh, H., Human growth hormone produced with recombinant DNA technology: Development and production. *Acta Paediatr.*, 75, 1, 1986.
19. Rajeev, L., Antibiotic discovery. *Mater. Methods*, 8, 2671, 2018.
20. Schatz, A., Bugie, E., Waksman, S.A., Hanssen, A.D., Patel, R., Osmon, D.R., The classic: Streptomycin, a substance exhibiting antibiotic activity against Gram-positive and Gram-negative bacteria. *Clin. Orthop. Relat. Res.*, 437, 3, 2005.
21. Waksman, S.A., Antibiotics—20 years later. *Bull. N. Y. Acad. Med.*, 37, 202, 1961.
22. Kapoor, D., Sharma, P., Sharma, M.M.M., Kumari, A., Kumar, R., Microbes in pharmaceutical industry, in: *Microbial Diversity, Interventions and Scope*, pp. 259–299, Springer, Singapore, 2020.
23. Rehman, K., Niaz, S., Tahir, A., Akash, M.S.H., Microorganisms and antibiotic production, in: *Antibiotics and Antimicrobial Resistance Genes in the Environment*, pp. 1–6, Elsevier, Cambridge, MA, 2020.
24. Subramaniyam, R. and Vimala, R., Solid state and submerged fermentation for the production of bioactive substances: A comparative study. *Int. J. Sci. Nat.*, 3, 480, 2012.
25. Hook, D.J., Production of antibiotics by fermentation, in: *Basic Biotechnololody*, p. 433, 2006.
26. Martin, Y.C. and Fischer, E.W., Semisynthetic antibiotics, in: *Medicinal Chemistry*, vol. 19, pp. 77–135, Elsevier, New York, 1983.
27. Bruggink, A. and Roy, P.D., Industrial synthesis of semisynthetic antibiotics, in: *Synthesis of β-Lactam Antibiotics*, pp. 12–54, Springer, Dordrecht, 1983.
28. Najafpour, G., Production of antibiotics, in: *Biochemical Engineering and Biotechnology*, pp. 263–279, Elsevier, Amsterdam, 2007.
29. Etebu, E. and Arikekpar, I., Antibiotics: Classification and mechanisms of action with emphasis on molecular perspectives. *Int. J. Appl. Microbiol. Biotechnol. Res.*, 4, 90, 2016.
30. Rolinson, G.N. and Sutherland, R., Semisynthetic penicillins. *Adv. Pharmacol.*, 11, 151, 1973.
31. Katz, L. and Mankin, A.S., Macrolides, in: *xPharm: The Comprehensive Pharmacology Reference*, pp. 1–4, Elsevier, Boston, 2009.
32. Dinos, G.P., The macrolide antibiotic renaissance. *Br. J. Pharmacol.*, 174, 2967, 2017.

33. Nelson, M.L. and Levy, S.B., The history of the tetracyclines. *Ann. N.Y. Acad. Sci.*, 1241, 17, 2011.
34. Mingeot-Leclercq, M.P., Glupczynski, Y., Tulkens, P.M., Aminoglycosides: Activity and resistance. *Antimicrob. Agents Chemother.*, 43, 727, 1999.
35. Kang, H.K. and Park, Y., Glycopeptide antibiotics: Structure and mechanisms of action. *J. Bacteriol. Virol.*, 45, 67, 2015.
36. Bui, T. and Preuss, C.V., *Cephalosporins*, StatPearls, StatPearls Publishing, Treasure Island, Florida, 2021.
37. Nguyen, R., Khanna, N.R., Safadi, A.O., Yan, S., *Bacitracin Topical*, StatPearls, StatPearls Publishing, Treasure Island, Florida, 2021.
38. Evans, M.E., Feola, D.J., Rapp, R.P., Polymyxin B sulfate and colistin: Old antibiotics for emerging multiresistant gram-negative bacteria. *Ann. Pharmacother.*, 33, 960, 1999.
39. Farzam, K., Nessel, T.A., Quick, J., *Erythromycin*, StatPearls, StatPearls Publishing, Treasure Island, Florida, 2021.
40. Veirup, N. and Kyriakopoulos, C., *Neomycin*, StatPearls, StatPearls Publishing, Treasure Island, Florida, 2021.
41. Chopra, I. and Roberts, M., Tetracycline antibiotics: Mode of action, applications, molecular biology, and epidemiology of bacterial resistance. *Microbiol. Mol. Biol. Rev.*, 65, 232, 2001.
42. Patel, S., Preuss, C.V., Bernice, F., *Vancomycin*, StatPearls, StatPearls Publishing, Treasure Island, Florida, 2021.
43. Chaves, B.J. and Tadi, P., *Gentamicin*, StatPearls, StatPearls Publishing, Treasure Island, Florida, 2021.
44. Floss, H.G. and Yu, T.W., Rifamycin-mode of action, resistance, and biosynthesis. *Chem. Rev.*, 105, 621, 2005.
45. Kausar, S., Said Khan, F., Ishaq Mujeeb Ur Rehman, M., Akram, M., Riaz, M., Rasool, G., Hamid Khan, A., Saleem, I., Shamim, S., Malik, A., A review: Mechanism of action of antiviral drugs. *Int. J. Immunopathol. Pharmacol.*, 1, 35, 2021.
46. Gertrud, U., Virology, Bacteria Make Antivirals, https://www.virology.ws/2021/02/04/bacteria-make-antivirals/, 2021.
47. Mulwa, L.S. and Stadler, M., Antiviral compounds from Myxobacteria. *Microorganisms*, 6, 73, 2018.
48. Yasuhara-Bell, J. and Lu, Y., Marine compounds and their antiviral activities. *Antiviral Res.*, 86, 231–240, 2010.
49. Teng, Y.F., Xu, L., Wei, M.Y., Wang, C.Y., Gu, Y.C., Shao, C.L., Recent progresses in marine microbial-derived antiviral natural products. *Arch. Pharm. Res.*, 43, 1215, 2020.
50. Moghadamtousi, S.Z., Nikzad, S., Kadir, H.A., Abubakar, S., Zandi, K., Potential antiviral agents from marine fungi: An overview. *Mar. Drugs*, 13, 7, 4520, 2015.

51. Ghannoum, M.A. and Rice, L.B., Antifungal agents: Mode of action, mechanisms of resistance, and correlation of these mechanisms with bacterial resistance. *Clin. Microbiol. Rev.*, 12, 501–517, 1999.
52. Chen, S.C.A. and Sorrell, T.C., Antifungal agents. *Med. J. Aust.*, 187, 404, 2007.
53. Wang, H., Yan, Y., Wang, J., Zhang, H., Qi, W., Production and characterization of antifungal compounds produced by *Lactobacillus plantarum* IMAU10014. *PloS One*, 7, e29452, 2012.
54. Leyva Salas, M., Mounier, J., Valence, F., Coton, M., Thierry, A., Coton, E., Antifungal microbial agents for food biopreservation: A review. *Microorganisms*, 5, 37, 2017.
55. Luciana, T., Abreu, P.A., Teixeira, V.L., Paixão, I.C.P., Pereira, R., Leal, B., Lourenço, A.L., Rampelotto, P.H., Castro, H.C., Mycoses and antifungals: Reviewing the basis of a current problem that still is a biotechnological target for marine products. *Front. Mar. Sci.*, 1, 12, 2014.
56. Muharram, M.M., New microbial source of the antifungal allylamine "Terbinafine". *Saudi Pharm. J.*, 25, 440, 2017.
57. Pham, J.V., Yilmam, M.A., Feliz, A., Majid, M.T., Maffetone, N., Walker, J.R., Kim, E., Cho, H.J., Reynolds, J.M., Song, M.C., Park, S.R., Yoon, Y.J., A review of the microbial production of bioactive natural products and biologics. *Front. Microbiol.*, 10, 1404, 2019.
58. Dixon, D.M. and Walsh, T.J., Antifungal agents, in: *Medical Microbiology*, 1996.
59. Risdian, C., Mozef, T., Wink, J., Biosynthesis of polyketides in *Streptomyces*. *Microorganisms*, 7, 124, 2019.
60. Moghadamtousi, S.Z., Nikzad, S., Kadir, H.A., Abubakar, S., Zandi, K., Potential antiviral agents from marine fungi: An overview. *Mar. Drugs*, 13, 4520–4538, 2015.
61. Singh, R., Tiwari, S., Rai, A., Mohapatra, T.M., Cyanobacteria: An emerging source for drug discovery. *J. Antibiot.*, 64, 401–412, 2011.
62. Santos, J.D., Vitorino, I., Reyes, F., Vicente, F., Lage, O.M., From ocean to medicine: Pharmaceutical applications of metabolites from marine bacteria. *Antibiotics*, 9, 455, 2020.
63. Raihan, T., Rabbee, M.F., Roy, P., Choudhury, S., Baek, K.H., Azad, A.K., Microbial metabolites: The emerging hotspot of antiviral compounds as potential candidates to avert viral pandemic alike COVID-19. *Front. Mol. Biosci.*, 8, 732256, 2021.
64. Minor, P.D., Live attenuated vaccines: Historical successes and current challenges. *Virology*, 479–480, 379–92, 2015.
65. Goldblatt, D., Conjugate vaccines. *Clin. Exp. Immunol.*, 119, 1–3, 290, 2000.
66. Iwasaki, A. and Omer, S.B., Why and how vaccines work. *Cell*, 183, 2020.
67. Badgett, M.R., Auer, A., Carmichael, L.E., Parrish, C.R., Bull, J.J., Evolutionary dynamics of viral attenuation. *J. Virol.*, 76, 20, 10524–10529, 2002.

68. Pulendran, B. and Ahmed, R., Immunological mechanisms of vaccination. *Nat. Immunol.*, 12, 6, 509, 552, 2011.
69. Baxter, D., Active and passive immunity, vaccine types, excipients and licensing. *Occup. Med.*, 57, 2007.
70. Herzog, C., Influence of parenteral administration routes and additional factors on vaccine safety and immunogenicity: A review of recent literature. *Expert Rev. Vaccines*, 13, 3, 399–415, 2014.
71. Gasparini, R., Amicizia, D., Lai, P.L., Panatto, D., Live attenuated influenza vaccine–A review. *J. Prev. Med. Hyg.*, 52, 3, 95–101, 2011.
72. Morrow, W.J.W., Sheikh, N.A., Schmidt, C.S., Davies, D.H., *Vaccinology: Principles and Practice*, John Wiley & Sons, Hoboken, 2012.
73. Levine, M.M., Chen, W.H., Kaper, J.B., Lock, M., Danzig, L., Gurwith, M., PaxVax CVD 103-HgR single-dose live oral cholera vaccine. *Expert Rev. Vaccines*, 16, 3, 197–213, Mar 2017.
74. Okafor, C.N., Rewane, A., Momodu, I. I., *Bacillus Calmette Guerin*, StatPearls, Treasure Island, Florida, 2021.
75. Amicizia, D., Arata, L., Zangrillo, F., Panatto, D., Gasparini, R., Overview of the impact of Typhoid and Paratyphoid fever. Utility of Ty21a vaccine (Vivotif). *J. Prev. Med. Hyg.*, 58, 1, E1–E8, 2017.
76. Demicheli, V., Rivetti, A., Debalini, M.G., Pietrantonj, C.D., Vaccines for measles, mumps and rubella in children. *Evid. Based Child. Helth A Cochrane Rev. J.*, 2, CD004407, 2013.
77. Stratton, K., Ford, A., Rusch, E., Clayton, E.W., Adverse effects of vaccines: Evidence and causality: Measles, mumps, and rubella vaccine. *Natl. Acad.*, 1, 1, 2011.
78. Kota, V. and Grella, M.J., *Varicella (Chickenpox) Vaccine*, StatPearls, Treasure Island, Florida, 2021.
79. Monath, T.P., Yellow fever vaccine. *Expert Rev. Vaccines*, 4, 4, 553–74, 2005.
80. Sridhar, S., Brokstad, K.A., Cox, R.J., Influenza vaccination strategies: Comparing inactivated and live attenuated influenza vaccines. *Vaccines*, 3, 373–389, 2015.
81. Bandyopadhyay, A.S., Garon, J., Seib, K., Orenstein, W.A., Polio vaccination: Past, present and future. *Future Microbiol.*, 10, 5, 791–808, 2015.
82. Folegatti, P.M., Ewer, K.J., Aley, P.K., Angus, B., Becker, S., Belij-Rammerstorfer, S., Bellamy, D., Bibi, S., Bittaye, M., Clutterbuck, E.A., Dold, C., Faust, S.N., Finn, A., Flaxman, A.L., Hallis, B., Heath, P., Jenkin, D., Lazarus, R., Makinson, R., Minassian, A.M., Safety and immunogenicity of the ChAdOx1 nCoV-19 vaccine against SARS-CoV-2: A preliminary report of a phase 1/2, single-blind, randomised controlled trial. *Lancet*, 396, 467, 2020.
83. Sanders, B., Koldijk, M., Schuitemaker, H., *Vaccine Analysis: Strategies, Principles, and Control*, Springer, Berlin, Heidelberg, pp. 45–80, 2015.
84. Kapil, P. and Merkel, T.J., Pertussis vaccines and protective immunity. *Curr. Opin. Immunol.*, 59, 72–78, 2019.

85. Ogden, S.A., Ludlow, J.T., Alsayouri, K., *Diphtheria Tetanus Pertussis (DTaP) Vaccine*, StatPearls, Treasure Island, Florida, 2021.
86. Kwon, S.Y. and Lee, C.H., Epidemiology and prevention of hepatitis B virus infection. *Korean J. Hepatol.*, 17, 2, 87–95, 2011.
87. Sapkal, G.N., Yadav, P.D., Ella, R., Deshpande, G.R., Sahay, R.R., Gupta, N., Vadrevu, K.M., Abraham, P., Panda, S., Balram, B., Inactivated COVID-19 vaccine BBV152/COVAXIN effectively neutralizes recently emerged B.1.1.7 variant of SARS-CoV-2. *J. Travel Med.*, 28, taab051, 2021.
88. Dewan, K.K., Linz, B., DeRocco, S.E., Harvill, E.T., Acellular pertussis vaccine components: Today and tomorrow. *Vaccines (Basel)*, 8, 2, 217, 2020.
89. Sociedade Brasileira de Pediatria, Vacina contra hepatite B [Hepatitis B vaccine]. *Rev. Assoc. Med. Bras.*, 52, 5, 288–9, 2006.
90. Verez-Bencomov, V., Fernández-Santanaeugenio, V., Hardy, E. *et al.*, A synthetic conjugate polysaccharide vaccine against *haemophilus influenzae* type b. *Science*, 305, 2004.
91. Ewer, K.J., Lambe, T., Rollier, C.S., Spencer, A.J., Hill, A.V., Dorrell, L., Viral vectors as vaccine platforms: From immunogenicity to impact. *Curr. Opin. Immunol.*, 41, 522, 47–54, 2016.
92. Turnbaugh, P.J., Ley, R.E., Hamady, M., FraserLiggett, C.M., Knight, R., Gordon, J., II, The human microbiome project. *Nature*, 449, 7164, 804–810, 2007.
93. Arumugam, M., Raes, J., Pelletier, E., Le Paslier, D., Yamada, T., Mende, D.R., Bork, P., Enterotypes of the human gut microbiome. *Nature*, 473, 7346, 174–180, 2011.
94. Morgan, X.C. and Huttenhower, C., Human microbiome analysis. *PloS Comput. Biol.*, 8, 12, e1002808, 2012.
95. Ursell, L.K., Metcalf, J.L., Parfrey, L.W., Knight, R., Defining the human microbiome. *Nutr. Rev.*, 70, suppl_1, S38–S4, 2012.
96. Gilbert, J.A., Blaser, M.J., Caporaso, J.G., Jansson, J.K., Lynch, S.V., Knight, R., Current understanding of the human microbiome. *Nat. Med.*, 24, 4, 392–400, 2018.
97. Behera, S.K. and Mulaba-Bafubiandi, A.F., Microbes assisted mineral flotation a future prospective for mineral processing industries: A review. *Miner. Process. Extr. M.*, 38, 2, 96–105, 2017.
98. Chapman, A.S., Stévant, P., Larssen, W.E., Stévant, P., Larssen, W.E., Food or fad? Challenges and opportunities for including seaweeds in a Nordic diet. *Botanica Marina*, 58, 423–433, 2015.
99. Cornish, M., Critchley, A., Mouritsen, O., Consumption of seaweeds and the human brain. *J. Appl. Phycol.*, 29, 2377–2398, 2017.
100. Tseng, C.K., Algal biotechnology industries and research activities in China. *J. Appl. Phycol.*, 13, 375, 2001.
101. Grobbelaar, J.U., Algal biotechnology: Real opportunities for Africa. *S. Afr. J. Bot.*, 70, 140–144, 2004.

102. Volkman, J.K., Sterols in microorganisms. *Appl. Microbiol. Biotechnol.*, 60, 495–506, 2003.
103. Inagaki, F., Nunoura, T., Nakagawa, S., Teske, A., Lever, M., Lauer, A. *et al.*, Biogeographical distribution and diversity of microbes in methane hydrate-bearing deep marine sediments on the Pacific Ocean margin. *Proc. Natl. Acad. Sci. U. S. A.*, 103, 2815–2820, 2006.
104. Bhadury, P., Mohammad, B.T., Wright, P.C., The current status of natural products from marine fungi and their potential as anti-infective agents. *J. Ind. Microbiol. Biotechnol.*, 33, 325–337, 2006.
105. Newman, D.J. and Hill, R.T., New drugs from marine microbes: The tide is turning. *J. Ind. Microbiol. Biotechnol.*, 33, 539–544, 2006.
106. Kjer, J., Debbab, A., Aly, A.H., Proksch, P., Methods for isolation of marine-derived endophytic fungi and their bioactive secondary products. *Nat. Protoc.*, 5, 479–490, 2010.
107. Du, L., Zhu, T., Fang, Y., Liu, H., Gu, Q., Zhu, W., Aspergiolide, A., A novel anthraquinone derivative with naphtho(1,2,3-de)chromene-2,7-dione skeleton isolated from a marine-derived fungus *Aspergillus glaucus*. *Tetrahedron*, 63, 1085–1088, 2007.
108. Du, L., Feng, T., Zhao, B., Li, D., Cai, S., Zhu, T. *et al.*, Alkaloids from a deep ocean sediment-derived fungus *Penicillium* sp. and their antitumor activities. *J. Antibiot.*, 63, 165–170, 2010.
109. Kwong, T.F., Miao, L., Li, X., Qian, P.Y., Novel antifouling and antimicrobial compound from a marine-derived fungus *Ampelomyces* sp. *Mar. Biotechnol.*, 8, 634–640, 2006.
110. Xiong, H., Qi, S., Xu, Y., Li, M., Qian, P.-Y., Antibiotic and antifouling compound production by the marine derived fungus *Cladosporium* sp. F14. *J. Hydro-Environ. Res.*, 2, 264–270, 2009.
111. Li, X., Kim, S.K., Nam, K.W., Kang, J.S., Choi, H.D., Son, B.W., A new antibacterial dioxopiperazine alkaloid related to gliotoxin from a marine isolate of the fungus Pseudallescheria. *J. Antibiot. (Tokyo)*, 59, 248–250, 2006.
112. Takahashi, Y. and Omura, S., Isolation of new actinomycete strains for the screening of new bioactive compounds. *J. Gen. Appl. Microbiol.*, 49, 141, 2003.
113. Olano, C., Méndez, C., Salas, J.A., Antitumor compounds from marine actinomycetes. *Mar. Drugs*, 7, 210–248, 2009.
114. Prudhomme, J., McDaniel, E., Ponts, N., Bertani, S., Fenical, W., Jensen, P. *et al.*, Marine actinomycetes: A new source of compounds against the human malaria parasite. *PloS One*, 3, e2335, 2008.
115. El-Gendy, M.M., Hawas, U.W., Jaspars, M., Novel bioactive metabolites from a marine derived bacterium Nocardia sp. ALAA 2000. *J. Antibiot. (Tokyo)*, 61, 379–386, 2008.
116. McArthur, K.A., Mitchell, S.S., Tsueng, G., Rheingold, A., White, D.J., Grodberg, J. *et al.*, Lynamicins A-E, chlorinated bisindole pyrrole antibiotics from a novel marine actinomycete. *J. Nat. Prod.*, 71, 10, 1732–7, 2008.

117. Cho, J.Y., Williams, P.G., Kwon, H.C., Jensen, P.R., Fenical, W., Lucentamycins A–D, cytotoxic peptides from the marine-derived actinomycete *Nocardiopsislucentensis. J. Nat. Prod.*, 70, 1321–1328, 2007.
118. Hawas, U.W., Shaaban, M., Shaaban, K.A., Speitling, M., Maier, A., Kelter, G. *et al.*, Cytotoxic isoquinoline quinones from a marine streptomycete. *J. Nat. Prod.*, 72, 2120–2124, 2009.
119. Pérez, M., Crespo, C., Schleissner, C., Rodríguez, P., Zúñiga, P., Reyes, F., Tartrolon, D., A cytotoxic macrodiolide from the marine-derived actinomycete *Streptomyces* sp. MDG-04-17-069. *J. Nat. Prod.*, 72, 2192–2194, 2009.
120. Blunt, J.W., Copp, B.R., Munro, M.H.G., Northcote, P.T., Prinsep, M.R., Marine natural products. *Nat. Prod. Rep.*, 21, 1–49, 2004.
121. Arnold, L.D. and Jose, L.A., Contributions of microorganisms to industrial biology. *Mol. Biotechnol.*, 38, 41, 2008.
122. Bhanu, P., Prem, P.S., Akshay, K., Somenath, D., Anand, K.C., Microbes as a novel source of secondary metabolite products of industrial significance, in: *Role of Plant Growth Promoting Microorganisms in Sustainable Agriculture and Nanotechnology*, K. Ajay, K.S. Amit, K.C. Krishna, (Eds.), pp. 21–37, Elsevier, Switzerland, 2019.
123. Cheptsov, V.S., Tsypina, S.I., Minaev, N.V., Yusupov, V.I., Chichkov, B.N., New microorganism isolation techniques with emphasis on laser printing. *Int. J. Bioprinting*, 5, 165, 2019.
124. Eric, A. and William, J.H., Grand challenges in microbiotechnology: Through the prism of microbiotechnology. *Front. Microbiol.*, 11, 430, 2020.
125. Luca, R.F., Anders, N., Andrijana, B., Daniela, Q.H., Timo, H., Anna-Lena, H., Hilde, L., Lisa, M., Miguel, M., Ulrich, K., Krist, V.G., Challenges in industrial fermentation technology research. *Biotechnol. J.*, 9, 1, 2014.

2

Emerging Nutraceutical Prospective of Microbes and their Therapeutic Aspects for Lifestyle Diseases

Richa Saxena[1*], Sana Ansari[1], Richa Sharma[2], Keshwanand Tripathi[1], Pankaj Kumar Rai[1], Naima Mirza[1], Mohd. Tahir[1], Anshdha Johari[1], Hemanreet Kaur[1], Aman Noor[1] and Ashal Ilyas[1]

[1]*Department of Biotechnology, Invertis University, Bareilly, U.P., India*
[2]*Department of Microbiology, Mahatma Gandhi University of Medical Science & Technology, Jaipur, Rajasthan, India*

Abstract

Humans have developed such food habits that restrict the intake of adequate beneficial nutrients. This nutrient deficiency resulted in diabetes, hypertension, and other lifestyle diseases. Researchers developed nutrients that act as medicines to supplement the deficient nutrients and are termed "nutraceuticals." Nutrition and pharmaceutical amalgamate to form a new hybrid word, "Nutraceutical." Nutraceuticals are considered food or part of food that provides medicinal or health benefits and their primary nutritional value. The current health trends and increasing population have been key reasons behind the rising sector of the nutraceutical market globally. The sayings of Hippocrates over 2500 years ago that "Let food be the medicine and medicine be the food," has become most popular in the field of nutraceuticals. This chapter focuses on the health benefits of nutraceuticals and their utilization for treating diseases occurring in humans. In addition to that, it covers challenges and opportunities that promote the production and regulation of nutraceuticals.

***Keywords*:** Nutraceuticals, prebiotics, probiotics, hypertension, diabetes

**Corresponding author*: saxena.richa50@gmail.com; richa.s@invertis.org

Sanjay Kumar, Narendra Kumar and Shahid-ul-Islam (eds.) Role of Microbes in Industrial Products and Processes, (35–58) © 2023 Scrivener Publishing LLC

2.1 Introduction

The term *Nutraceutical* has been coined by combining "nutrition" and "pharmaceutical." Stephen Defelice, in 1989, first fabricated the label 'Nutraceutical' and defined it as "any substance that is a food or a part of the food and provides medical or health benefits, including the avoidance and treatment of ailment" [1]. In general, they have a crucial part in dynamic and sustaining everyday physiological activity in humans. Nutraceuticals offer an advantage over pharmaceuticals because they have fewer side effects, are natural dietary supplements, and so forth. With the modern lifestyle, people have changed their eating habits too. Nowadays, the new generation population has been more prone to fast foods. These junk foods have low nutritional value and, in return, cause several diseases, such as cardiac arrest, which is the most common disease that ultimately leads to death in developing countries, followed by several other diseases like cancer, arthritis, and many more. Nutraceuticals are a natural product that blurs the barrier to interlink food and medicine [2]. Due to overcrowding and an unbalanced ecosystem, food has played a significant noteworthy role in the healthiness of the gut leading to several gut diseases like Crohn's disease, constipation, and many more. Many of the latest food components are included in a person's diet, like "nutraceuticals." These components have positively affected the gut's microbiota and help improve human health. Fungi, bacteria, and yeast play a vital role as producers and catalysts [3]. As there is a trend of using microbes to make probiotics and prebiotics, ingredients like microbial flavors are used to make healthy food. It has gained popularity recently because it has no harmful effect on human health [4]. For the manufacturing of vital nutraceuticals, lactic acid bacteria, namely *Lactococcus lactis* has become an ideal cell factory through manipulation or alteration in the gene of the lactic acid bacteria. The nutraceuticals developed by these bacteria will be more enhanced and valuable than the previous one as they can be easily achieved through genetic engineering [5]. Nutraceuticals have therapeutic properties and treat arthritic, cancer, constipation, diabetes, diarrhea, and several other diseases [6]. Recently, the market of nutraceuticals has been expanding rapidly because people have analyzed the potential of nutraceuticals as per research and scientific approaches. Nutraceuticals act as a food component and have a positive and significant health impact. Nutraceuticals are trustworthy for curing and preventing disease compared to pharmaceutical compounds, subject to government regulations [7].

2.2 Different Types of Nutraceuticals

Nutraceuticals can be divided into two categories which are potential nutraceuticals and established nutraceuticals. Potential nutraceuticals are those food components that provide health and medical benefits that do not have sufficient clinical data. In contrast, established nutraceuticals are the ones that have sufficient clinical data to show some significant health benefits. Several nutraceuticals are categorized as potential nutraceuticals until they are not actively utilized for commercial or clinical purposes with recorded clinical data [6]. Nutraceuticals are structured in various customs to make them easier to understand and apply, such as clinical testing, efficient food manufacture, or dietetic guidance. Nutraceuticals may also be classified in several different methods, including food source, method of exploitation, composition, and whatever the food substances are used as natural nutraceuticals. Some of them are discussed below (Figure 2.1) [8, 9].

2.2.1 Probiotics

A probiotic contains plenty of living good microbes; when provided in sufficient proportions, they positively impact the host by enhancing its

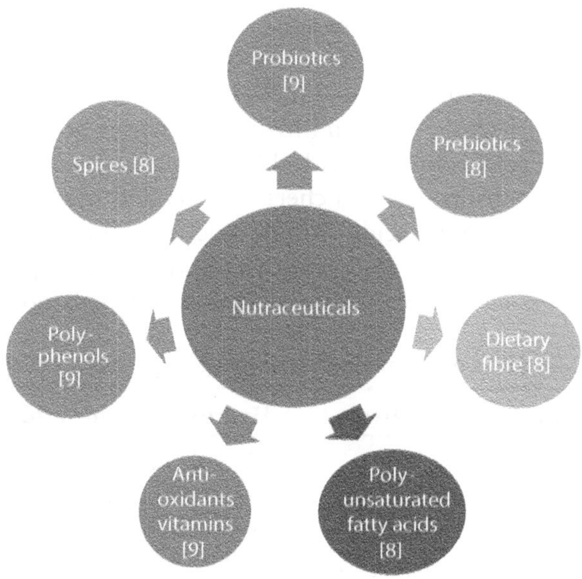

Figure 2.1 Representation of different roles of nutraceuticals [8, 9].

gut microbial stability. Generally, probiotics include bacteria like Gram-positive cocci such as *Lactococcus lactis*, Bifidobacteria such as *B. bifidun, and* Lactobacilli such as *L. acidophilus*. Probiotics can be taken in any form like gel, powder, or capsules [10]. Consumption of probiotics reduces the threat of several health-related issues such as allergy, asthma, problems related to the intestine, and various other such health issues [11].

2.2.2 Prebiotics

Prebiotics are related to dietary substances that significantly affect the host's gut microbiota by changing its environment [12]. Consuming prebiotics regularly helps increase the growth of lactibacillous and bifidobacterial in the gut so that the metabolism of the gut can function effectively [13].

2.2.3 Polyphenols

Polyphenols are natural phytochemicals that are mainly categorized as secondary metabolites. They exist in plant-based food items such as vegetables, fruits, cereals, tea, coffee, and other plant components. Polyphenols protect the plant from abiotic and biotic stress like pathogens, oxidants, and radiations [14]. Polyphenols can be categorized based on the numerical digit of phenol rings present and the element that binds these rings. Eight thousand polyphenols have been discovered, like flavonoids, flavones, and many more [15]. Phenolic acids are one-third of the polyphenols that are present in our diet and come under two categories: The derivatives of hydroxybenzoic acid include (gallic acid, parahydroxybenzoic acid, protocatechuic acid) and hydroxycinnamic acid derivative this includes (ferulic acid, coumaric acid, chlorogenic acid, sinapic acid, caffeic acid). Some other fruits, such as cherry, apple, and many more, contain high amounts of phenolic acids. To one side from this, polyphenols also perform as an anti-inflammatory, antioxidant they also help in the inhibition of diabetes mellitus [16]. The classification of polyphenols on the basis

Table 2.1 Classification of polyphenols into two groups with some examples [16].

Hydroxybenzoic acid derivatives	Hydroxycinnamic acid derivatives
Gallic acid, Parahydroxybenzoic acid, Protocatechuic acid [16].	Chlorogenic acid, Coumaric acid, Ferulic acid, Sinapic acid, Caffeic acid [16].

of hydrobenzoic acid derivatives and hydroxycinnamic acid derivatives with some examples has been mentioned in Table 2.1 [16].

2.2.4 Spices

Spices are fragment components in vegetables that serve as a spice rather than providing nutrition to the food. They can be in powdered or in pulverized form. They provide different flavors and pungency to the food. The pungency comes from the oil spices and is mainly responsible for aroma and taste [17]. Spices are mainly used in pesticides, pharmaceuticals, nutraceuticals, beverages, dental, and cosmetics. Spices play a prominent role in the country's economy [18]. These qualities are attributable to the spices and can create a wide range of compounds, for instance, turmeric, black pepper, ginger, red pepper, and clove. They have been confirmed to use its activity against neurodegenerative diseases [19, 20]. Chawanprash is one of India's leading marketing nutraceuticals, consisting of spices like ginger, clove, turmeric, etc. [21].

2.2.5 Polyunsaturated Fatty Acids

PUFAs are termed essential fatty acids; subsequently, they cannot be linked to the human figure, and the integrity of life is essential. Therefore, should be found in food. Linoleic acid (LA) is in the n-6 family, alpha-linolenic acid (LNA) belongs to the n-3 family. These parents are not essential compounds that can be transformed into the human figure for a long-time chain (LC) fatty acid, but people could not convert n-3 and n-6fatty acids [22]. High omega-3-fatty acids are eicosapentaenoic acid (EPA), Docosahexaenoic acid (DHA), alpha-linolenic acid (ALA) [23, 24]. ALA is a predecessor to the EPA and DHA. The essential origins of ALA are mainly soybeans, red/black currant seeds, and other nuts (e.g., walnuts) [25–29].

2.2.6 Antioxidant Vitamins

These vitamins work in a simultaneous and synergistic manner to prevent oxidative reactions that lead to several degradation diseases, including cancer, cardiovascular disease, cataracts, etc. [30, 31]. Tocotrienols travel more extensively within biological membranes than tocopherol due to presence of incompletechain and enters the tissue with layers of saturated fat, i.e., in the brain and liver. They have more digestive ability and are better liver antioxidants [32]. Vitamin E and Selenium have a synergistic function in contrast to lipid per-oxidation. Vitamin C, better known as ascorbic acid, provides hydrogen to the atom in lipid radicals, removes the singlet oxygen radical, and releases the oxygen of the cells [33].

Cell damage caused by free radicals is believed to play a vital role in the aging process and disease progression. Antioxidants are first line to protect from the damage of free radicals, and it is crucial to maintain high health and well-being. Oxygen is the most active atom capable of being part of potentially destructive molecules that are frequently referred to as "free radicals".

Free radicals can attack healthy immune cells, causing them to lose their structure and function. Antioxidants are known to stabilize, or shut down, free radicals before they attack cells. Antioxidants are vital for maintaining good cellular health and system and wellness [34].

2.2.7 Dietary Fibers

Dietary fiber is a food commodity, especially plant non-hydrolyzed substances by enzymes produced by digestion but digested by the intestinal microflora. Dietary fiber usually includes non-starch polysaccharides (NSP), such as cellulose, gums, hemicellulose, pectin, lignin, drug-resistant dextrins, and starch-resistant starch. Food loaded with soluble fiber includes oats, fruits, barley, and beans. The chemical dietary fiber is carbohydrate which has a polymerization rate of not less than 3, and are not digested or absorbed in small intestines. Based on their solubility in water, food fibers may segregate into two types:

1. Insoluble dietary fiber (IDF), which includes cellulose, some hemicellulose, and lignin, and are extracted from its limited expansion of the colon.
2. Soluble dietary fiber (SDF), which includes β-glucans, pectins, gums, mucilages, and hemicellulose, and are digested in the gut [35].

Fiber also promotes the growth of Intestinal bifidobacteria (especially fructosoligosacchar-ides). People who consume large amount of food fiber have lesser risk of CHR as compared to the low food fiber consumer [36], stroke [37], hypertension [38], diabetes [39], obesity [40], and definite gastrointestinal disorders [41]. Again, enlarging in the ingestion of high fiber food improves serum lipoprotein values [42] and decreases blood stress level.

2.3 Improving the Yields of Nutraceuticals Using Microbes

Improving the yields of nutraceuticals using microbes can be done via use of enzyme technology, and generation bio-active compounds obtained from seaweed for use as prebiotics and nutraceuticals.

2.3.1 Potential of Seaweed Bioactive Compounds as Functional Foods and Nutraceuticals

The bulk of seaweed biomass is made up of carbohydrates; many investigations looking at seaweed-derived chemicals have so focused on polysaccharides and oligosaccharides. Apart from it, phenolic chemicals and proteins found in seaweeds have also been studied. As potential functional components, they have been extensively researched. This review will primarily focus on three classes.

2.3.1.1 Polysaccharides

Polysaccharides are the most important products produced from seaweeds in terms of economy [43]. Polysaccharides are the most abundant components in seaweeds as 76% of the dry weight (DW) [44]. Seaweeds have a lot of nutrients and a high overall dietary fiber content [45]. The majority of these polysaccharides are fermentable and are found in gut microbiota, which may give people a health advantage [46]. Sulphated polysaccharides have also been demonstrated to have anti-inflammatory, antioxidant, antibacterial, and immunological properties [47].

2.3.1.2 Phenolic Compounds

Phlorotannins are extremely hydrophilic components produced by polymerization of phloroglucinol (1,3,5-trihydroxybenzene) monomer units with molecular weights ranging from 126 to 650 kDa: fucols (phenyl linkage), fucophloroethols (ether and phenyl linkage), fucols (phenyl linkage), and eckols (dibenzodioxin linkage) [48]. The use of phlorotannins as a functional food additive has been investigated; antioxidant, anti-inflammatory, and other biological activities anti-diabetic, anti-cancer, anti-hypertensive, and anti-allergic properties [49].

2.3.1.3 Proteins

Seaweed bioactive proteins and peptides have been shown to have antioxidant, anti-hypertensive, and anticoagulant properties. In general, green and red seaweeds (10–47 percent of DW) have higher protein content than brown seaweeds (3–15 percent of DW). Lectin and other bioactive proteins found in red and green seaweeds are essential. Furthermore, most seaweed species are a rich source of both necessary and acidic nutrients a mixture of amino acids [49].

2.4 Nutraceuticals as an Alternative for Pharmaceuticals

These nutraceuticals as well as the isolated nutrients, advantageous supplement (vitamins, minerals, amino acids, and sauces), precise diets to hereditarily finagled developer foods, and phytochemicals (flavonoids, polyphenols, phytosterols, tannins) promise to include beneficial healthy goods, in addition the usual nourishment values plant within food. There is an extensive involvement among the salutary custom and helps to reduce chances of disease similar as cataract, cardiovascular conditions, diabetes, gallstones, seditious conditions, neurodegenerative conditions (Alzheimer's/Huntington's complaint/ Parkinson's), rotundity, and frequent forms of cancers [50–52].

2.5 Therapeutic Aspects of Nutraceuticals for Lifestyle Diseases

Nutraceuticals plays an essential role in treating and inhibiting various diseases like diabetes, cancer, hypertension, obesity. These nutraceutical helps in decreasing the chances of various diseases in humans (Figure 2.2) [53]. In maintaining and elevating the health and well-being, diet and lifestyle is very important. It also averts the beginning of diseases arise because of wrong lifestyle and dietary habits. Mostly these diseases are cured through pharmacological therapy, but some can be restricted by using nutraceuticals by including them in our daily diet [54]. To decrease the occurrence

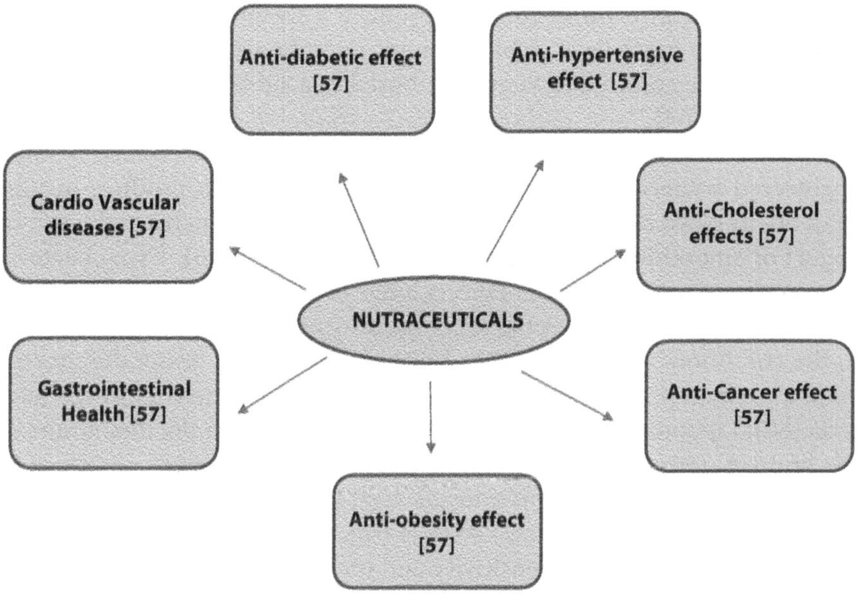

Figure 2.2 Nutraceuticals effect for the cure of several diseases [57].

of diseases, especially the health conditions that arise due to unsuitable lifestyle and wrong dietary, these nutraceuticals should be appropriately used [55]. To better effect these food-drugs, it should be included in our daily diet as a proactive mediator. A successful proactive medicine is the primary approach for prevention, which prevents and decreases the risk linked to particular lifestyle causing diseases [56]. Their inclusion should be beyond diet, but prior to use of medication for treating the symptom of arising diseases [57].

2.5.1 Anti-Diabetic Effect

Diabetes is a condition caused by disturbances in energy metabolism and defined by unbalanced glucose homeostasis. Insulin secretion is hampered, and resistance to insulin is developed [58, 59]. Insufficient insulin in the body to break down the sugar leads to diabetes, which is now very common and sometimes can be deadly. For managing glucose levels many drugs are in clinical use. But most of them have shown side effects like hypoglycemia and weight loss [60]. In the manufacture of insulin, biotechnology shows a significant role. A massive amount of human insulin is produced by genetically modified bacteria [61]. A famous brand of insulin, "Humulin," is also produced by genetically modifying the bacteria and used almost in the entire world [62]. To produce a specific protein that helps to regulate the sugar level in blood, a gut microbe was engineered. These bacteria can be cultured in yogurt and used as a substituent treatment for diabetics [63, 64]. Non-pathogenic *Escherichia coli* bacteria are used to establish a strain that releases the peptide "Glucagon-like peptide 1". This engineered bacterial cell could activate the pancreas to produce insulin [65]. So, in place of using insulin, diabetic people can increase the use of yogurt or smoothie to control their blood sugar level [66]. Using bacteria that produce protein in place of direct usage of protein is more beneficial as bacteria can create precisely the exact quantity of protein needed according to the condition, which cannot be managed by directly using the protein and has to be self-monitored. Hence it is better to use the engineered bacteria [67]. Lipopolysaccharide (LPS), which is present in the membrane of the gram-negative bacteria, can stimulate the inflammatory response in insulin resistance. By attaching to the CD14 toll-like receptor-4 (TLR-4) complex on innate leukocytes, LPS may trigger inflammation.

CD14 regulates insulin sensitivity and metabolic disorders in this context [68]. *Lactobacillus* is a genus of probiotic bacteria found to improve glucose metabolism in humans. These findings may look strange, given that several Lactobacillus species have been linked to an increase in type 2 diabetes patients. However, it has never been proven that a rise in Lactobacillus bacteria in particular type 2 diabetic patients directly impacts the condition [69]. *Akkermansia muciniphila* seems to be of specific importance among the bacteria that could potentially be utilized to treat type 2 diabetes [70]. In recent years, numerous nutraceuticals and clinically used dietary supplements have been verified and proven by science to improve type 2 diabetes and its problems by regulating pathological processes [71]. These compounds comprise omega-3 fatty acids,α-lipoic acid, berberine, phytoestrogens (Isoflavones), antioxidant vitamins, dietary fiber, i-flavonoids, minerals (chromium and magnesium) and conjugated linoleic acid [72], green tea, cinnamon tea, and ethyl esters of n-3 fatty acids can be helpful in treating diabetes issues [73] utilization of Omega-3 fatty acids, and soy isoflavones have been shown to reduce glucose forbearance to sugar, in addition, is linked with lesser and higher mortality rates for certain cancers, heart disease, and osteoporosis [74].

2.5.2 Anti-Hypertensive Effect

The most frequent, costly, and treatable CVD (cardiovascular disease) risk factor is hypertension [75]. As per the Joint National Committee Seven (JNC 7) blood pressure (BP) standards, approximately 70% of Americans suffer from excessive blood pressure (BP) [76]. Using JNC 7 blood pressure standards, the lifespan probability of developing high blood pressure is 90%, and one out of five individuals with pre-hypertension will suffer hypertension before four years [77]. Thekallikrein-kinin system (KKS) and the renin-angiotensin system (RAS) are essential for controlling salt homeostasis and blood pressure. Several bioactive peptides have been proven to have anti-hypertensive properties [78]. According to the investigation of Koyama and coworkers, bioactive substances from lactic-fermented buckwheat sprouts affect the blood pressure [79]. Six new bioactive peptides (VAE, FQ, VVG, FDART, WTFR, and DVWY) were discovered. Also, a solitary injection of these peptides injected to rats causes a considerable reduction in blood pressure [80].

2.5.3 Anti-Cholesterol Effects

A number of variables, including high cholesterol levels, contribute to coronary heart disease. Patients with hypercholesterolemia, or chronically raised blood cholesterol levels, are at a higher risk for cardiovascular disease, as well as coronary-related suffering and death [81]. Certain strains of lactic acid bacteria (LAB) were demonstrated in animal tests to reduce cholesterol levels in the blood, probably by disintegrating bile in the stomach and blocking their re-absorption [82]. Lower cholesterol levels are beneficial, as indicated by the 20–50% reduction in atherosclerosis in hypercholesterolemic patients who sustain a 10% cholesterol reduction [83]. The cholesterol levels of babies fed Lactobacillus acidophilus-fortified milk were shown to be lower, while the number of probiotics in their feces increased. In most of these investigations, a fall in blood cholesterol was only detected in humans and only when substantial doses of fermented dairy products were consumed. Other researchers that used "normal" dosages of the fermented milk product could not replicate the findings.

2.5.4 Anti-Cancer Effect

Cancer is one of the well-being difficulties that arise worldwide, with a continuous increase in cases and deaths. The unregulated development of aberrant cells that invade the tissues around them is termed cancer. The World Cancer Report says that the number of cancer patients is increasing drastically, and in a few years, 150 million young people will be diagnosed with the disease, an increase of 50%. To fulfill their biological demands, fast-growing cancer cells go through a number of changes in their lipid, fat, carbohydrate, and protein metabolism. According to information from different well-managed clinical trials, a healthy lifestyle that includes a diet rich in nutraceuticals/antioxidants can help prevent cancer and improve overall health. Nutraceuticals are having antioxidant properties were found to reduce DNA oxidative damage as well as modify numerous intracellular expression pathways such as COX-2, PI3K/Akt, AMPK-mTOR, NF-κB, and EGFR to reduce cancer risk and consequently lead to profitable profit results [84].

Several phytochemicals are antioxidants in the environment, including tetraterpenoids, flavonoids obtained from a fruit rich in citric acid, compounds containing multiple phenolic functionalities such as epigallocatechin-3-gallate, Vitamin E, *Curcuma longa*, Basidiomycetes Polysaccharide, and provitamin A exhibits chemopreventive properties by acting as antioxidants [85]. Foods that are rich in Vitamin C, lycopene (*Carica papaya*, tomato, Citrus × paradisi, *Citrullus lanatus*, *Psidium guajava*), glutathione, cysteine, Vitamin E, selenium, ellagic acid (pecans, Juglans, *Vaccinium spp*, *Punica granatum*), and so seemed to promote the anti-oxidative power. Though, more comprehensive knowledge is required to be estimated their therapeutic properties in cancer diagnosis and cure [86].

Uninhibited cell multiplication is the known property of cancer's growth and progression. Consequently, the great majority of anti-cancer medications are designed to target cells with a high rate of proliferation and regeneration [87]. Numerous bioactive peptides derived from a variety of resources have been found to have anti-cancer properties [88]. Probiotic microbes present in fermented food products have been shown to have beneficial impacts on mucosal damage, including reducing the impacts of cancer on the gastrointestinal tract [89]. Thorough research has revealed that the link between bowel cancer and probiotics appears to be primarily reliant on probiotic bacteria's bioactive substances that help in metabolism, resulting in the production of curative anti-carcinogenic components [90]. On adenocarcinoma cell lines, various probiotic microorganisms and their derivatives show a growth inhibition effect [91]. *Lactobacillus spp.* and *Bifidobacterium* divisions with high quantities of bacterial carbohydrates (polysaccharides, extracellular glycoproteins, and peptidoglycan) were discovered to have potent tumor-suppressing properties [92]. Nagaoka *et al.* reported that *B. breve* and *B. bifidum* constitute probiotic polysaccharides which have been studied to demonstrate the anti-ulcer activity by direct action on epidermal and fibroblast growth factors or by indirect action on immune system stimulation: which leads to the rapid production of 6-keto-prostaglandin F1 by macrophage [91].

2.5.5 Anti-Obesity Effect

Obesity is an extra and unwanted, and harmful amount of body fat causing preconception to health. In clinical care, it is measured by the body mass index (BMI), which is described as the fraction of body mass (kg) divided by tallness (m^2) (kg/m^2) [93, 94]. Chances of various diseases like coronary heart disease, hypertension, Type II diabetes, and certain forms of cancer increase because of obesity [95]. When energy intake is stored in the form of triglycerides overdose, energy spending will lead to obesity [96]. Obesity is a multifaceted feature that is impacted by various factors, including diet, developmental stage, age, physical activity, and genetics [97, 98].

A good nutraceutical to control obesity must be able to raise energy costs and/or reduce calorie consumption to lose mass. Numerous nutraceuticals similar like Curcumin, vitamins (Chromium, Magnesium, and vitamin E), Amla (*Emblica officinalis*), and Fenugreek (*Trigonella foenum-graecum*) are operative in combating rotundity as well as its metabolic disorders. Further nutraceuticals with capsaicin conjugated linoleic acid, *Momordica charantia, Vigna mungo, Psyllium, Murraya koenigii, and Lagenaria siceraria* fiber has anti-obesity possessions and hypolipidemic and hypoglycemic effects [99, 100]. In addition, herbal supplements similar to caffeine, chitosan, ephedrine, green tea (*Camellia sinensis*), and Huang-guarana may be helpful to ease the loss in weight [101].

Probiotic bacteria are utilized to make a natural medicine for weight loss, obesity prevention, increased satiety, prolonged satiation, decreased food intake, fat accumulation minimization, enhanced treatment power, energy metabolism and treating obesity, and strengthening insulin sensitivity [95, 96]. In both experimental models and humans, *Bifidobacterium* genus and Lactobacillus (*Lb. sporogenes* and *Lb. acidophilus* NCFB 1748) relatives have been seen to perform an essential part in weight maintenance as an anti-obesity effect [94]. Previous clinical investigations have not entirely defined the processes by which LAB lowers body fat; nonetheless, one of the anticipated mechanisms that work against obesity and metabolic disorders is the inhibition of chronic inflammation [95]. It was observed that consuming heat-treated bacteria can help to lower body fat. Heat-treated *Akkermansia muciniphila* cells were found to be more efficient than live bacterial cells at minimizing chronic inflammation, IR, and body fat formation [102]. Yogurt is proven effective for decreasing the risk of obesity and other metabolic disorders [103]. Other research on obesity after pregnancy found that women who took probiotics (Lactobacillus and Bifidobacterium strains) throughout the prenatal period were less expected to weight increment after giving birth [104].

2.5.6 Cardiovascular Diseases

The World Health Organization (WHO) reported [105] that the major cardiovascular diseases (CVD) like hypertension, congestive heart failure (CHF), angina pectoris, acute myocardial infarction, coronary heart diseases (CHD), and rheumatic heart diseases that kill over 17.5 million people per year. The primary factors that contribute to the occurrence of CVDs include aging, poor diet, excessive stress at the workplace, and an unhealthy lifestyle [106]. Cordycepin derived from fungi is known for combating against cardiovascular diseases [107]. TMAO — a gut metabolite synthesized from precursors like L-carnitine — and dietary phosphatidylcholine by gut bacteria. Both the precursors of TMAO can be found in high-fat foods such as cheese, eggs, and meat from gamey mammals. High TMAO levels have been associated with an increased risk of adverse cardiovascular events, showing the importance of TMAO as a marker for cardiovascular diseases [108, 109].

Nutraceutical molecules rich in vitamins, antioxidants, polyphenols, dietetic fiber, minerals, polyunsaturated fatty acids (PUFAs), and flavonoids (flavones/flavanones/flavonols) are attributed to prevent and treat CVDs [110]. Several nutraceuticals show beneficial effects on CVDs by altering endothelial function. Therefore, nutraceuticals will significantly impact the therapies of CVDs with little or no complexity [34, 111].

2.5.7 Nutraceuticals in Other Complications

Moringa oleifera Lam has an impressive list of remedial use and is a decent basis of countless phenolics, vitamins, acids, and amino acids proteins, kaempferol, caffeoylquinic acid, and beta carotene with high nourishing and healing values. Different portions of this flora such as leaves, flowers, seeds, bark, fruit, leaves, immature pods, and roots work as heart and anti-tumor circulatory stimulants, anti-hypertensive, anti-epileptic, antispasmodic, antioxidant, anti-inflammatory, diuretic, antipyretic, anti-diabetic, anti-ulcer, cholesterol reducing [112].

2.6 Current Market of Nutraceuticals

Sales and custom statistics are not accessible in most regions, subsist, and are naturally unreliable because of inadequate definition and variability of categories in nutraceuticals. Products containing polyphenols have high market value, because of their effectiveness, especially in cardiovascular conditions [113]. Nutraceuticals containing carotenoids and curcuminoids, (astaxanthin, turmeric) are well represented in the best-selling products, and herbs from TCM and Ayurveda are circulating the global market [114].

US sales statistics shows detailed information about the source of plants for each supplement and its use following the trend. The patterns of nutritional utilization in Western countries are similar to those in the US, but differ from those in China, India and other regions. In US sales of pharmaceutical ingredients exceeded $ 8 billion in 2017 [115, 116]. In contrast, in China, by the end of 2014, the whole sales value of Chinese material medica (CMM) exceeded up to $US120 billion. In China, the best-selling pharmaceutical products (nutritional or nutritious food products released in 2014 were reishi mushroom (*Ganoderma lucidum)*, American ginseng (*Panax quinquefolium*), goji (wolfberry, *Lycium chinense*), and Asian/Chinese/Korean ginseng (*Panax ginseng*) [115–118]. The nutraceutical industry of India is oppressed mainly by pharmaceutical companies and FMCG, which has fewer nutritional gaming establishments. Pharmacists and FMCG practitioners working in the nutrition field have a heterogeneous production range by introducing product extensions and different development under active trade names [119].

The Covered Area of Nutraceutical Products

In the treatments of arthritis and coughs, sleeping disorders, digestive disorders and cancer prevention, osteoporosis, high blood pressure, depression, diabetes, cholesterol, and cosmetics are combined with therapeutic medications [120–122].

2.7 Future Perspective of Nutraceutical Development

Companies currently working further how nutrients influence people from a health care outlook. Thus, they are looking for favorable sectors of medical treatments and nutrition incorporation to create holistic medical care. Nowadays, medical care is limited by manufacturing safe and effective drugs, but the other aspects of medicine are explored in the form of nutrition for creating a healthy, disease-free world to live for humans [123]. The Indian Nutraceutical markets are anticipated to grow at a compounded rate of 21% annually and reach $10 billion by 2022 and $18 billion in 2025 as consumers become more conscious of health and fitness. It is a usual belief that nutraceutical products are needful medicines for handling fitness issues with no other side effects. This faith is a result of the usage of nutraceuticals for lengthy periods with no severe toxicities. Although the faith is right in the case of biggie nutraceuticals, they usually have low side effects comparatively than medicines. However, they do have some kind of side effects. But if research continues on such Nutraceuticals, then the side effects which are there, will be tackled soon in future [124, 125].

2.8 Conclusion

Nutraceuticals are substances that provide physiological health benefits along with the protection and prevention against chronic disease. They have a beneficial effect on health that includes improved health, slow aging, prevention of chronic disease, increased expectancy of lifespan, and supports the body's structure or functions. In recent times, nutraceuticals have been a significant area of interest because of their potential in providing nutrition, safety against several microbes, and therapeutic effects for various chronic diseases. Many researchers have demonstrated the promising outcomes for the compounds found or isolated from the microbes as a nutraceutical against different complications. In this review, efforts have been made to provide the utilization of nutraceuticals for the protection or prevention of diseases related to allergy, cardiovascular, cancer, diabetes, cholesterol, digestion, and obesity. Sales and custom statistics are not accessible in most regions, and where they exist, they naturally may not be reliable because of their inadequate description and variability of categories in nutraceuticals, dietary supplements, and functional foods in several and varied regions markets. Companies currently understand how nutrients influence people from a health care outlook. Therefore, nutraceuticals have a multi-component nature and provide the potential for targeting areas along with a myriad of products such as probiotics, prebiotics, dietary supplements, etc., which requires holistic approaches for increasing its production and awareness of nutraceuticals.

References

1. Aronson, J.K., Defining 'nutraceuticals': Neither nutritious nor pharmaceutical. *Br. J. Clin. Pharmacol.*, 83, 1, 8–19, 2017.
2. Das, L., Bhaumik, E., Raychandhuri, U., Chakraborty, R., Role of nutraceuticals in human health. *J. Food Sci. Technol.*, 49, 173–183, 2012.
3. Hugenholtz, J. and Smid, E.J., Nutraceutical production with food-grade microorganisms. *Curr. Opin. Biotechnol.*, 13, 497–507, 2002.

4. Dufosse, L., Pigments, in: *Microbial Encyclopedia Microbiology*, vol. 4, pp. 457–471, 2009.
5. Bronzwaer, S., EFSA scientific forum 'from safe food to healthy diets'. EU risk assessment – Past, present and future. *Trends Food Sci. Technol.*, 19, S2–S8, 2008.
6. Pandey, M., Verma, R.K., Saraf, S.A., Nutraceuticals: New era of medicine and health. *Asian J. Pharm. Clin. Res.*, 3, 11–15, 2010.
7. Chauhan, V. and Kanwar, S.S., Chapter 4: Bioactive peptides: Synthesis, functions and biotechnological applications, in: *Biotechnological Production of Bioactive Compounds*, pp. 107–137, 2020.
8. Prabhakar, D.P., Nutraceuticals as therapeutic agent: A novel approach, in: *Recent Progress in Medicinal Plants*, J.N. Govil (ed.), pp. 141–156, Studium Press, India, Pvt. Ltd., 2015.
9. Kokate, C.K., Purohit, A.P., Gokhale, S.B., Nutraceutical and cosmaceutical, in: *Pharmacognosy*, 21st edition, pp. 542–549, Nirali Prakashan, Pune, India, 2002.
10. Lenoir-Wijnkoop, I., Sanders, M.E., Cabana, M.D., Caglar, E., Corthier, G., Rayes, N., Sherman, P.M., Timmerman, H.M., Vaneechoutte, M., Van Loo, J., Wolvers, D.A., Probiotic and prebiotic influence beyond the intestinal tract. *Nutr. Rev.*, 65, 11, 469–89, 2007.
11. Suvarna, V.C. and Boby, V.U., Probiotics in human health: A current assessment. *Curr. Sci.*, 88, 1744–1748, 2005.
12. Macfarlane, S., Macfarlane, G.T., Cummings, J.H., Prebiotics in the gastrointestinal tract. *Aliment. Pharmacol. Ther.*, 24, 5, 701–14, 2006.
13. Hord, N.G., Eukaryotic-microbiota crosstalk: Potential mechanisms for health benefits of prebiotics and probiotics. *Annu. Rev. Nutr.*, 28, 215–231, 2008.
14. Wang, L., Zhu, L.H., Jiang, H., Tang, Q.Z., Yan, L., Wang, D., Liu, C., Bian, Z.Y., Li, H., Grape seed proanthocyanidins attenuate vascular smooth muscle cell proliferation via blocking phosphatidylinositol 3-kinase-dependent signaling pathways. *J. Cell. Physiol.*, 223, 3, 713–726, 2010.
15. Duthie, G.G., Gardner, P.T., Kyle, J.A.M., Plant polyphenols: Are they the new magic bullet? *Proc. Nutr. Soc.*, 62, 599–603, 2003.
16. Scalbert, A., Johnson, I.T., Saltmarsh, M., Polyphenols: Antioxidants and beyond. *Am. J. Clin. Nutr.*, 81, 215S–217S, 2005.
17. Kochhar, K.P., Dietary spices in health and diseases: I. *Indian J. Physiol. Pharmacol.*, 52, 106–122, 2008.
18. Lampe, J.W., Spicing up a vegetarian diet: Chemopreventive effects of phytochemicals. *Am. J. Clin. Nutr.*, 78, 579S–583S, 2003.
19. Kunnumakkara, A.B., Sailo, B.L., Banik, K., Harsha, C., Prasad, S., Gupta, S.C., Bharti, A.C., Aggarwal, B.B., Chronic diseases, inflammation, and spices: How are they linked? *J. Transl. Med.*, 16, 1, 1–25, 2018.

20. Rao, B.N., Bioactive phytochemicals in Indian foods and their potential in health promotion and disease prevention. *Asia Pac. J. Clin. Nutr.*, 12, 9–22, 2003.
21. Shweta, K.G., Rajan, B.M., Urvashi, K.P., Blessy, M., Hitesh, N.J., Herbal plants: Used as cosmetic. *J. Nat. Prod. Plant Resour.*, 1, 1, 24–32, 2011.
22. Escott-Stump, E. and Mahan, L.K., *Krause's Food, Nutrition And Diet Therapy*, 10th edn, pp. 553–559, WB Saunders Company, Philadelphia, 2000.
23. Institute of Medicine, *Dietary Reference Intakes For Energy, Carbohydrate, Fibre, Fat, Fatty Acids, Cholesterol, Protein And Amino Acids*, National Academy Press, Washington DC, 2002.
24. Leray, C., Wiesel, M.L., Freund, M., Cazenave, J.P., Gachet, J., Long chain n-3 fatty acids specifically affect rat coagulation factors dependent on vitamin K: Relation to peroxidative stress. *Arterioscler. Thromb. Vasc. Biol.*, 21, 459–465, 2001.
25. Stoll, A.L., Severus, W.E., Freeman, M.P., Rueter, S., Zboyan, H.A., Diamond, E., Cress, K.K., Marangell, L.B., Omega 3 fatty acids in bipolar disorder. A preliminary double-blind, placebocontrolled trial. *Arch. Gen. Psychiatry*, 56, 407–412, 1999.
26. Buchner, H.C., Hengstler, P., Schindler, C., Meier, G., N-3 polyunsaturated fatty acids in coronary heart disease—a meta-analysis of randomized controlled trials. *Am. J. Med.*, 112, 298–304, 2002.
27. Nemets, B., Stahl, Z., Bemaker, R.H., Addition of omega-3 fatty acid to maintenance medication treatment for recurrent unipolar depressive disorder. *Am. J. Psychiatr.*, 159, 477–479, 2002.
28. Hiroyasu, I., Rexrode, K.M., Stampfer, M.J., Manson, J.E., Colditz, G.A., Speizer, F.E., Hennekens, C.H., Willett, W.C., Intake of fish and omega-3 fatty acids and risk of stroke in women. *J. Am. Med. Assoc.*, 285, 304–312, 2001.
29. Albert, C.M., Campos, H., Stampfer, M.J., Ridker, P.M., Manson, J.E., Willett, W.C., Ma, J., Blood levels of long-chain n-3 fatty acids and the risk of sudden death. *N. Engl. J. Med.*, 346, 1113–1118, 2002.
30. Elliot, J.G., Application of antioxidant vitamins in foods and beverages. *Food Technol.*, 53, 46–48, 1999.
31. Meydani, M., Effect of functional food ingredients: Vitamin E modulation of cardiovascular diseases and immune status in the elderly. *Am. J. Clin. Nutr.*, 71, 1665S–1668S, 2000.
32. Watkins, T.R., Bierenbaurn, M.I., Giampalala, A., Tocotrienols: Biological and health effects, in: *Antioxidant Status, Diet, Nutrition and Health*, A.M. Papas, (Ed.), pp. 479–496, CRC Press, Boca Raton, 1999.
33. Lee, J., Koo, N., Min, D.B., Reactive oxygen species, aging and anti-oxidative nutraceuticals. *Compr. Rev. Food Sci. Food Saf.*, 3, 21–33, 2004.
34. Li, S.Q. and Zhang, Q.H., Advances in the development of functional foods from buckwheat. *Crit. Rev. Food Sci. Nutr.*, 41, 6, 451–64, 2001.

35. Leclere, C.L., Champ, M., Boillot, J., Guille, G., Lecannu, G., Molis, C., Bornet, F., Krempff, M., Delort-Laval, J., Role of viscous guar guam in lowering the glycemic response after a solid meal. *Am. J. Clin. Nutr.*, 59, 914–921, 1994.
36. Liu, S., Stampfer, M.J., Hu, F.B., Whole-grain consumption and risk of coronary heart disease: Results from the Nurses' Health study. *Am. J. Clin. Nutr.*, 70, 412–419, 1999.
37. Steffen, L.M., Jacobs, D.R.J., Stevens, J., Shahar, E., Carithers, T., Folsom, A.R., Associations of whole-grain, refined grain, and fruit and vegetable consumption with risks of all-cause mortality and incident coronary artery disease and ischemic stroke: The atherosclerosis risk in communities (ARIC) study. *Am. J. Clin. Nutr.*, 78, 383-390, 2003.
38. Whelton, S.P., Hyre, A.D., Pedersen, B., Yi, Y., Whelton, P.K., He, J., Effect of dietary fiber intake on blood pressure: A metaanalysis of randomized, controlled clinical trials. *J. Hypertens.*, 23, 475–481, 2005.
39. Montonen, J., Knekt, P., Jarvinen, R., Aromaa, A., Reunanen, A., Whole-grain and fiber intake and the incidence of type 2 diabetes. *Am. J. Clin. Nutr.*, 77, 622–629, 2003.
40. Lairon, D., Arnault, N., Bertrais, S., Planells, R., Clero, E., Hercberg, S., Boutron-Ruault, M.-C., Dietary fiber intake and risk factors for cardiovascular disease in French adults. *Am. J. Clin. Nutr.*, 82, 1185–1194, 2005.
41. Petruzziello, L., Iacopini, F., Bulajic, M., Shah, S., Costamagna, G., Review article: Uncomplicated diverticular disease of the colon. *Aliment. Pharmacol. Ther.*, 23, 1379–1391, 2006.
42. Brown, L., Rosner, B., Willett, W.W., Sacks, F.M., Cholesterol lowering effects of dietary fiber: A meta-analysis. *Am. J. Clin. Nutr.*, 69, 30–42, 1999.
43. Michalak, I. and Chojnacka, K., Algae as production systems of bioactive compounds. *Eng. Life Sci.*, 15, 160–176, 2015.
44. Holdt, S.L. and Kraan, S., Bioactive compounds in seaweed: Functional food applications and legislation. *J. Appl. Phycol.*, *23*, 543–597, 2011.
45. de Jesus Raposo, M.F., de Morais, A.M.M.B., de Morais, R.M.S.C., Emergent sources of prebiotics: Seaweeds and microalgae. *Mar. Drugs*, 14, 1–27, 2016.
46. O'Sullivan, L., Murphy, B., McLoughlin, P., Duggan, P., Lawlor, P.G., Hughes, H., Gardiner, G.E., Prebiotics from marine macroalgae for human and animal health applications. *Mar. Drug*, 8, 2038–2064, 2010.
47. Synytsya, A., Čopíková, J., Kim, W.J., Park, Y., Cell wall polysaccharides of marine algae, in: *Springer Handbook of Marine Biotechnology*, S.K. Kim (Ed.), pp. 543–590, Springer-Verla Berlin Heidelberg, Heidelberg, 2015.
48. Li, Y.X., Wijesekara, I., Li, Y., Kim, S.K., Phlorotannins as bioactive agents from brown algae. *Process Biochem.*, 46, 2219–2224, 2011.
49. Freitas, A.C., Pereira, L., Rodrigues, D., Carvalho, A.P., Panteleitchouk, T., Gomes, A.M., Duarte, A.C., Marine functional foods, in: *Springer Handbook of Marine Biotechnology*, S.K. Kim (Ed.), pp. 969–994, Springer-Verla Berlin Heidelberg, Heidelberg, 2015.

50. Smith–Warner, S.A., Elmer, P.J., Tharp, T.M., Increasing vegetable and fruit intake: Randomized intervention and monitoring in an at–risk population. *Cancer Epidemiol. Biomarkers Prev.*, 9, 3, 307–317, 2000.
51. Andersen, V., Holst, R., Vogel, U., Systematic review: Diet–gene interactions and the risk of colorectal cancer. *Aliment. Pharmacol. Ther.*, 37, 4, 383–391, 2013.
52. Golla, U., Department of Biological Sciences, Indian Institute of Science Education and Research Bhopal (), 2018.
53. Ramnani, P., Chitarrari, R., Tuohy, K., Grant, J., Hotchkiss, S., Philp, K., *In vitro* fermentation and prebiotic potential of novel low molecular weight polysaccharides derived from agar and alginate seaweeds. *Anaerobe.*, 18, 1–6, 2012.
54. Santini, A., Tenore, G.C., Novellino, E., Nutraceuticals: A paradigm of proactive medicine. *Eur. J. Pharm. Sci.*, 96, 53–61, 2017.
55. Gallo, M., Ferrara, L., Calogeo, A., Montesano, D., Naviglio, D., Relationships between food and diseases: What to know to ensure food safety. *Int. Food Res. J.*, 137, 1–55, 2020.
56. Daliu, P., Santini, A., Novellino, E., A decade of nutraceutical patents: Where are we now in 2018? *Expert Opin, Ther. Pat.*, 28, 875–882, 2018.
57. Simmons, R.K., Unwin, N., Griffin., S.J., International diabetes dederation: An update of the evidence concerning the prevention of type 2 diabetes. *Diabetes Res. Clin. Pract.*, 87, 2, 143–149, 2010.
58. Whiting, D.R., Guariguata, L., Weil, C., Shaw, J., IDF diabetes atlas: Global estimates of the prevalence of diabetes for 2011 and 2030. *Diabetes Res. Clin. Pract.*, 94, 3, 311–321, 2011.
59. Modi, P., Diabetes beyond insulin: Review of new drugs for treatment of diabetes mellitus. *Curr. Drug Discovery Technol.*, 4, 39–47, 2007.
60. Vallianou, N.G., Stratigou, T., Tsagarakis, S., Microbiome and diabetes: Where are we now? *Diabetes Res. Clin. Pract.*, 146, 111–118, 2018.
61. Gupta, V., Sengupta, M., Prakash, J., Tripathy, B.C., Production of recombinant pharmaceutical proteins, in: *Basic And Applied Aspects of Biotechnology*, pp. 77–101, 2016.
62. Bajzer, M. and Seelay, R.J., Physiology: Obesity and gut flora. *Nature*, 444, 1009–1010, 2006.
63. Hossain, P., Kawar, B., El Nahas, M., Obesity and diabetes in the developing world–A growing challenge. *N. Engl. J. Med.*, 356, 213–215, 2007.
64. Gayathry, K.S. and John, J.A., Functional beverages: Special focus on antidiabetic potential. *J. Food Process. Preserv.*, 45, e15974, 2021.
65. Etemadian, Y., Ghaemi, V., Shaviklo, A.R., Pourashouri, P., Mahoonak, A.R.S., Rafipour, F., Development of animal/ plant-based protein hydrolysate and its application in food, feed and nutraceutical industries: State of the art. *J. Clean. Prod.*, 278, 123219, 2021.

66. Shi, H., Kokoeva, M.V., Inouye, K., Tzameli, I., Yin, H., Flier, J.S., TLR4 links innate immunity and fatty acid-induced insulin resistance. *J. Clin. Investig.*, 116, 3015–3025, 2006.
67. Zaborska, K.E. and Cummings, B.P., Rethinking bile acid metabolism and signaling for type 2 diabetes treatment. *Curr. Diab. Rep.*, 18, 11, 109, 2018.
68. Gurung, M., Li, Z., You, H., Rodriguoes, R., Jump, D.B., Morgun, A., Shulzhenko, N., Role of gut microbiota in type 2 diabetes pathophysiology. *EBioMedicine*, 51, 1–9, 2020.
69. Rafieian-Kopaei, M. and Nasri, H., Ginger and diabetic nephropathy. *J. Renal Inj. Prev.*, 2, 1, 9–10, 2013.
70. Davì, G., Santilli, F., Patrono, C., Nutraceuticals in diabetes and metabolic syndrome. *Cardiovasc. Ther.*, 28, 4, 216–26, 2010.
71. Sirtori, C.R. and Galli, C., N-3 fatty acids and diabetes. *Biomed. Pharmacother.*, 56, 8, 397–406, 2002.
72. Tavafi, M., Diabetic nephropathy and antioxidants. *J. Nephropathol.*, 2, 1, 20–27, 2013.
73. Basu, S., Wagner, R.G., Sewpaul, R., Reddy, P., Davies, J., Implications of scaling up cardiovascular disease treatment in South Africa: A microsimulation and cost-effectiveness analysis. *Lancet Glob. Health*, 7, 2, e270–e280, 2019.
74. Phillips, R.A., Arnold, R.M., Peterson, L.E., Hypertension guidelines: The threads that bind them. *J. Am. Coll. Cardiol.*, 72, 11, 1246–1251, 2018.
75. Psara, E., Pentieva, K., Ward, M., McNulty, H., Critical review of nutrition, blood pressure and risk of hypertension through the lifecycle: Do B vitamins play a role? *Biochimie*, 173, 76–90, 2020.
76. Toldrá, F., Gallego, M., Reig, M., Aristoy, M.C., Mora, L., Bioactive peptides generated in the processing of dry-cured ham. *Food Chem.*, 321, 1–9, 2020.
77. Williams, B., Poulter, N., Brown, M. *et al.*, Guidelines for management of hypertension: Report of the fourth working party of the British hypertension society, 2004—BHS IV. *J. Hum. Hypertens.*, 18, 139–185, 2004.
78. Koyama, M., Naramoto, K., Nakajima, T., Aoyama, T., Watanabe, M., Nakamura, K., Purification and identification of anti-hypertensive peptides from fermented buckwheat sprouts. *J. Agric. Food Chem.*, 61, 12, 3013–3021, 2013.
79. Boffa, M.B. and Koschinsky, M.L., Oxidized phospholipids as a unifying theory for lipoprotein (a) and cardiovascular disease. *Nat. Rev. Cardiol.*, 16, 305–318, 2019.
80. Tsafrakidou, P., Michaelidou, A.M., Biliaderis, C.G., Fermented cereal-based products: Nutritional aspects, possible impact on gut microbiota and health implications. *Foods*, 9, 6, 734–759, 2019.
81. Parham, J.S. and Goldberg, A.C., Mipomersen and its use in familial hypercholesterolemia. *Expert Opin. Pharmacother.*, 20, 127–131, 2019.
82. Li, Y., Go, V.L., Sarkar, F.H., The role of nutraceuticals in pancreatic cancer prevention and therapy: Targeting cellular signaling, microRNAs, and eigenome. *Pancreas*, 44, 1, 1–10, 2015.

83. Thomasset, S.C., Berry, D.P., Garcea, G., Dietary polyphenolic phytochemicals—promising cancer chemopreventive agents in humans? A review of their clinical properties. *Int. J. Cancer*, 120, 3, 451–458, 2007.
84. Ardalan, M.R. and Rafieian-Kopaei, M., Antioxidant supplementation in hypertension. *J. Renal Inj. Prev.*, 3, 2, 39–40, 2014.
85. DeBerardinis, R.J., Lum, J.J., Hatzivassiliou, G., Thompson, C.B., The biology of cancer: Metabolic reprogramming fuels cell growth and proliferation. *Cell Metab.*, 7, 1, 11–20, 2008.
86. Chauhan, B., Kumar, G., Kalam, N., Ansari, S.H., The hindu newspaper; Current concepts and prospects of herbal nutraceutical: A review. *J. Adv. Pharm. Technol. Res.*, 4, 1, 4–8, 2013.
87. Selhub, E., Logan, A., Bested, A., Fermented foods, microbiota, and mental health: Ancient practice meets nutritional psychiatry. *J. Physiol. Anthropol.*, 33, 1–12, 2014.
88. Kumar, M., Nagpal, R., Verma, V., Kumar, A., Kaur, N., Hemalatha, R., Gautam, S., Singh, B., Probiotic metabolites as epigenetic targets in the prevention of colon cancer. *Nutr. Rev.*, 71, 23–34, 2013.
89. Barzegari, A., Kheyrolahzadeh, K., Khatibi, S.M.H., Sharifi, S., Memar, M.Y., Vahed, S.Z., The battle of probiotics and their derivatives against biofilms. *Infect. Drug Resist.*, 13, 659–672, 2020.
90. Raskov, H., Burcharth, J., Pommergaard, H.C., Linking Gut Microbiota to Colorectal Cancer. *J. Cancer*, 8, 17, 3378–3395, 2017.
91. Nagaoka, M., Hashimoto, S., Watanabe, T., Yokokura, T., Mori, Y., Anti-ulcer effects of lactic acid bacteria and their cell wall polysaccharides. *Biol. Pharm. Bull.*, 17, 1012–1017, 1994.
92. Ku, S., Park, M.S., Ji, G.E., You, H.J., Review on bifidobacterium bifidum BGN4: Functionality and nutraceutical applications as a probiotic microorganism. *Int. J. Mol. Sci.*, 17, 9, 1544, 2016.
93. González-Muniesa, P., Mártinez-González, M.A., Hu, F.B., Després, J.P., Matsuzawa, Y., Loos, R.J.F., Moreno, L.A., Bray, G.A., Martinez, J.A., Obesity. *Nat. Rev. Dis. Primers*, 3, 1–18, 2017.
94. Silveira, E.A., Kliemann, N., Noll, M., Sarrafzadegan, N., Oliveira, C.D., Visceral obesity and incident cancer and cardiovascular disease: An integrative review of the epidemiological evidence. *Obes. Rev.*, 22, 1–17, 2021.
95. Yao, J., Wu, D., Zhang, C., Yan, T., Zhao, Y., Shen, H., Xue, K., Huang, X., Wang, Z., Qiu, Y., Macrophage IRX3 promotes diet-induced obesity and metabolic inflammation. *Nat. Immunol.*, 22, 1268–1279, 2021.
96. Aoun, A., Darwish, F., Hamod, N., The influence of the gut microbiome on obesity in adults and the role of probiotics, prebiotics, and synbiotics for weight loss. *Prev. Nutr. Food Sci.*, 25, 2, 113–123, 2020.
97. Cerdó, T., García-Santos, J.A., Bermúdez, M.G., Campoy, C., The role of probiotics and prebiotics in the prevention and treatment of obesity. *Nutrients*, 11, 3, 636–667, 2019.

98. Kaur, G., Mukundan, S., Wani, V., Nutraceuticals in the management and prevention of metabolic syndrome. *Austin J. Pharmacol. Ther.*, 3, 1, 1063, 2015.
99. Rubin, S.A. and Levin, E.R., Clinical review 53: The endocrinology of vasoactive peptides: Synthesis to function. *J. Clin. Endocrinol. Metab.*, 78, 1, 6–10, 1994.
100. Boozer, C.N., Nasser, J.A., Heymsfield, S.B., An herbal supplement containing Ma Huang–Guarana for weight loss: A randomized, double-blind trial. *Int. J. Obes.*, 25, 3, 316–324, 2001.
101. Toshimitsu, T., Gotou, A., Sashihara, T., Furuichi, K., Hachimura, S., Shioya, N., Suzuki, S., Asami, Y., Ingesting yogurt containing lactobacillus plantarum OLL2712 reduces abdominal fat accumulation and chronic inflammation in overweight adults in a randomized placebo-controlled trial. *Curr. Dev. Nutr.*, 5, 2, 1–11, 2021.
102. Depommier, C., Everard, A., Druart, C., Plovier, H., Hul, M.V., Vieira-Silva, S., Falony, G., Raes, J., Maiter, D., Delzenne, N.M., Barsey, M.D., Loumaye, A., Hermans, H.P., Thissen, J.P., De Vos, V.M., Cani, P.D., Supplementation with *Akkermansia muciniphila* in overweight and obese human volunteers: A proof-of-concept exploratory study. *Nat. Med.*, 25, 1096–103, 2019.
103. Chen, Y., Feng., R., Yang, X., Dai, J., Huang, M., Ji, X., Li, Y., Okekunle, A.P., Gao, G., Onwuka, J.U., Pang, X., Wang, C., Li, C., Li, W., Sun, C., Yogurt improves insulin resistance and liver fat in obese women with nonalcoholic fatty liver disease and metabolic syndrome: A randomized controlled trial. *Am. J. Clin. Nutr.*, 109, 1611–1619, 2019.
104. Sousa, R., Halper, J., Zhang, J., Lewis, S.J., Li, W., Effect of *lactobacillus acidophilus* supernatants on body weight and leptin expression in rats. *BMC Complement. Altern. Med.*, 8, 5–12, 2008.
105. Mehta, D., Integrative medicine and cardiovascular disorders. *Prim. Care: Clinics Office Pract.*, 44, 351–367, 2017.
106. Harris, L.K., Diane, C., Berry, D.C., Cortés, Y.I., Psychosocial factors related to cardiovascular disease risk in young african american women: A systematic review. *Ethn. Health*, 20, 1–19, 2021.
107. Ashraf, S.A., Elkhalifa, A.E.O., Siddiqui, A.J., Patel, M., Awadelkareem, A.M., Snoussi, M., Snoussi, M., Ashraf, M.S., Adnan, M., Hadi, S., Cordycepin for health and wellbeing: A potent bioactive metabolite of an entomopathogenic medicinal fungus cordyceps with its nutraceutical and therapeutic potential. *Molecule*, 25, 12, 2735, 2020.
108. Lau, K., Srivatsav, V., Rizwan, A., Nashed, A., Liu, R., Shen, R., Akhtar, M., Bridging the Gap between Gut Microbial Dysbiosis and Cardiovascular Diseases. *Nutrients*, 9, 1–16, 2017.
109. Brown, J.M. and Hazen, S.L., The gut microbial endocrine organ: Bacterially derived signals driving cardiometabolic diseases. *Annu. Rev. Med.*, 66, 343–359, 2015.
110. Shahbazian, H., World diabetes day. *J. Renal Inj. Prev.*, 2, 4, 123–4, 2013.

111. Khosravi–Boroujeni, H., Sarrafzadegan, N., Mohammadifard, N., White rice consumption and CVD risk factors among Iranian population. *J. Health Popul. Nutr.*, 31, 2, 252–261, 2013.
112. Baradaran, A., Madihi, Y., Merrikhi, A., Rafieian-Kopaei, M., Nematbakhsh, M., Asgari, A., Nephrotoxicity of hydroalcoholic extract of teucrium polium in Wistar rats. *Pak. J. Med. Sci.*, 29, 329–333, 2013.
113. Andrew, R. and Izzo, A.A., Principles of pharmacological research of nutraceuticals. *Br. J. Pharmacol.*, 174, 1177–1194, 2017.
114. Stephen, D., A Report of National Nutraceutical Centre, Nutraceuticals India 2012 Webinar, p. 22, 2018.
115. Son, M.J., Hwi–Jin, I., Kim, Y.E., Ku, B., Lee, J.H., Son, C.G., Evaluation of the anti-fatigue effects of a traditional herbal drug, Gongjin-dan, under insufficient sleep conditions: Study protocol for a randomised controlled trial. *Trials*, 17, 418, 2016.
116. Smith, T., Kawa, K., Ecklb, V., Mortonc, C., Stredneyd, R., Herbal supplement sales in US increased 8.5% in 2017, topping $8 billion. *HerbalGram*, 119, 62–71, 2018.
117. Shinde, N., Bangar, B., Deshmukh, S., Kumbhar, P., Nutraceuticals: A review on current status. *Res. J. Pharm. Technol.*, 7, 1, 110–113, 2014.
118. Williamson, E.M., Liu, X., Izzo, A.A., Trends in use, pharmacology, and clinical applications of emerging herbal nutraceuticals. *B. J. Pharmaco.*, 177, 6, 1227–1240, 2020.
119. Gupta, S., Chauhan, D., Mehla, K., Sood, P., Nair, A., An overview of nutraceuticals: Current scenario. *JBCP*, 1, 2, 55–62, 2010.
120. Ramnani, P., Chitarrari, R., Tuohy, K., Grant, J., Hotchkiss, S., Philp, K., Campbell, R., Gill, C., Rowland, I., *In vitro* fermentation and prebiotic potential of novel low molecular weight polysaccharides derived from agar and alginate seaweeds. *Anaerobe*, 18, 1–6, 2012.
121. Santini, A., Tenore, G.C., Novellino, E., Nutraceuticals: A paradigm of proactive medicine. *Eur. J. Pharm. Sci.*, 96, 53–61, 2017.
122. Van Der Zanden, L.D.T., Van Kleef, E., De Vijk Rene, A., Van Trijp Hens, C.M., Knowledge, perceptions and preferences of elderly regarding protein-enriched functional food. *Appetite*, 80, 16–22, 2014.
123. Waliszewski, K.N. and Belasco, G., Neutraceutical properties of lycopene. *Salud Pública Méx.*, 52, 3, 254–265, 2010.
124. Sutter, M.E., Thomas, J.D., Brown, J., Morgan, B., Selenium toxicity: A case of selenosis caused by a nutritional supplement. *Ann. Intern. Med.*, 148, 12, 970–971, 2008.
125. Maurya, A.P., Chauhan, J., Yadav, D.K., Gangwar, R., Maurya, V.K., Chapter-11: Nutraceuticals and their impact on human health, in: *Preparation of Phytopharmaceuticals for the Management of Disorders: The Development of Nutraceuticals and Traditional Medicine (Part-II: Health Benefits of Nutraceuticals)*, C. Egbuna, A.P. Mishra, M. Goyal, (Eds.), pp. 229–254, Academic Press, Elsevier, Amsterdam, 2020.

3

Role of Microbes in the Food Industry

Ranjana Kumari[1]*, Lakshmi E. Jayachandran[2] and Sanjay Kumar[3]

[1]*Department of Biotechnology, Indian Institute of Technology Kharagpur, West Bengal, India*
[2]*Agricultural & Food Engineering Department, Indian Institute of Technology Kharagpur, West Bengal, India*
[3]*Department of Biochemical Engineering and Biotechnology, Indian Institute of Technology Delhi, New Delhi, India*

Abstract
Microbes play a very crucial role in a wide array of food applications. They are the main component for producing several food products and are also responsible for the food spoilage that causes huge economic roles. Several bacteria and fungi have been in use from ancient times to produce different kinds of fermented food products and enzymes. These days the use of probiotics and prebiotics has improved normal gut flora after drug treatments. On the other hand, the microbes can also lead to undesirable changes in food quality affecting food safety. The growth and survival of microbes in food depend on several factors, *viz.*, intrinsic properties (nutrient density, water activity, pH, food components, biological structures, oxidoreduction potential) and extrinsic factors (temperature, relative humidity, and presence of competitive microorganisms). These microbes can be monitored either by following traditional methods or rapid methods. Therefore, rapid detection or real-time monitoring of microbial load in the food product is a major concern for maintaining food quality and safety. Controlled use of these microbes can benefit humankind in several ways, and this calls for careful examination of the interaction between food and its microflora.

Keywords: Microbes, food products, pathogens, detection

**Corresponding author*: ranjanabt@gmail.com

Sanjay Kumar, Narendra Kumar and Shahid-ul-Islam (eds.) Role of Microbes in Industrial Products and Processes, (59–80) © 2023 Scrivener Publishing LLC

3.1 Introduction

Food processing and preservation using microorganisms have been common for decades. Microbes are employed at various stages of postharvest operations such as ripening, fermentation, etc. Microbes render taste, texture, and smell to the foods and produce certain inhibitory compounds that aid in food preservation. These organisms are also employed to develop functional and novel foods such as cheese, yogurt, prebiotics, and probiotics. A wide range of foods and beverages are produced with the help of microbial actions. The application of microorganisms in food processing and preservation is a slow process built on natural processes and experimentation observations. This chapter summarizes the applications of microbes in different food industries.

3.2 Microflora of Food Products

3.2.1 Bacteria

Bacteria are unicellular prokaryotes and are present almost everywhere in different shapes, sizes, and arrangements like coccus (spherical), bacillus (rod-shaped), vibrio (comma-shaped), and spirilla (spiral). Bacteria can be classified based on the cell wall composition as gram-positive bacteria containing peptidoglycans and gram-negative bacteria containing lipopolysaccharide in the cell wall. It can also be classified based on the mode of nutrition as autotrophs or heterotrophs and based on respiration as aerobic or anaerobic bacteria. Bacteria are beneficial for food industries in several ways. Several strains of bacteria such as *Lactobacillus, Streptococcus, Bifidobacterium, Erwinia*, etc., have been utilized extensively to produce different types of food products such as fermented food and alcoholic beverages.

3.2.2 Yeast

Yeast is a unicellular fungus, and it can convert sugar into carbon dioxide and ethanol. It has been in wide use in the food industry to produce baked products and make alcoholic beverages. There is evidence of its use in bread making in ancient Egyptians. Yeast species Saccharomyces cerevisiae has been used extensively in the food industry. It is commonly known as baker's yeast used in bakery products and brewer's yeast used in beverage production. Yeast is produced industrially in a different form for its specific application. Yeast is grown by the aerobic fermentation process in a large tank on sugar-rich substrates like molasses. They are collected from the tank and concentrated by centrifugation. After that, they are processed and available in different forms like freeze-dried powder, active dried granules, and vacuum-packed fine powder. Some of the commonly used yeast species of the food industry are *Saccharomyces carlsbergensis, Kluyveromyces marxianus, Brettanomyces, Yarrowia lipolytica Zygosaccharomyces, Cryptococcus, Galactomyces candidus, Debaryomyces hansenii*, etc. [1, 2]. They play a major role in developing texture and flavor in food products. In bread making, the yeast starts the fermentation process and releases carbon dioxides, giving a fluffy texture to the bread.

3.2.3 Molds

Molds are filamentous fungi. They have also been used to produce different foods and food products. They have been used to produce several types of

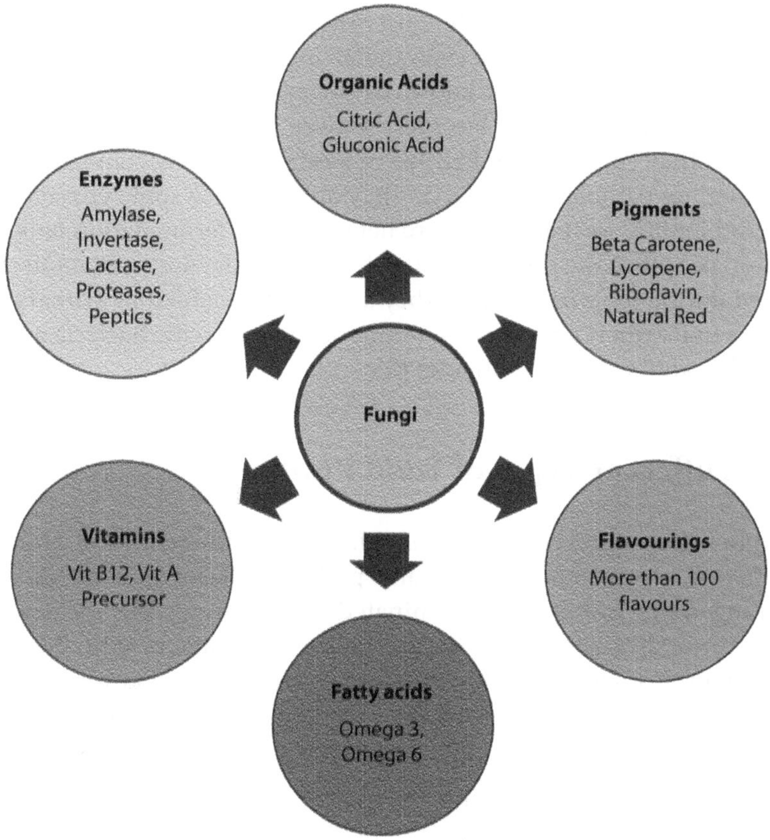

Figure 3.1 Application of fungi in the food industry. (Adapted from [3]).

enzymes like cellulases, amylases, proteases, etc. They are also directly used as feed and food. They have been extensively used for the production of oriental foods like Soy sauce (*Aspergillus oryzae* and *Aspergillus sojae*), Cheese (*Roquefort* and *Camembert species*), Miso (*Aspergillus oryzae*), Salami (*Penicillium species*), Tempeh (*Rhizopus oligosporo*), wine (*Bothrytiscinerea species*), etc. A fermented milk product Viili is also produced by lactic acid fermentation using the mold. Their extensive application in the production of different biomolecules is shown in Figure 3.1. They are also known for causing food spoilage.

3.2.4 Viruses

Viruses are acellular microorganisms and cannot grow independently. They first invade another living organism called a host and then multiply within host cells. Thus, viruses are parasites. A particular strain of virus generally infects a specific host. They cannot multiply on food but remain viable and infectious even under highly adverse conditions. The viruses infecting bacteria are known as bacteriophages. In the food industry, they have been used as bio preservatives and for decontamination of livestock. They enhance the shelf-life of food products by combating some food spoiling pathogenic bacteria. They also produce some enzymes that prevent biofilms' formation and are hence used in fermentation tanks [4]. They do not affect the sensory properties of food and are resistant to stress created during food processing. They are generally added in primary production units where they act as antimicrobial for specific bacteria and can be used as antibiotics or pesticides [5].

3.3 Different Types of Food Products

Microbes are employed for the manufacture of different food products. In 7000 B.C., ancient Babylonians preferred fermented drinks over drinking water (often be contaminated) to minimize the incidence of water-borne diseases. First records of cheese manufacture by Egyptians and Sumerians dates back to 3000 B.C. Fermented meats were popular among Romans as early as 1000 B.C. The application of microorganisms in food industries calls for strict legislation to target process development and food safety.

3.3.1 Fermented Dairy Products

Fermented dairy products such as yogurt and cheese are produced by lactic acid fermentation. The lactose present in milk (pasteurized or unpasteurized) is converted into lactic acid by useful bacteria (lactic acid bacteria) such as *Streptococcus salivarius, S. thermophilus*, and *Lactobacillus bulgaricus*. They feed on the lactose and release lactic acid as a waste product, which in turn causes the curdling of the milk protein, casein, into the curd. The liquid formed along with curd is known as whey. The fresh milk is boiled to eliminate pathogenic microbes and then cooled to about 40°C. This is followed by the addition of starter culture and thorough mixing. It is left undisturbed for about 6 hours. The increase in acidity during microbial fermentation prevents the growth of other pathogenic microbes. Several variants of fermented milk or yogurt are available across the globe. It is known as lassi in India and ayran in Turkey.

Cheese is solid food prepared by inoculating milk (from goat, sheep, reindeer, and water buffalo) with a starter culture (bacteria or fungi). There are several types of cheese based on the type of fermentation used.

a. Cheddar cheese – addition of bacteria and characterized by enhanced flavor and texture.
b. Roquefort cheese and blue cheese – addition of mold and fungi.
c. Camembert cheese – combination of bacteria and fungi.
d. Swiss cheese – addition of *Propionibacterium sharmanii.*

These microbes are known for their ability to produce large amounts of carbon dioxide resulting in large holes.

3.3.2 Alcoholic Beverages

Wine is produced by fermentation of fruit juices, employing autochthonous microbiota, or inoculating with *Saccharomyces cerevisiae*. Fermentation of fruit juices by *Saccharomyces* and *non-Saccharomyces* strains can increase the bioactive components, ferment musts with high sugar concentration (osmophilic yeast), and withstand high or low fermentation temperatures survive in wines with a high concentration of alcohol. Beer is manufactured by yeast fermentation transforming sugary substances to alcohol with a resultant low pH and characteristic flavor.

3.3.3 Sourdough

Sourdough technology is popular in producing bread and cakes, characterized by a complex microbial ecosystem comprising of lactic acid bacteria and yeasts. The fermented sourdough is metabolically active and characterized by palatability and high sensory quality. In addition to Lactobacillus strains, species such as *Leuconostoc, Enterococcus Weissella, Lactococcus, Pediococcus,* and *Streptococcus* have been isolated from sourdough.

3.3.4 Fermented Pickles

Fermented pickles are prepared in a medium of dilute brine (2–5% salt), which encourages the growth of naturally found fermentation bacteria (particularly lactic acid bacteria) over a period of 1–2 weeks. These bacterial strains produce lactic acid, which further prevents the growth of food poisoning bacteria and other spoilage microorganisms. Further, these strains of bacteria can survive well under high acidic conditions prevalent in most pickled products. The fermentation type and rate are dependent on the amount of salt added. The microorganisms involved in traditionally fermented pickles are restricted to a few categories of bacteria, yeasts, and fungi. Several yeast species have been reported from traditional pickles, which mainly include *Brettanomyces spp., Candida spp., Hyphopichia spp., Cryptococcus spp., Issatchenkia spp., Debaryomyces spp., Dekkera spp., Galactomyces spp., Geotrichum spp., Hansenula spp., Hanseniaspora spp., Kazachstania spp., Kluyveromyces spp., Torulaspora spp., Metschnikowia spp., Pichia spp., Rhodotorula spp., Saccharomyces spp., Saccharomycodes spp., Schizosaccharomyces spp., Torulopsis spp., Saccharomycopsis spp., Trichosporon spp., Yarrowia spp.,* and *Zygosaccharomyces spp.* The use of fungal species (*Mucor, Monilla,* and *Rhizopus*) as fermenting agents enables the release of exocellular amylase capable of hydrolyzing starchy materials, eliminating previous steps of hydrolysis.

3.4 Effect of Food Microbes on Human Gut Microflora

The animal gut is an inhabitant of different microbes' bacteria, archaea, and fungi, collectively known as gut microflora. They play several roles in nutrient metabolism and maintain homeostasis in health conditions (Figure 3.2). The diversity and dominance of gut microbiota get severely affected by food nutrient and microbial composition. The consumption

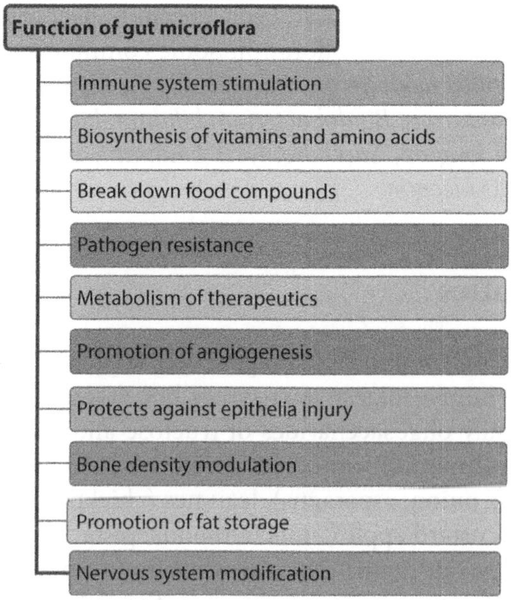

Figure 3.2 Depiction of major metabolic function played by gut microbiota.

of food material containing harmful bacteria can lead to colonization of bad bacteria in the human gut and negatively influences the healthy gut microbiota homeostasis. Similarly, uptake of high sugar and high-fat food can restrict insulin release. The intake of antibiotics and pesticides causes severe disturbance in gut microbiota by favoring the growth of only some species of microbes. The dominance of harmful bacteria in gut microbiota has also been responsible for increasing body mass index.

Therefore, it is necessary to intake food of good nutrient composition which carries a specific load of beneficial microbes. Probiotics and prebiotics are food supplements containing beneficial microbes that help in the recovery and maintenance of healthy gut microflora.

3.5 Probiotics

This food supplement contains live inoculum of beneficial bacteria. Many food products are naturally rich in beneficial microbes, such as yogurt, kefir, sauerkraut, tempeh, kimchi, miso, kombucha, natto, buttermilk, and pickles. The synthesized probiotic food supplements are available in the form of powder, capsule, and liquid drinks. Different probiotic supplements vary with bacterial strain types and concentration. Generally, it is taken to recover the normal gut microflora [6]. Some common strains of probiotics are *Lactobacillus acidophilus, Lactobacillus gasseri, Lactobacillus casei, Lactobacillus plantarum, Bifidobacterium bifidum, Lactobacillus bulgaricus, Bifidobacterium longum, Saccharomyces boulardii, Bifidobacterium lactis, and Enterococcus faecium.*

3.6 Prebiotics

These are fiber or some natural sugar food supplements which cannot be digested by human digestive enzymes but digestive to gut microbes. They are generally oligosaccharides of fructose and galactose molecules. There are several natural sources of prebiotics such as vegetables (chicory roots, garlic, onion, asparagus), legumes (chickpeas, lentil, soybeans), fruits (banana, custard apple), nuts (almonds, pistachio), and seeds (flaxseed, oats, barley, and bran). The concentration of prebiotics is very low in natural sources, so they are produced in the industry on a large scale. Prebiotics are the energy sources of gut microbes, so they can modulate the composition and function of gut microflora [7]. Thus, intake of the right amount of probiotics and prebiotics and general meals help keep homeostasis between harmful and beneficial gut microbiota.

3.7 Factors Affecting the Microbial Load in Food Products

The food spoilage due to microbial invasion results in dry matter loss due to carbohydrate consumption and the generation of a range of undesirable volatile organic compounds responsible for off-flavor and nutritional losses. It can also produce potentially harmful toxins, which could be harmful to both human beings and livestock.

The factors affecting the growth of microorganisms in food can be classified as intrinsic and extrinsic factors, briefly discussed in subsequent sections [8].

3.7.1 Intrinsic Factors

3.7.1.1 pH

The majority of the microorganisms grow and proliferate at the neutral pH range (6.6–7.5). The bacteria are more sensitive to pH changes than other microbe species among the microorganisms. However, mold and yeasts can survive over a wide range of pH. The yeast and molds are the main spoilage agents of fruits as they can survive well at pH <3.5. Vegetables having a higher pH is more prone to fungal and bacterial spoilage, while bacterial strains commonly infest meat and meat products (pH >5.6). While most of the food items possess an inherent acidity, few food products (fermented milk, sauerkraut, pickles) owe their pH values to the action of microorganisms. The letter is termed biological acidity.

Foods that can resist any change in their pH are referred to as buffered. For instance, the protein contents in meat offer good buffering capacity, thus resisting unwanted changes in pH of meat. A prolonged lag phase is noted when microorganisms grow beyond the optimum pH. This lag could be of longer duration in a highly buffered substrate. Thus, the length of the lag phase reflects the time required for the organisms to bring the external environment within their optimum growth conditions. Unfavorable pH conditions tend to affect the enzyme function of the microbial cells and their nutrient transport mechanism.

3.7.1.2 Water Activity (a_w)

The water activity of a food refers to the amount of water available for the growth and proliferation of microorganisms. Removal of moisture content/binding of moisture results in reduced $a_{w,}$ resulting in retarded growth of microbes. The water activity is related to the food environment's relative humidity (R.H.) as R.H. = $a_w \times 100$. Most of the fresh food has a water activity of above 0.99. The microbial species, based on the requirement of water activity, are classified as hygrophillic ($a_w = 1.00$), mesophilic ($0.95 < a_w < 1.00$), and xerophilic ($a_w < 0.95$). The water activity requirement of different classes of microorganisms varies as *bacteria > fungi > molds.* The gram-negative strains require higher a_w in comparison to the gram-positive strains. Bacteria often grow at $a_w > 0.91$. However, a few strains, such as *Clostridium botulinum*, survive at a_w as less as 0.86. Few halophilic bacteria can grow well at a_w in the range of 0.75. Yeast and mold can thrive over a wide range of a_w. Molds withstand $a_w > 0.80$, while xerophilic mold can withstand a_w as low as 0.65. Similarly, osmophilic yeasts can survive well at a_w as less as 0.61.

The lowering of a_w below the optimum range adversely affects the metabolic mechanisms delaying the microbial growth, growth rate, and final microbial load. The a_w is also affected by the other interacting factors such as pH, incubation temperature, and oxidoreduction potential of the food product. For instance, studies have shown that when the food media's pH and storage temperature are reduced, the minima a_w for microbial growth is higher. The water activity of the food products can be reduced by techniques such as drying, the addition of preservatives, osmosis, etc.

3.7.1.3 Oxidation-Reduction (O/R) Potential

The O/R potential of the food is the ease with which a food gains or loses electrons. A potential difference develops when the electron transfer occurs within a food system, expressed in millivolts (mV). An oxidized substance exhibits a positive potential, and a reduced substance is characterized by negative potential. The O/R potential of a food is determined by several factors such as the O/R potential of the original food, poising capacity, oxygen tension of the atmosphere around the food, and exposure of the food to the atmosphere.

The anaerobic microbes (e.g., *Clostridium*) tend to grow well under reduced conditions, whereas the anaerobic microbes (e.g., *Bacillus* genus) prefer oxidized conditions. Microaerophiles are aerobes that can grow well in slightly reduced conditions (e.g., *lactobacillus, campylobacter*). Few microbes can grow well under aerobic and anaerobic conditions and are called facultative anaerobes.

Most plant-based food products have an O/R potential of 200-300 mV, while meat products have in the range of -200 mV. The cheese products have O/R potential in the range of -20 to -200 mV, which indicates that aerobic microorganisms (bacteria and yeast) are the common cause of food deterioration.

3.7.1.4 Nutrient Content

Various components of foods like water, carbon source, nitrogen source, minerals, vitamins, and some growth factors are essential for the growth and proliferation of microorganisms.

The water requirement of different classes of microorganisms has been discussed in detail in previous sections. The requirements of these components, excluding water, are least for moldy fungi the other microbes like yeasts, gram-positive bacteria, and gram-negative bacteria.

The major energy sources for microbial growth include alcohols, sugars, and amino acids. However, few microbes can utilize starches, celluloses, and fats by breaking them into simpler forms and subsequent assimilation.

The main sources of nitrogen for microorganisms are amino acids. Microbes can also utilize proteins, nucleotides, and peptides as a nitrogen source.

Microorganisms require trace quantities of vitamins for growth and survival. The molds can synthesize almost all of their required food components. This unique ability of molds combined with conducive growth conditions (less pH and positive O/R potential) makes them the major spoilage factors in fruit-based food products.

3.7.1.5 Antimicrobial Components

Antimicrobial constituents are food components that render stability against microbial invasion. Eggs and milk contain lysozyme, and these products also contain conalbumin, making them a fairly efficient antimicrobial system. The derivatives obtained from hydroxycinnamic acid, ferulic acid, coumaric acid, etc., are a group of antifungal and antibacterial constituents mainly present in fruits, vegetables, tea, molasses, and other plant-based food sources. Glucosinolates in cruciferous plants render antifungal properties. The lactoperoxidase is a natural inhibition agent to microbial growth in bovine milk.

3.7.1.6 Biological Structures

These include the structural components (nutshell, fruit peels, etc.) of foods that provide remarkable protection against damage by spoilage organisms. In the case of nuts with damaged shells, nutmeats are readily susceptible to mold spoilage. Eggs stored under proper temperature and humidity are well preserved due to outer shells and inner membranes. Spoilage microorganisms easily attack fruits with damaged surfaces. Similarly, the skin of meat and fish products serves as a physical barrier to the entry of microbes.

3.7.2 Extrinsic Factors

These factors include different properties of the storage conditions affecting both the foods' stability and their microbial infestation level.

3.7.2.1 Temperature of Storage

The foodborne microorganisms can survive over an extended temperature range (-34°C to 100°C). The microbes can be classified based on the requirement of the growth temperature as psychrotrophs, mesophiles, and thermophiles. Psychrotrophs grow well at or below 7°C with optima between 20°C and 30°C (e.g., *Flavobacterium, Lactobacillus, Micrococcus, Pseudomonas*). They are responsible for the deterioration of fish, meats, eggs, poultry, and other refrigerated foods. Mesophiles can grow well between 20°C and 45°C with optima temperature between 30°C and 40°C (e.g., yeasts). Thermophiles can also grow above 45°C with optima between 55°C and 65°C (e.g., *Bacillus* and *Clostridium*). Among microbes, molds can survive over a large range of temperatures than bacteria.

3.7.2.2 Relative Humidity of Environment

Perishable foods should be kept under low R.H. conditions. High R.H. causes rapid surface spoilage of foods such as meat, further leading to deep spoilage. Although it is possible to minimize the possibilities of spoilage in foods matrixes by maintaining under low R.H., care should be taken to prevent food dehydration.

3.7.2.3 Level of Gases in the Storage Environment

Carbon dioxide can be used to prolong the shelf life of a different variety of food products. The CO_2-rich environment under hyperbaric conditions has been found to be more effective in stabilizing food products. The use of CO_2 under low-temperature conditions have been found to reduce the pH of food product, further increasing the number of hurdles involved. The presence of CO_2 has been found to inhibit the metabolic pathways of the microbial cells, eventually leading to cell death.

3.7.2.4 Competitive Microorganisms

The ability of a microbe to inhibit/destroy closely related food spoilage organisms in the mixed media accounts for an interference relationship. The lactic antagonism (lactic acid bacteria-killing closely related species) is one of the best examples of the same. The killing can take place due to various reasons such as the production of antimicrobial compounds, decrease in pH, hydrogen peroxide, lack of nutrients, diacetyl, and bacteriocins or bacteriocin-like factors. *Propionibacterium freudenreichii* sub sp. *shermanii* in pasteurized skim is associated with a multicomponent inhibitory system (e.g., 3-hydroxypropionalaldehyde), effective against used gram-negative bacteria and molds that may invade cottage cheese.

3.8 Food Spoilage

It is said that the dried food product must be kept dry or heat-processed food must be heated enough to kill all the microbes, and the concentration of preservative should also be high enough to inhibit the microbial growth. Even after following all these attributes of food storage, some microbes like *Xeromyces bisporus*, and *Zygosaccharomyces rouxii* cannot be processed out. Since these are extreme xenophiles. Other microbes like *Byssochlamys spp.*, *Neosartorya fischeri*, and *Talaromyces spp.* are highly heat resistant. *Zygosaccharomyces bailii* is yeast that can grow at a very high concentration of preservatives. Therefore, it is impossible to achieve food products free from microbes, and it is due to changing climatic conditions and improper handling [9]. Thus, improper handling and storage of food product may lead to infestation with harmful microbes that causes food spoilage. Sometimes food products are kept in storage for more than their self-life, leading to enzymatic degradation of food and making it unfit for human consumption. Microbial growth can deteriorate the food nutrient level and decreases its self-life. It may cause decoloration, foul smells, gas, and slime which causes huge wastage of consumable food and economic losses. The World Health Organization (WHO) and Food and Agricultural Association (FAO) of the United Nations have reported that one-third of the total consumable food produced worldwide is wasted annually [10]. In India, wastage is around 40 percent of total production, and its monetary value is around 14 billion US dollars. The food spoilage is mainly caused by the bacterial species belonging to the genus *Pseudomonas, Acinetobacter, Erwini, Brochothrix, Bacillus, Shewanella,* and *Clostridium* and fungal species of genus *Aspergillus, Penicillium, Phytophthora, Botrytis,* and *Alternaria* [11].

3.9 Foodborne Disease

Intake of food contaminated with microbes or with toxic compounds causes food poisoning as well as foodborne diseases. As per the world health organization, 600 million people fall sick due to consuming contaminated food products, out of which 33 million die every year [12]. The common symptoms of foodborne disease are nausea, headache, vomiting, abdominal pain, and diarrhea, but sometimes it affects severely, causing organ failure and lethal. These diseases are mainly caused by the bacteria belonging to the genus *Escherichia*, *Brucella*, *Campylobacter*, *Clostridium*, *Shigella*, *Salmonella*, *Mycobacterium*, and *Yersinia* [13–15]. Similarly, fungi belonging to the genus *Alternaria*, *Fusarium*, *Aspergillus*, *Candida*, and *Mucoromycetes*, and mycotoxins produced by them are also responsible for several foodborne diseases [16]. Parasites such as Toxoplasma are also one of the major culprits which cause food poisoning. Some virus-mediated food poisoning is also very commonly reported, mainly caused by Norovirus or Norwalk virus [17]. The World health organization (WTO) has reported that around 200 diseases are caused by eating contaminated food.

3.10 Analysis of Microbial Contaminants of Food

Management of food safety is one of the major concerns in the developed and the undeveloped country. In India, the government has established a food safety and standard authority, an autonomous body that regulates food safety standards. The major food contaminants and toxicants are adulterants, pesticides, toxins, preservatives, and microbial agents. Rapid detection of microbial contaminants in food products can save losses incurred due to food spoilage [18]. Several tools and techniques are available for the detection and quantification of microbial load in food products. The use of a particular technique depends upon microbial types, their load, the complexity of the food matrix, risk level, and the availability of technology platforms. Food products are highly heterogeneous and thus need to follow an extensive pre-treatment procedure for using any analytical methods for microbial analysis.

3.10.1 Traditional Method

Traditionally microbial contaminants are identified based on the morphological or generic features of the bacterial or fungal colony, such as size, shape, appearance, color, and spore formation. It starts with sample preparation, serial dilution, plating, enumeration, and isolation of pure bacterial colonies. The microbial loads are enumerated by counting CFU (colony forming unit) per gram or ml of food sample. The pure bacterial or fungal colonies were isolated from the food samples following serial dilution and plating methods. The use of selective or differential media is based on identifying the specific class of microbes [19]. The pure isolated culture of microbes is identified using testes like gram staining of bacteria to identify gram-positive or gram-negative bacteria. There are several disadvantages associated with this method. Counting colony-forming units is a slow process and does not indicate actual microbial activity. Moreover, preparation of culture medium, plate inoculation, counting colonies, and biochemical characterization is laborious [20]. These limitations have encouraged the focus of researchers towards the development of rapid isolation and identification methods.

3.10.2 Microscopic Method

Bacteria can be identified and enumerated directly by using a high-resolution microscope with an image analysis system such as epifluorescence and flow cytometry. In the epifluorescence technique, the microbes are trapped on polycarbonate members and then placed in the microscope to analyze viable and unviable bacteria [21]. This method has been used to detect and quantify bacteria in milk [22]. The detection limit of this method is 10^4 to 10^5 bacteria per ml of sample. Flow cytometry is a very powerful technique that works on the principle of scattered light by the cells. It differentiates the microbes based on their optical characteristics. The analysis can be performed by making a suspension of the cell. The cells can also be stained with fluorescent stains to analyze the viability and their metabolic conditions [23]. This technique has been used for the analysis of microbial load in milk [24], wine [25], and water [26].

3.10.3 Biochemical Method

The biochemical tests available for analysis of microbial contamination in food products are based on phenotypic and biochemical characteristics. Most of the available commercial microbial detection kits are based on the biochemical method. Some bacteria produce specific granules in the cytoplasm, which can be used as an identification marker for those bacteria. Some bacteria of genus *Pseudomonas* specifically produce a carbon source poly beta-hydroxybutyrate (PHB). The biochemical characteristics such as the source of carbon utilization, enzymes production such as catalases or oxidases, pH specificity, chemical sensitivity, and other specific metabolic activities such as the production of volatiles or non-volatiles fatty acids have also been targeted for the identification of particular bacterial species. These techniques can be automated fully or partially.

The most advanced biochemical analysis technique is mass spectrometry (matric assisted laser desorption/ionization-time of flight, MALDI-TOF). In this technique, the microbes are plated on mass with a special matrix, irradiated with high-intensity laser light, which results in the production of ions from the various biochemical compound of the microbes, and the ions are further accelerated through a mass spectrometer and determines their mass to charge ration (m/z) reaching at a different time to the detector. The signals are plotted as intensity versus m/z ratio yields a mass spectrum. A particular mass spectrum is uniquely related to a specific biochemical composition of microbes. The unknown microbes can be identified by analysis resulting mass spectrum with respect to the reported mass spectrum of specific microbes. Barreiro *et al.* [27] have used MALDI-TOF for identification of *Staphylococcus aureus*, *E. coli*, and *Streptococcus spp.* in milk. Similarly, the MALDI-TOF method has been used to detect Lactic acid bacteria in fermented vegetables [28], and *Streptococcus parauberis* is in seafood [29].

3.10.4 Immunological Method

Advances in biotechnology have led to the development of highly specific antibody-based tests for rapid (<10 min) identification and quantification of microbes and toxins in food. The affinities of monoclonal or polyclonal antibodies towards antigens (proteins) form the basis of these immunochemical tests. Quantification of the antibody-antigen reaction is done by measuring the absorbance spectrophotometrically. However, in some instances, the binding events and the resulting complexes are amplified with the aid of enzymes, fluorophores, and radioisotopes for enhanced signal recognition. The antigen-antibody-based techniques have gained popularity owing to their high specificity and sensitivity even in the presence of contaminants. In addition, these methods are less time-consuming and do not require a skilled operator for troubleshooting. The primary immunochemical methods used for the analysis of microbial and toxins contamination in food are Enzyme-Linked Immune Sorbent Assay (ELISA), Radioimmunoassay (RIA), Fluorescence *in situ* hybridization (FISH), immune-biosensor, etc. ELISA detection system has been used to detect E. coli and *Salmonella* in milk, vegetable, and ground beef [30, 31].

3.10.5 Molecular Methods

The molecular techniques fundamentally depend on polymerase chain reaction (PCR) technology. The three fundamental steps of PCR include (i) selection of a specific region of DNA/RNA for primer development; (ii) extraction of DNA/RNA; and (iii) amplification of target DNA/RNA region in the sample. The selection of target DNA can be attained through two ways, i.e., by using specific sequence information from databases, thereby enabling the primers to be designed across conserved and variable regions or through cloning and sequencing of the arbitrary regions of the bacterial or fungal genome. Several detection methods have been developed for the detection of foodborne microorganisms based on PCR principle, such as Fluorescence *in situ* hybridization (FISH), real-time PCR [32], multiplex PCR [33, 34], microarray [35], and next-generation sequencing

(NGS). The PCR-based detection method is a more preferred method than ELISA for detecting *Salmonella* in seafood [36].

The introduction of next-generation sequencing (NGS) has led to the analysis exceedingly higher number of nucleic acid sequences through shotgun sequencing (sequencing of the complete microbial nucleic acids) and targeted sequencing (gene-specific sequencing), permitting a more in-depth description of the microbial constituents. Many different NGS platforms as 454 (Roche), Illumina, SOLiD and Ion Torrent (Life technologies), and PacBio (Pacific Biosciences) systems are currently available [37]. The major disadvantages of NGS include high cost, laborious procedure, and biased results for dominant microbial species [38].

3.10.6 Chromatographic Methods

Chromatographic methods are the most widely and routinely used technique for analyzing toxins produced by a microbial contaminant in food products. This technique works on the principle of interaction between a stationary phase and a mobile phase. The constituents separate between the stationary and mobile phases. The stationary bed can be both solid and liquid, and the mobile phase (liquid, gas, or supercritical fluid) usually penetrates through or along the stationary bed. The nature of the mobile phase decides the name of any particular chromatographic method (Liquid/Gas/Supercritical fluid). The mobile phase carries the analyte along with it. The different constituents in the analytes travel at varying speeds leading to differential partitioning of the constituents between the mobile and stationary phases. The popular chromatographic techniques used for the detection of different volatile secondary metabolites and toxins are thin-layer chromatography (TLC), gas chromatography (G.C.), and high-performance liquid chromatography (HPLC). Alexandrakis *et al.* [39] have detected metabolites produced by the microbial agent responsible for the spoilage of Irish chicken using HPLC.

3.10.7 Biosensor

Biosensors detect the binding of complementary components coupled with a signal transducer like graphite, gold, and carbon. The recognition element selectively reacts with the target molecule, such as enzymes, antigens, and antibodies. The change in the physical variable resulting from the reaction is converted into a measurable signal by the transducing element. The biosensors are categorized based on the type of receptor molecules such as enzyme, DNA, cell, phage, antibody, and based on signal transducer such as piezoelectric, optical, and electrochemical [40]. Pal *et al.* [41] have used an electrochemical biosensor to detect *B. cereus* in strawberries, alfalfa sprouts, lettuce, tomatoes, cooked corn and fried rice. Similarly, biosensor has been used for the detection of *Listeria monocytogenes* in milk, lettuce, ground beef [42]; *E. coli* in milk, apple juice, and ground beef extract [43]; and *Cronobacter spp* in infant milk powder [44].

3.11 Conclusion

Microbes, especially bacteria and fungi, have benefited humans from ancient times to produce beverages and have also been used as food or drugs. Many microbes have been employed directly or indirectly in food and drug applications. A large pool of food products is also manufactured with the help of microbes. A section of microbes is also found to be harmful to food quality and human safety. They are responsible for food spoilage and foodborne diseases in humans and animals. Hence detection of microbes using different methods is of utmost importance. Several rapid detection methods have now evolved and replaced the traditional techniques. Proper understanding of the interaction between different microbes and the food environment is essential for effective use in the food industry.

References

1. Tofalo, R., Perpetuini, G., Di Gianvito, P., Arfelli, G., Schirone, M., Corsetti, A., Suzzi, G., Characterization of specialized flocculent yeasts to improve sparkling wine fermentation. *J. Appl. Microbiol.*, 120, 6, 1574–1584, 2016.
2. Demain, A.L., Phaff, H.J., Kurtzman, C.P., The industrial and agricultural significance of yeasts, in: *The Yeasts*, pp. 13–19, Elsevier, Amsterdam, Netherlands, 1998.
3. Copetti, M.V., Fungi as industrial producers of food ingredients. *Curr. Opin. Food Sci.*, 25, 52–56, 2019.
4. Połaska, M. and Sokołowska, B., Bacteriophages – A new hope or a huge problem in the food industry. *AIMS Microbiol.*, 5, 4, 324, 2019.
5. Fernández, L., Gutiérrez, D., Rodríguez, A., García, P., Application of bacteriophages in the agro-food sector: A long way toward approval. *Front. Cell. Infect. Microbiol.*, 8, 296, 2018.
6. Nagpal, R., Kumar, A., Kumar, M., Behare, P.V., Jain, S., Yadav, H., Probiotics, their health benefits and applications for developing healthier foods: A review. *FEMS Microbiol. Lett.*, 334, 1, 1–15, 2012.

7. Davani-Davari, D., Negahdaripour, M., Karimzadeh, I., Seifan, M., Mohkam, M., Masoumi, S.J., Berenjian, A., Ghasemi, Y., Prebiotics: Definition, types, sources, mechanisms, and clinical applications. *Foods*, 8, 3, 92, 2019.
8. Jay, J.M., Intrinsic and extrinsic parameters of foods that affect microbial growth, in: *Modern Food Microbiology*, pp. 38–66, Springer, Boston, MA, 1998.
9. Pitt, J., II and Hocking, A.D., Spoilage of stored, processed and preserved foods, in: *Fungi And Food Spoilage*, pp. 401–421, Springer, Boston, MA, 2009.
10. Food and Agricultural Association, *Global Food Losses and Food Waste*, Düsseldorf, Germany, 2011, https://www.fao.org/food-loss-and-food-waste/flw-data.
11. Rawat, S., Food spoilage: Microorganisms and their prevention. *Asian J. Plant Sci. Res.*, 5, 4, 47–56, 2005.
12. World Health Organization, Fact sheets on food safety, 2017. https://www.who.int/news-room/fact-sheets.
13. Adley, C.C. and Ryan, M.P., The nature and extent of foodborne disease, in: *Antimicrobial Food Packaging*, pp. 1–10, Academic Press, Limerick, Ireland, 2016.
14. Lamas, A., Regal, P., Vázquez, B., Miranda, J.M., Cepeda, A., Franco, C.M., Salmonella and campylobacter biofilm formation: A comparative assessment from farm to fork. *J. Sci. Food Agric.*, 98, 4014–4032, 2018.
15. McCabe, E., Burgess, C.M., Lawal, D., Whyte, P., Duffy, G., An investigation of shedding and super-shedding of Shiga toxigenic Escherichia coli O157 and E. coli O26 in cattle presented for slaughter in the Republic of Ireland. *Zoonoses Public Health*, 66, 83–91, 2019.
16. Benedict, K., Chiller, T.M., Mody, R.K., Invasive fungal infections acquired from contaminated food or nutritional supplements: A review of the literature. *Foodborne Pathog. Dis.*, 13, 7, 343–349, 2016.
17. Gupta, R., Gupta, A., Singh, R.P., Singh, P.K., Singh, R.L., Food allergies, in: *Food Safety and Human Health*, pp. 99–125, Academic Press, India, 2019.
18. Srey, S., Jahid, I.K., Ha, S.D., Biofilm formation in food industries: A food safety concern. *Food Control*, 31, 572–585, 2013.
19. Kuria, J.K.N., Wahome, R.G., Jobalamin, M., Kariuki, S.M., Profile of bacteria and fungi on money coins. *East Afr. Med. J.*, 86, 4, 151–155, 2009.
20. Zhao, X., Lin, C., Wang, J., Oh, D., Advances in rapid detection methods for foodborne pathogens. *J. Microbiol. Biotechnol.*, 24, 3, 297–312, 2014.
21. Jasson, V., Jacxsens, L., Luning, P., Rajkovic, A., Uyttendaele, M., Alternative microbial methods: An overview and selection criteria. *Food Microbiol.*, 27, 6, 710–730, 2010.
22. Rosmini, M.R., Signorini, M.L., Schneider, R., Bonazza, J.C., Evaluation of two alternative techniques for counting mesophilic aerobic bacteria in raw milk. *Food Control*, 15, 1, 39–44, 2004.
23. Flint, S., Drocourt, J.L., Walker, K., Stevenson, B., Dwyer, M., Clarke, I., McGill, D., A rapid, two-hour method for the enumeration of total viable

bacteria in samples from commercial milk powder and whey protein concentrate powder manufacturing plants. *Int. Dairy J.*, 16, 4, 379–384, 2006.
24. Gunasekera, T.S., Veal, D.A., Attfield, P.V., Potential for broad applications of flow cytometry and fluorescence techniques in microbiological and somatic cell analyses of milk. *Int. J. Food Microbiol.*, 85, 3, 269–279, 2003.
25. Malacrinò, P., Zapparoli, G., Torriani, S., Dellaglio, F., Rapid detection of viable yeasts and bacteria in wine by flow cytometry. *J. Microbiol. Methods*, 45, 2, 127–134, 2001.
26. Delgado-Viscogliosi, P., Simonart, T., Parent, V., Marchand, G., Dobbelaere, M., Pierlot, E., Pierzo, V., Menard-Szczebara, F., Gaudard-Ferveur, E., Delabre, K., Delattre, J.M., Rapid method for enumeration of viable *Legionella pneumophila* and other *Legionella* spp. in water. *Appl. Environ. Microbiol.*, 71, 7, 4086–4096, 2005.
27. Barreiro, J.R., Gonçalves, J.L., Braga, P.A.C., Dibbern, A.G., Eberlin, M.N., Veiga dos Santos, M., Non-culture-based identification of mastitis-causing bacteria by MALDI-TOF mass spectrometry. *J. Dairy Sci.*, 100, 2928–2934, 2017.
28. Nguyen, D.T.L., Van Hoorde, K., Cnockaert, M., De Brandt, E., Aerts, M., Vandamme, P., A description of the lactic acid bacteria microbiota associated with the production of traditional fermented vegetables in Vietnam. *Int. J. Food Microbiol.*, 163, 1, 19–27, 2013.
29. Fernández-No, I.C., Böhme, K., Calo-Mata, P., Cañas, B., Gallardo, J.M., Barros-Velázquez, J., Isolation and characterization of Streptococcus parauberis from vacuum-packaging refrigerated seafood products. *Food Microbiol.*, 30, 91–97, 2012.
30. Shen, Z., Hou, N., Jin, M., Qiu, Z., Wang, J., Zhang, B., Wang, X., Wang, J., Zhou, D., Li, J., A novel enzyme-linked immunosorbent assay for detection of Escherichia coli O157: H7 using immunomagnetic and beacon gold nanoparticles. *Gut Pathog.*, 6, 1, 1–8, 2014.
31. Bolton, F.J., Fritz, E., Poynton, S., Jensen, T., Rapid enzyme linked immunoassay for detection of Salmonella in food and feed products: Performance testing program. *J. AOAC Int.*, 83, 2, 299–303, 2000.
32. Ma, K., Deng, Y., Bai, Y., Xu, D., Chen, E., Wu, H., Li, B., Gao, L., Rapid and simultaneous detection of Salmonella, Shigella, and Staphylococcus aureus in fresh pork using a multiplex real-time PCR assay based on immunomagnetic separation. *Food Control*, 42, 87–93, 2014.
33. Ruan, J., Li, M., Liu, Y.P., Li, Y.Q., Li, Y.X., Rapid and sensitive detection of Cronobacter spp. (previously *Enterobacter sakazakii*) in food by duplex PCR combined with capillary electrophoresis-laser-induced fluorescence detector. *J. Chromatogr. B*, 921, 15–20, 2013.
34. Zhang, W., Yu, Y., Hertwig, F., Thierry-Mieg, J., Zhang, W., Thierry-Mieg, D., Wang, J., Furlanello, C., Devanarayan, V., Cheng, J., Deng, Y., Comparison of RNA-seq and microarray-based models for clinical endpoint prediction. *Genome Biol.*, 16, 1, 1–12, 2015.

35. Ehrenreich, A., DNA microarray technology for the microbiologist: An overview. *Appl. Microbiol. Biotechnol.*, 73, 255–273, 2006.
36. Kumar, R., Surendran, P.K., Thampuran, N., Evaluation of culture, ELISA and PCR assays for the detection of salmonella in seafood. *Lett. Appl. Microbiol.*, 46, 2, 221–226, 2008.
37. Glenn, T.C., Field guide to next-generation DNA sequencers. *Mol. Ecol. Resour.*, 11, 5, 759–769, 2011.
38. den Besten, H.M., Amézquita, A., Bover-Cid, S., Dagnas, S., Ellouze, M., Guillou, S., Nychas, G., O'Mahony, C., Pérez-Rodriguez, F., Membré, J.M., Next generation of microbiological risk assessment: Potential of omics data for exposure assessment. *Int. J. Food Microbiol.*, 287, 18–27, 2018.
39. Alexandrakis, D., Brunton, N.P., Downey, G., Scannell, A.G., Identification of spoilage marker metabolites in Irish chicken breast muscle using HPLC, G.C.–M.S. coupled with SPME and traditional chemical techniques. *Food Bioproc. Tech.*, 5, 5, 1917–1923, 2012.
40. Velusamy, V., Arshak, K., Korostynska, O., Oliwa, K., Adley, C., An overview of foodborne pathogen detection: In the perspective of biosensors. *Biotechnol. Adv.*, 28, 2, 232–254, 2010.
41. Pal, S., Ying, W., Alocilja, E.C., Downes, F.P., Sensitivity and specificity performance of a direct-charge transfer biosensor for detecting Bacillus cereus in selected food matrices. *Biosyst. Eng.*, 99, 461–468, 2008.
42. Kanayeva, D.A., Wang, R., Rhoads, D., Erf, G.F., Slavik, M.F., Tung, S., Li, Y., Efficient separation and sensitive detection of Listeria monocytogenes using an impedance immunosensor based on magnetic nanoparticles, a microfluidic chip, and an interdigitated microelectrode. *J. Food Prot.*, 75, 11, 1951–1959, 2012.
43. Waswa, J., Irudayaraj, J., DebRoy, C., Direct detection of E. coli O157:H7 in selected food systems by a surface plasmon resonance biosensor. *LWT–Food Sci. Technol.*, 40, 187–192, 2007.
44. Xu, F., Li, P., Ming, X., Yang, D., Xu, H., Wu, X., Shah, N.P., Wei, H., Detection of cronobacter species in powdered infant formula by probe-magnetic separation PCR. *Int. J. Dairy Sci.*, 97, 10, 6067–6075, 2014.

4

Food Preservatives From Microbial Origin: Industrial Perspectives

Lakshmayya N.S.V.[1†], Swarna Lekhya Y.[1†], Yugal Kishore Mohanta[2], Sanjeeb Kumar Mandal[1], Dinesh Chand Agrawal[3,4] and Bishwambhar Mishra[1*]

[1]*Department of Biotechnology, Chaitanya Bharathi Institute of Technology, Hyderabad, India*
[2]*Department of Applied Biology, University of Science and Technology, Meghalaya, India*
[3]*Department of Biotechnology, Faculty of Engineering and Technology, RAMA University Kanpur, India*
[4]*Department of Biotechnology, School of Life Sciences and Technology, IIMT University, Meerut, India*

Abstract

Biological preservatives derived from plants, animals, and bacteria have gained favor because they are more dependable, ethical, and safe. The food industry is looking for natural antimicrobials due to consumer preferences for the additives derived from nature and worries about the assurance of manufactured preservatives. Such preservatives are more stable, but they are also cheaper and more widely available in nature. Bacteriocins, enzymes, plant extracts, peptides, and fermented substances are examples of natural antimicrobials. Numerous *LAB* bacteriocins have promising implementations in the food market preservation, and their application in the food market can aid in alleviating the use of additives and preservatives along with the severity of thermal processing, culminating in food products that are effectively preserved naturally and have more organoleptic as well as nutritional properties. Bacteriocins are adjoined to foods in various ways like food flavoring agents, additives, mean life prolongers, or innards, perhaps can be manufactured in situ using bacteriocin forming initiators, add-ons, or conservational factors. Bacteriocins have received huge attention in recent years as part of hurdle technology, and they show enhancing or combinational effects when

**Corresponding author*: mishra.bishwambhar@gmail.com
†These authors contributed equally to this work

Sanjay Kumar, Narendra Kumar and Shahid-ul-Islam (eds.) Role of Microbes in Industrial Products and Processes, (81–106) © 2023 Scrivener Publishing LLC

blended with many more microbicidal substances. High-pressure processing or pulsed electric fields are physical treatments that allow for more effective food preservation. This chapter explores the vital functions and corresponding action mechanisms of natural preservatives obtained from various microbiological sources in recent years. This will provide thorough information for food additives and preservatives in use.

Keywords: Bacteriocins, preservatives, food, nisin, LAB, microorganisms

4.1 Introduction

Food preservation involves incorporating science-based insights and a range of innovative techniques and approaches to prevent food degradation and spoilage, extend shelf life, and assure customers of a pathogen-free product. The shelf-life of a product is the amount of time it takes for it to decay to an unacceptable level. Qualitative characteristics such as aroma, appearance, color, and many other organoleptic characteristics will be lost in damaged foodstuffs. Nutritional advantages are also impaired by food decay. Physical, biological, microbiological, chemical, and biochemical factors can contribute to food deterioration [1]. Preservation strategies, including effective postharvest care before the processing of both plant and animal meals, should be implemented as early as possible in the food production pipeline [21]. The estimated big world population in 2030 and 2050 has gotten a lot of attention to influence global food consumption. Apart from population expansion, income growth is also a major driver of food demand [43].

The Food Loss and Waste Protocol (FLW Protocol), a global multi-stakeholder coalition, provides worldwide standard nomenclature for defining, measuring, and reporting FLW concerning key stakeholder goals (e.g., businesses, NGOs, etc.) [32]. According to the FLW protocol, FLW stands for "the weight of food and/or associated inedible items removed from the food supply chain" [32]. Food losses (FL) are decreases in edible food mass in the FSC during the production, postharvest, and processing stages, even before it reaches the market.

Food waste (FW) is defined as the food of acceptable quality and suitability for human eating that is discarded at the end of the food supply chain (distribution and consumption) [36].

Owing to the overwhelming market for safe, eco-friendly preservatives for fragile storage, hurdle technology serves to safeguard foodstuffs from microorganism-caused spoiling to a certain degree. An antibacterial chemical synthesized by food-grade microbes as a preservative for commercial use is outlined in this chapter.

4.2 The Need for Food Biopreservation

Animal-derived consumables are highly perishable due to their high nutritional content, moisture, and neutral pH. These foods should be adequately stored to ensure quality and safety. Human illness and disease epidemics manifest from failure to do so. Foodborne diseases are indeed a serious and costly public health problem that affects people worldwide. Consequently, superior processing parameters, appropriate standards of hygiene, and technologies are commonly utilized in the food market to assure the quality and safety of goods.

On the other hand, food preservation is critical for food quality and safety. Low-temperature preservation procedures like refrigeration and freezing are available, as are high-temperature preservation techniques such as pasteurization, sterilization, and chemical preservation. Modern preservation procedures are now widely used, such as bio-preservation, irradiation, and hurdle technologies.

When food is maintained using conventional practice, it alters its condition and loses some nutrients. As a logical consequence, contemporary methods are more effective in ensuring food safety and quality. It is also significant in today's world due to the globalization of the food market, the introduction of innovative foods, technological breakthroughs, and the necessity for sparsely processed products with ready-to-eat quality and prolonged shelf life [51]. From a 'farm to fork' perspective, bio-preservation is the most reliable of all the preservation methods employed today. It can retain excellent quality, hygienic conditions, and minimal nutritional and organoleptic losses while extending shelf life [35]. Bio-preservation is a method of food preservation that relies on the antibacterial characteristics of naturally occurring organisms and their metabolites. It has the potential to harmonize and rationalize the necessary safety standards with ancient preservation methods and modern food safety and quality demands [14].

4.3 Antimicrobial Peptides (AMPs) and Their Mode of Action

Bacteriocins affect various vital operations in microorganisms, including cell wall formation, replication, translation, and transcription of living cells, due to the wide range of chemical structures or types. Most bacteriocins work by generating pores or channels with a diameter of 0.2–1.2 nm that obliterate the energy potential of bacteria [46].

Responsive cells, such as the mode of action of Class I bacteriocins such as nisin, subtilin, and lacticin 3147, remove the proton motive force that obstructs amino acid transport.

Nisin is a commonly accepted and hypothesized surface-active bacteriocin that binds to the lipid II component of the bacterial cell membrane, stabilizes the pore complex, and causes cell rupture by degrading the sulfhydryl group. Both nisin and subtilin are Class Lantibiotics that form a compound with lipid I and lipid II and interfere with cell wall production. Short fragments of nisin at the N-terminus are accountable for the formation of pores inhibition. The lacticin molecule (which is made up of lac1 and lac 2) is coupled with the cytoplasmic membrane along the peptide named A1 before connecting to the cell wall's lipid II component. The above step helps trigger a cascade of events, resulting in a location for the A2 molecule to attach with high affinity, which causes holes to develop in the cell membrane. Figure 4.1 depicts mechanism of an antimicrobial peptide to stop the growth of food spoilage.

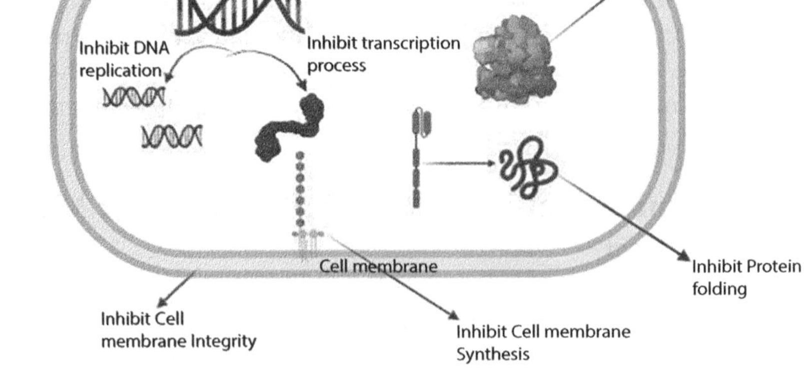

Figure 4.1 Mechanism of an antimicrobial peptide to stop the growth of food spoilage microorganisms.

4.3.1 Bacteriocin

AMPs and proteins are synthesized in huge quantities by various species, particularly bacteria. One significant type of antibacterial chemical is bacteriocins, which are proteinaceous antibacterial compounds performed by bacteria and show effect against species similar to the generating entities [4]. Initially, bacteriocins were generally characterized in colicins, which are large proteins with a molecular weight of up to 80 kDa initially generated from *E. coli* [65]. They could kill genetically agnate bacteria when they are attached to the internal membrane or other cellular targets. Bacteriocins are now known as *LAB*, which are small, thermodynamically favorable cationic peptides produced through gram-positive microorganisms and feature a much broader variety of inhibitory properties [7]. LAB-derived bacteriocins are often employed in food bio-preservation since LAB is historically regarded to be harmless for use in food. Bacteriocins have a wide range of antimicrobial activity against a wide spectrum of bacteria [65]. This clause implies that bacteriocins own a limited spectrum of action, which is fallacious. It also suggests that the parent strain may be naturally susceptible to the bacteriocin, necessitating immunization or inherent protection against it. Food-grade bacteria produce many bacteriocins during microbial fermentation, but only a few have been characterized [37].

Anomalous amino acid residues are found in many, but not all, bacteriocins. Many bacteriocins, including well-known examples like valinomycin and Gramicidin S, contain these atypical amino acids produced by non-ribosomal mechanisms. Even though bacteriocins are fascinating molecules, they have not received the same level of research as other antibiotics. One rationale for this is that their limited range of action limits their utility.

4.3.2 Nisin

Nisin is a ribosomal-derived peptide with broad antibacterial capabilities, including efficacy against many microorganisms that cause food spoilage [17]. Nisin's performance as a food preservative has led to its humongous use worldwide. Nisin is made from a food-grade bacterium's fermentation product [10]. Because nisin is still the only bacteriocin approved to be used as a preservative, it is becoming more commonly used in the food industry. Nisin has been THE GRAS (Generally Recognized as Safe) by the US Food and Drug Administration (FDA) since 1988 since it has long been used in food preservation (E234) without creating health problems. According to FDA regulations, this bacteriocin can only be used in cheese products at a maximum concentration of 250 ppm to inhibit *Clostridium botulinum* growth [20]. Nisin is efficacious against many pathogenic Gram-positive bacteria, including Listeria monocytogenes and *Clostridium botulinum*, and perhaps some Gram-negative pathogens, such as *Escherichia coli* and *Salmonella spp.*, when combined with chelating agents such as acid (EDTA), heat treatment, freezing, or any other treatment that causes the cell wall to become permeable. Nisin looks a lot like the antimicrobials known as lantibiotics, named after the unusual amino acid lanthionine. Lantibiotics have a lot of potential as food preservatives, but nisin has been studied intensively to see if it can be used for that purpose [2]. For various reasons, antimicrobial peptides produced by ribosomes are intriguing research topics. For one thing, because they're proteinaceous, they're physiologically safe because their core amino acids are digested by the digestive system, making them typical dietary suppliers like other proteins.

Another argument is that antimicrobial peptides produced by ribosomes are formed via transcription and translation routes, implying that their amino acid arrangements, and thus their forms and properties, are dictated by genomic sequences. The architecture of peptides can be changed by changing their codons. The antibacterial activity of nisin is primarily determined by its solubility in water and mechanical stability,

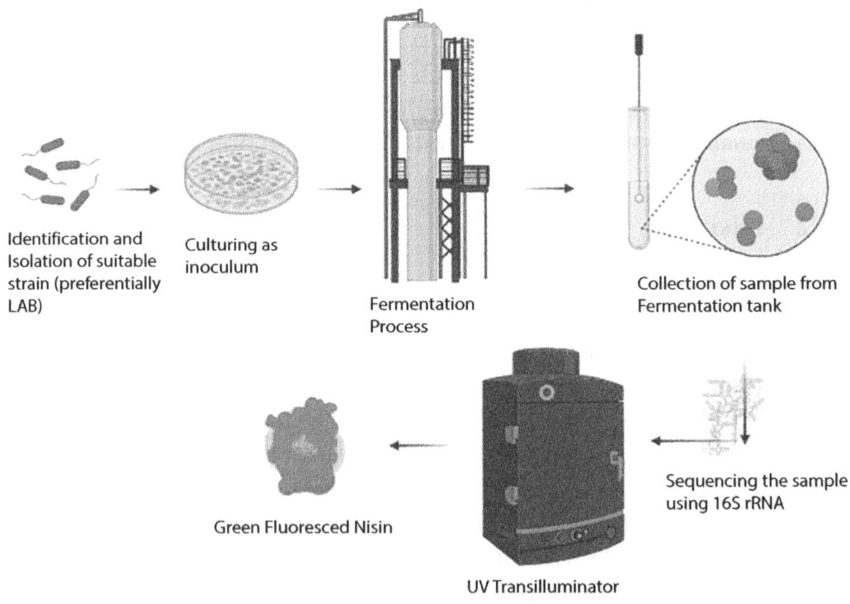

Figure 4.2 The mass production and characterization of nisin.

both of which are affected by ph. Indeed, nisin is more soluble and stable in acidic environments, with a solubility of 12 weight percent at pH 2.5 and 4 weight percent at pH 5.0. This solubility is approaching nil when the pH reaches and exceeds neutrality. Antibacterial activity is also stronger at acidic pH and decreases with increasing pH, which can be explained by an irreversible shift in the molecular structure of nisin [22]. Nisin activity is relatively stable at low temperatures (such as when the peptide is frozen), but it can be lost when heated over a lengthy period. The mass production and characterization of nisin has been illustrated in Figure 4.2.

4.3.3 Biosynthesis

Antimicrobial peptides can be derived from bacteria and fungi, and some well-known peptides include nisin, gramicidin, *Bacillus subtilis*, and *Bacillus brevis* from *Lactococcus lactis*, *Bacillus subtilis*, and *Bacillus brevis* [21, 48]. The biological expression has received increased attention due to the expensive cost of the chemical manufacture of AMPs. For expression systems, *Pichia pastoris*, *Saccharomyces cerevisiae*, and bacteria such as *Escherichia coli*, *B. subtilis*, and plants have been used, but it should be noted that AMPs are difficult to produce in *E. coli* due to toxicity, proteolytic degradation, and purification, which is required to take advantage of fusion tags [24].

Furthermore, various AMPs have been recovered and separated from plant stems, seeds, and leaves, and they are divided into different classes, including thionins, defensins, and snakins [29]. Increased marine-derived AMPs have been claimed to have contributed to people's choices increasing value for marine biodiversity. Several of the identified marine AMPs have shown promising outcomes *in vivo*; for example, As-CATH4 has an immunity-stimulating impact *in vivo* and can boost the anti-infective power of medications taken in combination with it. Myticusin-beta is an immune-related AMP produced by *Mytilus coruscus* that can be a viable alternative to antibiotics [30]. Furthermore, GE33, also known as pardaxin, is a marine AMP, and a vaccination based on it has been proven to improve anticancer immunity in mice [27].

4.3.4 Mode of Action

Antimicrobial peptides are unique molecules whose underlying mechanisms (MOA) have been extensively studied since their inception. Establishing the MOA of these AMPs is crucial for their potential use as therapeutics and preservatives. AMPs were originally thought to have only one method of action: membrane targeting, but new evidence is accumulating, indicating they have multiple ways of action [52]. The two main types of MOA are direct killing and immunological change.

4.3.4.1 Membrane Permeabilizing Mechanism of Action: Immediate Execution

It's feasible for membrane homing AMPs to interact with receptors or non-receptors. The receptor-mediated route is predominantly mediated by AMPs produced by bacterial species and effectual even if artificially synthesized in the sub micromolecular, like nisin. The Nisin peptide is composed of two zones, one which contains an excessive affinity for the cell wall precursor molecule, which is a membrane-anchored, that's lipid II molecule [31]. Membrane-rooted is another zone and aperture-fabricating domain. Mesentericin comes as one more example of a receptor-arbitrated membrane attacking AMP. In vertebrates and invertebrates, most AMPs target the membrane rather than receptors. These AMPs interact with membrane components and are often active against bacteria *in vitro* at micromolar doses. On the outside surfaces of Gram-positive and Gram-negative microscopic organisms, lipopolysaccharide and teichoic acids confer an add-up to negative charge on the fringe, allowing for the essential electrostatic interaction with cationic AMPs.

Much significantly, AMPs exploit a structural difference between bacterial membranes and multicellular eukaryotic membranes. The external thin layer of the phospholipids in the bacterial cell membrane is generally built of lipids with head groups of a negative charge, like cardiolipin, although in animal membranes, the external outlet is often composed of several zwitterionic membrane lipids; one of them is sphingomyelin and cholesterol is a neutral component. The preponderance of lipids with negatively charged head groups in animal membranes is found on the inner leaflet towards the cytoplasm. The positively charged AMP interacts strongly with the negatively charged lipids on the outer leaflet of the bacterial surface. Certain AMPs, as previously mentioned, have anticancer effects [15]. In cancer cells, the equilibrium of negatively charged phospholipids between the inner and outer membranes is disrupted, which enhances AMP interlinkage due to the phosphatidylserine (PS) present on the outer leaflet.

Furthermore, the increased membrane-spanning potential and membrane flexibility, as well as the overexpression of additional negatively charged biomolecules like O-ubiquitinated mucins, sialylated gangliosides, and heparan sulfate, allows AMPs to target cancer cells more effectively [26].

Aurein shows a high action against 50 cancer cell lines while inflicting minimal harm. Furthermore, certain AMPs are sensitive to lipid characteristics other than charge, i.e., they aren't just sensitive to charge in liposomes framed of phosphatidyl glycine, an anionic phospholipid tracked down generally in cell layers of microbes. Magainins can prompt leakage more efficiently than liposomes originating from negatively charged PS, the main phospholipid in animal membranes.

Plant cyclotides and other AMPs can bind to zwitterionic lipids like phosphatidylethanolamine (PE), prevalent inside the interior cytoplasmic leaflet of animal membranes and on the surface of bacterial membranes. These results indicate that transmembrane charge isn't the only factor impacting the initial connection; other characteristics, including curvature of the membrane, may be critical as well. Following the early hydrophobic and electrostatic interactions, the AMPs concentrate at the periphery and flat-pack within the membrane of bacterial cells once they reach a distinct convergence. To this point, several paradigms were illustrated to describe the operation of AMPs [5]. There are two types of models: transmembrane pore models and non-pore models. Barrel-stave and toroidal-pore types are categorized under transmembrane pore models. The AMPs are originally transverse to the membrane in the model of the barrel-stave, even though they are subsequently implanted straight down into the lipid bilayer. This improves oblique peptide-peptide interactions exactly as the membrane protein ion channel does. The amphipathic structure of the peptide (and/or sheet) is crucial in creating the pore because the water-repelling regions interface with the lipids on the membrane, and the water-attracting residues construct the channel lumen. A distinguishing characteristic is that AMPs under this group have a core size of 22 (helical) or 8 residues (sheet) to bridge the lipid bilayer. Only a few AMPs have been demonstrated to create barrel stave channels, including Pardaxin, Alamethicin, and Protegrin.

The peptides in the toroidal pore model also penetrate right angles in the lipid bilayer; however, no distinctive peptide-peptide interactions exist. Rather, the peptides produce a limited bending of the lipid multilayer, resulting in openings partially caused by the peptides and in part by the head groups of phospholipids [37]. The "toroidal pore" is a lipid-peptide supramolecule that is dynamic and ephemeral. The gross configuration of the bilayer differentiates this paradigm from the Barrel-stave pore: The hydrophilic and hydrophobic organization of the lipids is conserved in barrel-stave pores, but the hydrophilic and hydrophobic organization of the multilayer is disrupted in toroidal pores. This provides distinct surfaces for the lipid tail and lipid head groups to engage with. Several peptides begin to migrate to the inner cytoplasmic layer, wherein they access the cytoplasm

Table 4.1 Summary of AMPs synthesized by microorganisms and their activity towards killing pathogenic organisms.

S. no	Microorganisms responsible to produce AMPs	Physicochemical properties of AMP	Target organism	References
1.	*Leuconostoc gelidium*	• 37, 3930.3 Da Mol. wt	LAB	[53]
2.	*Brochothrix campestris*	• stable at 121°C for 15 min and inactivated by enzyme treatment	*C. botulinum*	[54]
3.	*E. faecium*	• 71 Da Mol. wt	Active against Gm + and Gm- bacteria	[54]
4.	*Streptomyces scopuliridis*	• 27,000 Da Mol. wt, • Stable at −20°C to 60°C and pH 3–9, Active against bacteria at pH 7	Inhibits Gm + and Gm- bacteria	[55]
5.	*Bifidobacterium longum*		Active against Gm + and Gm- bacteria	[56]
6.	*Lactobacillus casei*	• Optimum temperature 38°C	Active against Gm + and Gm- bacteria	[57]
7.	*Bacillus coagulans*	• 8000 Da Mol. wt	*S. aureus*	[58]

(Continued)

Table 4.1 Summary of AMPs synthesized by microorganisms and their activity towards killing pathogenic organisms. (*Continued*)

S. no	Microorganisms responsible to produce AMPs	Physicochemical properties of AMP	Target organism	References
8.	*Haloferax larsenii HA1*	• 14,000 Da Mol. wt, • Stable at 100°C, pH 5–9, organic solvents, surfactants, detergents	*Hfx. larsenii HA10*	[59, 66]
9.	*Pediococcus pentosaceous* NCDC 273	• 4600 Da Mol. wt		[60, 67]
10.	*Lactobacillus gasseri K7*		*Lactobacillus*	[61, 68]
11.	*Bacillus subtilis SN7*	• 3400 Da Mol. wt, • Stable at wide range of temp (50–121°C), pH 3–9, resistant to lipase and amylase	Gm+ *sps*	[62, 69]
12.	*Lactobacillus paracasei HD1–7*	• 10,000 Da Mol. wt, • Stable at 121°C, 20 min, pH 3–8	Gm + and Gm- bacteria	[63, 70]
13.	*Weissella confusa A3*	• 2700 Da Mol. wt, • Stable at 100°C, 30 min, pH 2–6	*B. cereus*	[64, 71]

and may address intracellular components, so apertures collapse quickly. Ion choice and discontinuous diameter are two further toroidal pore traits. Both these illustrations can cause cell membrane depolarization and death.

AMPs can also act without causing membrane holes to form. One of these models is referred to as Carpet. The AMPs bind to the bilayer membrane linearly, in this case, and reach a concentration high enough to blanket the membrane's surface, forming a "carpet." This causes unfavorable reciprocities on the membrane surface. As a result, the integrity of the cell membrane is harmed, causing a detergent-like action that causes the membrane to break down by forming micelles. The detergent-like concept describes how the membrane bilayer structure eventually collapses into micelles. The carpet model does away with the need for particular peptide-peptide interactions between peptide monomers that are membrane-bound and the peptide insertions to generate transmembrane channels in the hydrophobic core or specialized peptide structures. Many peptides work as antibacterial agents, regardless of their nucleic acid makeup or sequence length. Due to their amphiphilic nature, such AMPs frequently function in a carpet-like fashion and at high concentrations. AMPs that act through the carpet model include cecropin, aurein, and indolizidine.

The MOA of AMPs can be defined using several different models. In addition to the models stated above, the Shai-Huang-Matsazuki illustration, electroporation, and interfacial activity models are all comparable. Some models don't distinguish between them at all.

For example, a carpet-like process has been proposed as a pre-processing sequence for the toroidal pore model. In most studies, model membranes are utilized to determine the MOA of AMPs. The mechanism of merely one or two AMPs was investigated in complete bacterial cells using various imaging techniques. Other membrane models or test conditions are likely to produce different results, meaning that the illustrations presented here whether or not apply to what occurs in bacteria. Table 4.1 depicts comprehensive information regarding the AMPs synthesized by microorganisms and their activity towards killing pathogenic organisms.

4.3.4.2 Direct Killing: Non-Membrane Targeting Mechanism of Action

AMPs that do not target the membrane are classified into two groups: those that attack the bacterial cell wall and those that earmark the intracellular targets. Traditional antibiotics like penicillin can prevent the production of cell walls, and AMPs can do the same. Most classical antibiotics connect to fixed proteins that are involved in the development of cell wall components, whereas AMPs typically interact with a variety of antecedents that are required for cell wall production [16]. A prominent target is the lipid II molecule, which is very conserved. AMPs, like defensins, adhere to the lipid II molecule's negatively charged pyrophosphate sugar moiety. This imperative event can cause membrane breakdown and pore formation. AMPs like human defensin 1 and defensin 3 depend on preferential binding to provide bactericidal action. Since the advent of AMPs, it was presumed that they couldn't possess intracellular targets. Research with the original helical peptides, cecropin, magainin and melittin, demonstrated that the whole-d variant was just as effective as the natural all-l peptides, proving that stereospecific targets like RNA/DNA or proteins aren't required for antibacterial activity and that AMPs target the membrane. According to a subsequent study, different AMPs containing all-d or all-l amino acids did not have the same effect. Several AMPs are now recognized as intracellular targets, as they boost bacterial mortality despite not producing membrane permeabilization at the lowest effective dose. The above AMPs initially engage with the cell membranes subsequently aggregate within the cell, prompting vital functioning to be disrupted.

Protein/nucleic acid synthesis inhibition and enzymatic/protein activity disruption are among the novel intracellular targets discovered. Buforin II, a frog-derived histone-derived AMP, for example, binds to *E. coli* DNA and RNA without permeabilizing the bacterial membrane.

Human defensin 5 translocates into *Escherichia coli* cytoplasm and amass at the cell division plate and odd poles, implying that cytoplasmic targets may be responsible for some of the antibacterial activity. Indolizidine, Human Defensin 4,1, and PR-39 have all been discovered to attack intracellular bacterial components.

4.3.4.3 Immune Modulating Mechanism

Apart from directly killing bacteria, AMPs can also engage and stimulate immune cells, resulting in greater microbial death and/or inflammation management. Because AMPs are fabricated by a variety of immune cells, including macrophages and neutrophils, they are one of the pioneer molecules to interact with incoming bacteria. It is required to trigger an immunological response to attract other immune cells and control inflammation during infection. Surprisingly, some AMPs can trigger a series of immunological responses, including leukocyte attraction, activation, and differentiation; inflammation reduction by lowering proinflammatory chemokine production; angiogenesis stimulation; and chemokine and reactive oxygen/nitrogen species regulation. Human AMPs such as defensins and LL-37 use a chemo attractive method to lure immune cells such as leukocytes, dendritic cells, and mast cells. In addition, innate defense regulators (IDR), which are chemical replicas of native AMPs like IDR-1018 and IDR-1, subdue proinflammatory cytokines in mouse bioassays. IDR-1018 has proved potential in decreasing the inflammatory response in severe malaria cases, despite its direct anti-malaria activity. Mice given a combination of anti-malarial medications and IDR-1018 had less harmful brain inflammation, which would have resulted in death otherwise. Though majority AMPs have been demonstrated to have synergy with innate immune system components like macrophages and neutrophils, there is an attestation that they also affect the adaptive immune system, specifically B and T lymphocytes.

Although the exact mechanisms are uncertain, a recent study suggests that AMPs could be employed to improve penetration.

The above-mentioned studies imply that AMPs function in a range of maverick or cooperative "faceted" routes, making them promising candidates for further research.

4.4 Applications of AMPs for Food Preservation

4.4.1 Fruits and Vegetables

There are many different forms of rot bacterial pathogens in fruits and vegetables, and many traditional sterilization treatments, such as applying severe temperatures and extremely high voltage, can destroy tissues and cause nutritional loss, decreasing the food's quality [19]. Bioderived antibacterial peptides are unusual in that they have a broad antibacterial spectrum, can preserve product flavor and aroma at room temperature, may increase the shelf life of vegetables and fruits, and ensure food safety [34].

Consequently, they now have the option of replacing chemical preservatives in the preservation of fruits and vegetables. The identified RPT-0001, an antimicrobial peptide that inhibits Gram-negative and Gram-positive bacteria's function, used it to make silver nanoparticles. When compared to citrate-coated silver nanoparticles, RPT-0001 and SNPs showed good antibacterial additive effects, and so might be used in food preservation, packaging materials, and the treatment of foodborne illnesses [39, 45]. Nisin significantly delayed the deterioration of white jade at 4°C [38, 39]. Nisin not only extends the storage life of carrots by 15 to 25 days, but it also reduces freon emissions, which helps to protect the environment [25], combined 1% pullulan with an antibiotic peptide from Potassium *Bacillus subtilis* R21-4 and coated broccoli with the mixture at 10°C, finding that the peptide effectively delayed vegetable deterioration. Studied the effects of varied amounts of gluconolactone, chitosan, and antimicrobial peptides (produced by *Bacillus licheniformis* ES-2-4) on the retention of cherry tomatoes at 81°C [30]. The optimal combination was found to be 0.5 percent gluconolactone + 4 percent chitosan + antibacterial peptide standard solution [40, 42].

4.4.2 Animal Foods

For more than 50 years, antibiotics have been utilized in the animal business. Although its use has benefits, it has also generated issues, such as the 241 rise of antibiotic-resistant bacteria and drug residues in livestock products [50].

Enteric infections from animals are an immediate source of food contamination. The prohibition of antibiotics as growth promoters (AGPs) has animal nutrition to test, necessitating new ways to control and prevent harmful bacterial colonization. Modifying the gut microbiome with new feed additives like nutraceuticals to improve animal health is a current topic in animal breeding, and it has a lot of promise. Although our understanding of the impacts of such feed additives has grown, crucial information on their influence on the host is still lacking. The term "probiotic" has been defined in a variety of ways. The more frequently recognized definition is "live microorganisms that impart a health benefit on the host when delivered in suitable doses" [22]. This definition suggests that a probiotic's health benefit must be established. Intestinal microbial homeostasis management, gastrointestinal barrier function stability, bacteriocinogenesis, enzymatic activity boosting uptake and nourishment, immunostimulatory effects [49], procarcinogenesis suppression [3]. Competitive exclusion (CE), also known as the "Nurmi concept," stems from the discovery that feeding freshly hatched chicks with a suspension of gut content obtained from healthy adult chickens could protect them from *Salmonella* invasion of the stomach CE bacteria from gut material should be introduced early in childhood so that they can establish themselves in the gastrointestinal system and become competitive or antagonistic against opportunistic pathogens.

Prebiotics are "nondigestible food ingredients that beneficially affect the host by selectively stimulating the growth and/or activity of one or a limited number of bacteria in the colon." At least three requirements must be fulfilled for a nutritional substrate to be classified as a prebiotic: (1) The substrate must not be hydrolyzed or absorbed in the stomach or small intestine; (2) it had to be picky for favorable synergistic bacteria in the large intestine, such as bifidobacteria; and (3) fermentation of the substrate should result in advantageous luminal/systemic effects within the host [2].

Bacillus species are Gram-positive, spore-forming microbes found in water, soil, and the atmosphere. *Bacillus* species are allochthonous bacteria that enter the digestive tract through the inadvertent consumption of contaminated foodstuff. The use of live *Bacillus* spores as a probiotic supplement has prompted concerns about their safety: numerous *Bacillus* species are reported to create food poisoning when used as livestock feed supplementation, nutraceuticals, crop protection products, or seed coating agents [3]. Feed additives have been utilized in animal nutrition over the past couple of years to enhance nutrient uptake, metabolic parameters, and livestock productivity. AMP is part of animals' natural innate immune systems, and in plants, they constitute a defense mechanism that is analogous to animals' innate immunity, defending them from infections and pests [8]. The existence of even-numbered cysteine residues coupled by disulfide bridges, which provides AMP its exceptional stability, is a typical trait. Furthermore, their amino acid makeup, amphipathicity, helicity, cationic, and size allow them to enter lipid membranes and kill the target bacterium [2, 21].

Synthetic AMP A3 and P5 are two more kinds of AMP that have been employed [54] tested varied rising doses of AMP A3 as an alternative to conventional antibiotics, including 0, 60, and 90 mg/kg of diet. The study looked at the growth performance of weanling pigs and the coefficient of total tract apparent digestibility (CTTAD) of nutrients and intestinal characteristics. There was a linear increase in final body weight and average daily growth with no influence on average daily feed intake or feed efficiency. In addition, they observed a linear improvement on CTTAD of dry matter and crude protein, and no differences were observed between ileal total anaerobic bacteria, *Clostridium spp*., and coliform populations of pigs feed with AMP A3 compared to conventional antibiotics. Therefore, the results showed that AMP A3 is a potential molecule to replace antibiotics used as growth promoters in pigs.

4.5 Future Outlook and Limitations

4.5.1 AMPs and Nanotechnology

Nisin pertains to the lantibiotics class and is a bacteriocin with peculiar amino acids produced post-translational modification that is enzyme-mediated. This peptide comprises dehydrated threonine and serine residues, resulting in the amino acids dehydroalanine (DHA) and dehydrobutyrine (DHB) [23]. More unsaturated amino acids and cysteine residues generate lanthionine and B methyllanthionine residues, whose intermolecular crosslinking causes rings to develop in mature bacteriocin [20]. Furthermore, *Lactococcus lactis* produces nisin, a 34-amino-acid polypeptide. Nisin has two variations (A and Z), with histidine at position 27 in the A version and asparagine in the Z form [44]. The change in structure benefits nisin Z because it has an ability to diffuse and solubility, both of which are important for food [15]. Nisin is effective against a wide range of Gram-positive bacteria and is generally non-toxic to mammals [33]. As a result, it can help increase the shelf life of food while also protecting it from contamination. The following are some examples of nisin being put into a nanodelivery system to boost bioavailability.

Nisin has two variations (A and Z), with histidine at position 27 in the A version and asparagine in the Z form. The change in structure benefits nisin Z because it has a solubility and ability to diffuse, which are important for food [15, 18]. Nisin is effective against a wide range of Gram-positive bacteria and is generally non-cytotoxic to mammals [33]. As a result, it can help increase the shelf life of food while also protecting it from contamination. The following are some examples of nisin being put into a nanodelivery system to boost bioavailability. This peptide has been encapsulated into polymeric nanoparticles made of polycaprolactone for use in the food industry and as an alternative antifungal treatment. Nanoparticles were produced, described, and then examined *in vitro* using Franz diffusion cells and mucosa from pig vagina for topical treatment of nisin against either acute or chronic Candidiasis. Encapsulated nisin had antibacterial efficacy against *Candida albicans* strains comparable to free nisin, indicating that it could be used as a preventive against Candidiasis and other gynecological infections. Following permeation investigations, the encapsulated nisin was found to have penetration potential as well as the ability to gradually release in the pig vaginal mucosa [11, 47].

Moreover, once contained in nanofibers containing wheat gluten, zirconia, and polyvinyl alcohol nisin has also been evaluated against *Staphylococcus aureus* [47]. After 7 hours, these nanofibers released nisin gradually and inhibited *S. aureus*. These findings suggested that nanofibers containing nisin could be effective for wound treatment and active food packaging.

Nanotechnology's application in the food business offers a lot of potential, especially in the food packaging industry. Transpiring food packaging materials using antimicrobial nanoparticles have the prospective to address considerable world food grant concerns [9]. These new antimicrobial packaging options can extend the shelf life and reduce food waste due to rotting. However, the usage of food-packaging materials incorporating nanoparticles has raised concerns about the risk to humans from nanoparticle exposure, which could move from nanocomposite food packaging into food and drink.

Nanoparticle migration from nanocomposite food packaging has recently been a source of concern. Regulatory criteria for the use of nanotechnology in the food industry are also being developed [13]. During storage, the movement of nanoparticles from packaging material into food/food chiral molecules has been investigated. The safety of nanocomposite packaging materials used for food goods must be evaluated.

In general, there are two release paradigms for nanoparticles released from polymer nanocomposite materials: (1) energy independent diffusion, dissipation, and desorption into external liquid media, and (2) emancipation of nanoparticles aided by matrix degradation [12]. Numerous research has been carried out to see how different nanoparticle qualities (including morphology, dimension, distribution properties, and penetration enhancer kind) or host polymer qualities (such as degree of polymerization, structural properties, fluidity, and compatibilizers/additives) alter dispersion.

The releasing characteristics of nanoparticles are also likely to be affected by the creation process of polymer nanocomposite films. Furthermore, other parameters such as storage temperature, media PH, and even UV radiation may significantly impact release characterizations. Furthermore, research into the post-release phase is required to determine if the released particles disintegrate, agglomerate, or change composition or shape. It is significant from the standpoints of human health and environmental safety since such parameters are expected to influence the pace of retrograde redistribution through diverse biological systems. Associations before full-scale deployment of nanotechnology in the food packaging industry, characterization of implanted nanoparticles (such as geometry, dimension, or surface morphology pdi), and consequent deployments of released nanoparticles must be established.

In food packaging, nanotechnology is becoming increasingly relevant, particularly for antimicrobial packaging. Antimicrobial nanoparticles can be included in a food-packaging system to suppress the growth of spoilage and pathogenic microorganisms, improve food safety, and dramatically lengthen food shelf life. Many organic and inorganic nanomaterials, such as copper nanoparticles, zinc oxide nanoparticles, silver nanoparticles, chitosan, and antimicrobial peptides, have been investigated as options for creating antimicrobial nanocomposite packaging materials [28].

4.5.2 AMPs and Hurdle Technology

Although 2012 was a successful year for novel peptide-based medicines, what are the barriers to their substantial development and commercialization after thirty years of promising research data? The most significant stumbling block may be the high costs of production (estimated at US $300–500 per gram). Even though AMPs can be chemically changed to produce more stable molecules, proteolytic degradation is a problem to consider. Peptides can be confined to precise conformations to produce high-affinity pharmaceuticals structurally stable concerning degradation, thereby enhancing their pharmacological characteristics [47]. Bioinformatic methods may also aid in correctly anticipating the 3-Dimensional structure of these molecules, enabling the development of products with superior cell permeability properties and targeting certain intracellular processes. Despite this, molecular knowledge of these compounds' mechanisms of action is still absent, and elucidating these mechanisms is critical for the successful design and manufacture of novel peptide-based treatments. Decrease of potency and degeneration all through stockpiling are also major shortcomings that have been transcended using a variety of techniques, such as loading these peptides into nano-sized molecules, self-assemblage them as nanoscopic arrangements, entrapment in lipid nanovesicles, and complexation in electrospun nanoparticles. A huge number of potentially helpful peptide-based medications have been rejected by the FDA, resulting from poor physicochemical characteristics [3], highlighting the need for a complete investigation of production processes and action mechanisms to achieve success. The relevance of incorporating known structural and functional data on AMPs into the building of models for predicting their effectiveness is critical. The combination of computational biology and experimental data is also likely to lead to a better understanding of the interactions between these molecules and their suspected targets and, as a result, to improvements in the properties of these compounds [27].

4.5.3 Limitations and Drawbacks

The emergence of multi-resistant bacterial strains in the mid-1980s highlighted the need to replace synthetic antibiotics. Peptide-based antibiotics are now widely regarded as a viable solution to the rising resistance to traditional antibiotics [41]. Defensins have been proposed as therapeutic agents for a variety of infectious disorders, but their development has been delayed by the difficulties and costs associated with large-scale manufacture. Human neutrophil defensins and analogs have recently been shown to be produced in high quantities. Though reports of the synthesis of modest amounts of defensins have surfaced, recombinant technology for the development of defensins has yet to materialize. It has also been established that transforming the defensin gene and expressing it in plants confers disease resistance, which is equal to traditional fumigation [2]. Similar human research will be crucial in controlling or preventing viral and inflammatory disorders. Despite their potential in therapeutic development, AMPs have a number of drawbacks that make them unsuitable for clinical use. Naturally occurring peptides are often unstable in the GI tract and other bodily serums, with constricted absorption and distribution, as well as rapid metabolic breakdown and excretion [6, 7]. As a result, they have low bioavailability. Their adaptable architecture may potentially cause unintentional interactions with other components, resulting in unwanted side effects. In recent decades, antimicrobial resistance has emerged as one of the most serious risks to global health security, according to the World Health Organization's (WHO) global report on surveillance, owing to the widespread use of traditional antibiotics in health care systems, animal production, and in the community.

4.5.4 Scope

Several cationic and non-cationic peptides have been shown to be expressed throughout many vertebrate and invertebrate species in recent investigations, and it's probable that these molecules work together to boost the immune response to infections. Peptides from a variety of creatures, as well as their synthetic equivalents, will almost certainly play a key role in the development of AMPs as human medicines. Despite the fact that more work remains to be done before these molecules may be used in therapeutic settings, it is important to note that tremendous progress has been achieved in this regard. Furthermore, examining these peptides has provided a new basic understanding of biochemistry and biology, as well as the discovery of novel peptides that could be used to construct analogs to prevent or treat infections.

4.6 Conclusions

Antibacterial peptides (AMPs) have been investigated extensively as potential bio-preservatives; however, their interactions with specific dietary components (proteins and lipids), as well as their stability, such as proteolytic degradation, may result in diminished antimicrobial activity. To address these issues, chemical and bioengineering solutions have recently been proposed to generate powerful, selective, and metabolically stable peptide formulations at a lower cost and with a lower likelihood of causing unwanted side effects. Antimicrobial peptides are becoming an emerging category for clinical applications due to these innovative approaches and discoveries in biological roles. AMPs are intriguing alternatives to traditional antibiotics because of their broad-spectrum (antibacterial, antifungal, and antiviral) actions and low levels of induced resistance [1]. These peptides have a lot of potential as feed additives in the swine and poultry industries. In piglets and broilers, antimicrobial peptides have been shown to boost growth performance, stimulate food digestibility and gut health, modify intestinal microbiota, and enhance immune function. The antibacterial and immunomodulating activity of AMPs contributes to their favorable impacts on growth performance, increasing nutrient digestibility and health. Antimicrobial agents known as AMPs are commonly utilized in food packaging. They have potent antibacterial properties in foods and can help extend the shelf life of various goods. AMPs, their structural qualities, method of action, kinds, and fabrication for food packaging, and how they promote food preservation, were discussed in this review. Although AMPs are a viable food preservation option, their application in the packaging industry has several limits. AMPs are challenging to use on an industrial scale due to their high cost, in addition to their antifungal and antibacterial capabilities. Purified AMPs are too expensive in commercial use. Another significant impediment to the industrial usage of AMPs like bacteriocins is the lack of recognized rules.

On the other hand, nisin has been employed on a large scale in the industry. This suggests that bacteriocins can overcome these constraints, which is good news for future applications. Furthermore, active packaging is a good approach to keep AMPs contained. It guarantees that the encapsulated AMPs are released at the right timing and concentration. This improves the shelf life of certain perishable foods by extending the antibacterial action of AMPs.

Furthermore, *in vitro* and *in vivo* performance of active packaging for AMP encapsulation must be thoroughly evaluated. It's also a potential method for food preservation and protection against microbial contamination and degradation. To thoroughly analyze the benefits of AMPs, several tools and techniques must be developed.

References

1. Agarwal, G., Gabrani, R., Antiviral peptides: Identification and validation. *Int. J. Pept. Res. Ther.*, 27, 1, 1–20, 2020.
2. Bierbaum, G. and Sahl, H.G., Lantibiotics: Mode of action, biosynthesis and bioengineering. *Curr. Pharm. Biotechnol.*, 10, 1, 2–18, 2009.
3. Boman, H.G., Antibacterial peptides: Basic facts and emerging concepts. *J. Intern. Med.*, 254, 3, 197–215, 2003.
4. Boparai, J.K. and Sharma, P.K., Mini-review on antimicrobial peptides, sources, mechanism and recent applications. *Protein Pept. Lett.*, 27, 1, 4–16, 2020.
5. Chan, D., II, Prenner, E.J., Vogel, H.J., Tryptophan-and arginine-rich antimicrobial peptides: Structures and mechanisms of action. *BBA-Biomembranes*, 1758, 9, 1184–1202, 2006.
6. Cheng, Q., Shi, X., Liu, Y., Liu, X., Dou, S., Ning, C., Liu, Z.Q., Sun, S., Chen, X., Ren, X., Production of nisin and lactic acid from corn stover through simultaneous saccharification and fermentation. *Biotechnol. Biotechnol. Equip.*, 32, 2, 420–426, 2018.
7. Das, D. and Goyal, A., Antioxidant activity and γ-aminobutyric acid (GABA) producing ability of probiotic Lactobacillus plantarum DM5 isolated from Marcha of Sikkim. *LWT-Food Sci. Technol.*, 61, 1, 263–268, 2015.
8. Da Costa, J.P., Cova, M., Ferreira, R., Vitorino, R., Antimicrobial peptides: An alternative for innovative medicines? *Appl. Microbiol. Biotechnol.*, 99, 5, 2023–2040, 2015.
9. de Arauz, L.J., Jozala, A.F., Mazzola, P.G., Penna, T.C.V., Nisin biotechnological production and application: A review. *Trends Food Sci. Technol.*, 20, 3–4, 146–154, 2009.
10. Delves-Broughton, J., Nisin and its application as a food preservative. *Int. J. Dairy Technol.*, 43, 3, 73–76, 1990.
11. Deng, X., Nikiforov, A.Y., Leys, C., Antimicrobial nanocomposites for food packaging, in: *Food Preservation*, pp. 1–34, Academic Press, Ghent, Belgium, 2017.
12. Duncan, T.V. and Pillai, K., Release of engineered nanomaterials from polymer nanocomposites: Diffusion, dissolution, and desorption. *ACS Appl. Mater. Interfaces*, 7, 1, 2–19, 2015.
13. Farhoodi, M., Nanocomposite materials for food packaging applications: Characterization and safety evaluation. *Food Eng. Rev.*, 8, 1, 35–51, 2016.
14. Field, D., Ross, R.P., Hill, C., Developing bacteriocins of lactic acid bacteria into next generation biopreservatives. *Curr. Opin. Food Sci.*, 20, 1–6, 2018.
15. Fu, W. and Kapila, Y.L., Nisin, a probiotic bacteriocin, combined with limited chemoradiation therapy in head and neck cancer. *Arch. Clin. Med. Case Rep.*, 05, 04, 531–536, 2021.

16. García-Bayona, L., Guo, M.S., Laub, M.T., Contact-dependent killing by *Caulobacter crescentus* via cell surface-associated, glycine zipper proteins. *Elife*, 6, e24869, 2017.
17. Gharsallaoui, A., Oulahal, N., Joly, C., Degraeve, P., Nisin as a food preservative: Part 1: Physicochemical properties, antimicrobial activity, and main uses. *Crit. Rev. Food Sci. Nutr.*, 56, 8, 1262–1274, 2016.
18. Gujarathi, S.S., Bankar, S.B., Ananthanarayan, L.A., Fermentative production, purification and characterization of nisin. *Int. J. Food Eng.*, 4, 5, 2–3, 2008.
19. Han, J., Tao, W., Hao, H., Zhang, B., Jiang, W., Niu, T., Li, Q., Cai, T., Physiology and quality responses of fresh-cut broccoli florets pretreated with ethanol vapor. *J. Food Sci.*, 71, 5, S385–S389, 2006.
20. Hansen, J.N., Nisin as a model food preservative. *Crit. Rev. Food Sci. Nutr.*, 34, 1, 69–93, 1994.
21. Hurst, A., Nisin, in: *Advances in Applied Microbiology*, pp. 85–123, Elsevier, Microbiology Research Division, Health Protection Branch, Health and Welfare Canada, Ottawa, Ontario, Canad, 1981.
22. Hurst, A. and Paterson, G.M., Observations on the conversion of an inactive precursor protein to the antibiotic nisin. *Can. J. Microbiol.*, 17, 11, 1379–1384, 1971.
23. Jayant, D. and Halami, P.M., Industrial perspective of food preservatives from microbial origin, in: *Current Developments in Biotechnology and Bioengineering*, pp. 243–261, Elsevier, CSIR-Central Food Technological Research Institute, Mysore, Karnataka, India, 2020.
24. Joint FAO/WHO Expert Committee on Food Additives, Joint FAO/WHO Expert Committee on Food Additives. Meeting, and World Health Organization, *Evaluation of Certain Food Additives and Contaminants: Fifty-Seventh Report of the Joint FAO/WHO Expert Committee on Food Additives*, vol. 57, World Health Organization, International Programme on Chemical Safety World Health Organization, Geneva, 2002.
25. Joo, N.E., Ritchie, K., Kamarajan, P., Miao, D., Kapila, Y.L., Nisin, an apoptogenic bacteriocin and food preservative, attenuates HNSCC tumorigenesis via CHAC 1. *Cancer Med.*, 1, 3, 295–305, 2012.
26. Kamarajan, P., Hayami, T., Matte, B., Liu, Y., Danciu, T., Ramamoorthy, A., Worden, F., Kapila, S., Kapila, Y., Nisin ZP, a bacteriocin and food preservative, inhibits head and neck cancer tumorigenesis and prolongs survival. *PloS One*, 10, 7, e0131008, 2015.
27. Kaur, S. and Kaur, S., Bacteriocins as potential anticancer agents. *Front. Pharmacol.*, 6, 272, 2015.
28. Kny, E., Polymer nanocomposite materials used for food packaging, in: *Ecosustainble Polymer Nanomaterials for Food Packaging. Innovative Solutions, Characterisation Needs, Safety and Environmental Issues*, pp. 337–375, 2013.

29. Kellner, R., Jung, G., Hörner, T., Zähner, H., Schnell, N., Entian, K.-D., Götz, F., Gallidermin: A new lanthionine-containing polypeptide antibiotic. *Eur. J. Biochem.*, 177, 1, 53–59, 1988.
30. Keymanesh, K., Soltani, S., Sardari, S., Application of antimicrobial peptides in agriculture and food industry. *World J. Microbiol. Biotechnol.*, 25, 6, 933–944, 2009.
31. Kumar, P., Kizhakkedathu, J.N., Straus, S.K., Antimicrobial peptides: Diversity, mechanism of action and strategies to improve the activity and biocompatibility *in vivo*. *Biomolecules*, 8, 1, 4, 2018.
32. Laridi, R., Kheadr, E.E., Benech, R.-O., Vuillemard, J.C., Lacroix, C., Fliss, I., Liposome encapsulated nisin Z: Optimization, stability and release during milk fermentation. *Int. Dairy J.*, 13, 4, 325–336, 2003.
33. Luo, X., Wang, M., Dessie, W., Yang, J., Hou, Y., He, N., Research progress on antimicrobial peptides within the food industry. *Open Access J. Biomed. Eng. Biosci.*, 3, 368–369, 2019.
34. Li, Y., Xiang, Q., Zhang, Huang, Y., Su, Z., Overview on the recent study of antimicrobial peptides: Origins, functions, relative mechanisms and application. *Peptides*, 37, 2, 207–215, 2012.
35. Liu, Y., Sameen, D.E., Ahmed, S., Dai, J., Qin, W., Antimicrobial peptides and their application in food packaging. *Trends Food Sci. Technol.*, 112, 471–483, 2021.
36. Luo, Y. and Song, Y., Mechanism of antimicrobial peptides: Antimicrobial, anti-inflammatory and antibiofilm activities. *Int. J. Mol. Sci.*, 22, 21, 11401, 2021.
37. Cabrele, C., Martinek, T.A., Reiser, O., Berlicki, Ł., Peptides containing β-amino acid patterns: Challenges and successes in medicinal chemistry. *J. Med. Chem.*, 57, 23, 9718–9739, 2014.
38. Maher, S. and McClean, S., Investigation of the cytotoxicity of eukaryotic and prokaryotic antimicrobial peptides in intestinal epithelial cells *in vitro*. *Biochem. Pharmacol.*, 71, 9, 1289–1298, 2006.
39. Małaczewska, J. and Kaczorek-Łukowska, E., Nisin-A lantibiotic with immunomodulatory properties: A review. *Peptides*, 137, 170479, 170479, 2021.
40. Matsusaki, H., Endo, N., Sonomoto, K., Ishizaki, A., Lantibiotic nisin Z fermentative production by Lactococcus lactis IO-1: Relationship between production of the lantibiotic and lactate and cell growth. *Appl. Microbiol. Biotechnol.*, 45, 1, 36–40, 1996.
41. Papagianni, M., Ribosomally synthesized peptides with antimicrobial properties: Biosynthesis, structure, function, and applications. *Biotechnol. Adv.*, 21, 6, 465–499, 2003.
42. Meng, S., Xu, H., Wang, F., Research advances of antimicrobial peptides and applications in food industry and agriculture. *Curr. Protein Pept. Sci.*, 11, 4, 264–273, 2010.

43. Mokoena, M.P., Lactic acid bacteria and their bacteriocins: Classification, biosynthesis and applications against uropathogens: A mini-review. *Molecules (Basel, Switzerland)*, 22, 8, 1255, 2017.
44. Mulders, J.W.M., Boerrigter, I.J., Rollema, H.S., Siezen, R.J., de Vos, W.M., Identification and characterization of the lantibiotic nisin Z, a natural nisin variant. *Eur. J. Biochem.*, 201, 3, 581–584, 1991.
45. Mythili, R.S. and Sathiavelu, A., Recovery of bacteriocin (NISIN) from lactococcus lactis and testing its ability to increase the shelf life of vegetables (carrot and beans). *Res. J. Biol. Sci.*, 5, 11, 727–730, 2010.
46. Özel, B., Şimşek, Ö., Akçelik, M., Saris, P.E.J., Innovative approaches to nisin production. *Appl. Microbiol. Biotechnol.*, 102, 15, 6299–6307, 2018.
47. Papagianni, M., Ribosomally synthesized peptides with antimicrobial properties: Biosynthesis, structure, function, and applications. *Biotechnol. Adv.*, 21, 6, 465–499, 2003.
48. Penna, T.C.V., Jozala, A.F., De Lencastre Novaes, L.C., Pessoa, A., Cholewa, O., Production of nisin by Lactococcus lactis in media with skimmed milk. *Appl. Biochem. Biotechnol.*, 122, 1, 619–637, 2005.
49. Rai, M., Pandit, R., Gaikwad, S., Kövics, G., Antimicrobial peptides as natural bio-preservative to enhance the shelf-life of food. *J. Food Sci. Technol.*, 53, 9, 3381–3394, 2016.
50. Rodrigues, G., Maximiano, M.R., Franco, O.L., Antimicrobial peptides used as growth promoters in livestock production. *Appl. Microbiol. Biotechnol.*, 105, 19, 7115–7121, 2021.
51. Saeed, F., Afzaal, M., Tufail, T., Ahmad, A., Use of natural antimicrobial agents: A safe preservation approach, in: *Active Antimicrobial Food Packaging*, vol. 18, 2019.
52. Shai, Y., Mode of action of membrane active antimicrobial peptides. *Biopolymers*, 66, 4, 236–248, 2002.
53. Tang, Z., Ma, Q., Chen, X., Chen, T., Ying, Y., Xi, X., Wang, L., Ma, C., Shaw, C., Zhou, M., Recent advances and challenges in nanodelivery systems for antimicrobial peptides (AMPs). *Antibiotics*, 10, 8, 990, 2021.
54. Tegginamath, A. and Novák, I.O., *Antimicrobial Treatments for Textile Materials*, Tul.Cz, Accessed Jan. 16 2022, https://dspace.tul.cz/bitstream/handle/15240/161344/Akshat_Tegginamath_Diploma_Thesis.pdf?sequence=1.
55. Fathi, M., Mohebbi, M., Razavi, S.M.A., Application of image analysis and artificial neural network to predict mass transfer kinetics and color changes of osmotically dehydrated kiwifruit. *Food Bioprocess Tech.*, 4, 8, 1357–1366, 2011.
56. Wang, S., Zeng, X., Yang, Q., Qiao, S., Antimicrobial peptides as potential alternatives to antibiotics in food animal industry. *Int. J. Mol. Sci.*, 17, 5, 603, 2016.
57. Yoon, J.H., Ingale, S.L., Kim, J.S., Kim, K.H., Lee, S.H., Park, Y.K., Kwon, I.K., Chae, B.J., Effects of dietary supplementation of antimicrobial peptide-A3 on growth performance, nutrient digestibility, intestinal and fecal microflora

and intestinal morphology in weanling pigs. *Anim. Feed Sci. Technol.*, 177, 1–2, 98–107, 2012.

58. Zainodini, N., Hassanshahi, G., Hajizadeh, M., Falahati-Pour, S.K., Mahmoodi, M., Mirzaei, M.R., Nisin induces cytotoxicity and apoptosis in human asterocytoma cell line (SW1088). *Asian Pac. J. Cancer Prev.*, 19, 8, 2217, 2018.
59. Vilas Boas, L.C.P., Campos, M.L., Berlanda, R.L.A., de Carvalho Neves, N., Franco, O.L., Antiviral peptides as promising therapeutic drugs. *Cell. Mol. Life Sci.*, 76, 18, 3525–3542, 2019.
60. M.E. Stiles, J.C. Vederas, M.J. van Belkum, R.W. Worobo, R.J. Worobo, J.K. McCormick, G.G. Greer, L.M. McMullen, J.J. Leisner, A. Poon, C.M.A.P. Franz, Bacteriocins, transport and vector system and method of use thereof. US Patent. US6403082 B1, 2002.
61. M.E. Stiles, J.C. Vederas, M.J. van Belkum, R.W. Worobo, R.J. Worobo, J.K. McCormick, G.G. Greer, L.M. McMullen, J.J. Leisner, A. Poon, C.M.A.P. Franz, Novel bacteriocins, transport and vector system and method of use thereof. US Patent. 20100129864, A1, 2010.
62. M.H. Farris, P.F. Churchill, J.B. Olson, Novel bacteriocin from a new Streptomyces species. US Patent US20100286030 A1, 2010.
63. D.J. O'Sullivan and J.H. Lee, Lantibiotics and uses thereof. US Patent. 7960505 B2, 2011.
64. Sharma, D. and Saharan, B.S., Simultaneous production of biosurfactants and bacteriocins by probiotic Lactobacillus casei MRTL3. *Int. J. Microbiol.*, 2014, 2–4, 2014.
65. Ranganath, B.S., Sharmila, T., Balasubramanyam, B.V., Optimisation of media for the growth and production of bacteriocin from Bacillus coagulans. *Int. J. Innov. Res. Technol. Sci. Eng.*, 1, 109–114, 2015.
66. Kumar, V. and Tiwari, S.K., Halocin HA1: An archaeocin produced by the haloarchaeon Haloferax larsenii HA1. *Process Biochem.*, 61, 202–208, 2017.
67. Simha, V.B., Sood, S.K., Kumariya, R., Garsa, A.K., Simple and rapid purification of pediocin PA-1 from Pediococcus pentosaceous NCDC 273 suitable for industrial application. *Microbiol. Res.*, 167, 544–549, 2012.
68. Mavric, A., Tompa, G., Trmcic, A., Rogelj, I., Matijasic, B.B., Bacteriocins of lactobacillus gasseri K7 – monitoring of gassericin K7 A and B genes' expression and isolation of an active component. *Process Biochem.*, 49, 1251–1259, 2014.
69. Lee, S.G. and Chang, H.C., Purification and characterization of mejucin, a new bacteriocin produced by Bacillus subtilis SN7. *LWT Food. Sci. Technol.*, 87, 8–15, 2018.
70. Ge, J., Sun, Y., Xin, X., Wang, Y., Ping, W., Purification and partial characterization of a novel bacteriocin synthesized by Lactobacillus paracasei HD1-7 isolated from Chinese sauerkraut juice. *Sci. Rep.*, 6, 19366, 2016.
71. Goh, H.F. and Philip, K., Purification and characterization of bacteriocin produced by Weissella confusa A3 of dairy origin. *PloS One*, 10, 3–5, 2015.

5

Marine Microbes as a Resource for Novel Enzymes

Kanagaraj Suganya[1], Balraj Sudha[1], Bishwambhar Mishra[2], Bapatla Sumithra[2], Sanjeeb Kumar Mandal[2*] and Sundaravadivelu Sumathi[1†]

[1]Department of Biochemistry, Biotechnology and Bioinformatics, Avinashilingam Institute for Home Science and Higher Education for Women, Coimbatore, Tamil Nadu, India
[2]Department of Biotechnology, Chaitanya Bharathi Institute of Technology, Hyderabad, India

Abstract

Microbial enzymes have many benefits over the enzymes obtained from plant and animal sources because of their chemical action activities, cheaper, abundant supplies, and relatively a lot of stability. A marine setting, which includes seventy-one of the surfaces, is not only wealthy with variety but also an enormous resource for potential microorganisms for helpful applications. Microorganisms and fungi in marine ecosystems secrete various enzymes that support their environment and ecological roles. Marine microbic enzymes became the main target of attention, and various enzymes had drawn the eye of microbic explorers. Many enzymes are extracted from water or marine sediments, refined and characterized for their properties and potential applications. Primary targets of recent protein technology still working on the maintenance of foods and food parts, economical use of raw resources, the development of food quality like flavor and style, manufacture of dietic foods, eliminating opposing alimentary substances from sure biological process raw materials, utilization of raw materials for the preparation of animal feed. Enzymes are used as cost-efficient and environmentally sensitive substitutes for chemical processes in many industries together with prescription drugs. Marine metagenomics could be a rapidly growing and exciting field of genomic research; by that, we tend to examine the microbic populations in marine habitats. Marine metagenomics is typically used instead while not culturing microorganisms to

Corresponding author: sanjeeb.vit@gmail.com; ORCID; 0000-0002-3920-9456
†*Corresponding author*: sumathi_bc@avinuty.ac.in

Sanjay Kumar, Narendra Kumar and Shahid-ul-Islam (eds.) Role of Microbes in Industrial Products and Processes, (107–144) © 2023 Scrivener Publishing LLC

induce distinctive biocatalysts for completely new functions applied in biotechnological applications. The event of metagenomics has raised the invention of biocatalysts with several unique properties. This review is focused on marine-derived microbic enzymes for industrial applications.

Keywords: Drugs, enzymes, marine environment, metagenomics, xylanase

5.1 Introduction

More than seventieth of the surface of our planet contains the world's oceans. The seas and oceans still stay undiscovered treasures of incalculable biological and chemical entities that require knowing and controlling for the prolonged property. Microbic species play a key role in the organic carbon and biogeochemical cycle in aquatic environments. Marine microorganisms play a full-of-life role in the mineralizing of advanced organic matter within the marine setting via degrading pathways of their metabolism and contribute to secondary marine development [1]. Additionally, to dead plants and animal residues, marine microorganisms are degraded by advanced polysaccharides like polysaccharides, pectin, xylan, starch, lignin, fats, sugar, urea, proteins, aromatic and acyclic hydrocarbons, and plenty of different organic substances that enter the aquatic environment. As an expensive supply of hydrolytic enzymes of business price, their presence within the breakdown of organic substances and also the retting of yarns and fibers testify to their prospective [2].

For medicine applications, marine microbes represent a prospective supply of bioactive compounds. The emerging scientific literature describes novel metabolites with enticing medicine properties isolated from marine microbes. Economic and enticing thanks to getting novel natural merchandise are provided by the marine setting. Marine microorganisms are the supply of the latest compounds and extracts with medicinal potential in the treatment of chronic inflammatory disorders [3]. They represent an eco-friendly and possible alternative for multiple mechanisms of action and essential medicine applications to get drug candidates [4].

Nowadays, microbic enzymes are cheap and less harmful various to artificial medication. Natural merchandise became the main target of exploring new pharmacologically active compounds. Especially, attributable to their abundance and variety, marine microorganisms are an excellent supply of bioactive substances for medicine applications [5]. The key organism categories embody microorganisms, fungi, microalgae, and dependent microorganisms. Over 20,000 bioactive compounds are produced by microbial sources, with almost 10,000 obtained primarily from actinomycetes species. Streptomycetes became a significant source of natural antibiotics for medicinal and industrial applications because most of those metabolites are potent antibiotics [6]. Many novel compounds with entirely different biological activity have been discovered due to more profound genetic research of actinomycetes and the use of high-throughput screenings. The producers of innovative and unique chemical structures with high medicine prospects also are different categories of marine microorganisms like fungi and eubacterium. Additionally, like aquatic plants and animals, macroorganism-related microbes are rising as a plentiful supply of the latest bioactive compounds [7].

There has been an increasing interest and demand for enzymes with unique properties since the introduction of biotechnology. Compared to the terrestrial environment, the marine environment offers marine microorganisms, unique genetic structures, and living habitats. The marine system varies from nutrient-rich areas to nutritionally distributed places where only a couple of species will live [8]. The efficiency of the marine environment includes high salinity, air, mass, vasoconstrictive, and special lighting conditions. These will contribute to the many differences between the enzymes produced by microbial sources and the sequence of homology enzymes produced by terrestrial microorganisms, which has led to the emergence of marine microbes in recent years. These enzymes are used as prescription drugs, fine chemicals, and food additives [9]. Researchers have extracted a variety of enzymes with specific activity from marine microorganisms, actinomycetes, fungus, and other marine microbes in recent years, and a few products have already been used in commercial processes. Particularly, a significant variety of drug targets have yielded some marine microbic enzymes. Marine microorganisms, the large genetic and organic chemistry diversity that continues to be in infancy are currently attractive as a potential new source of enzymes with numerous application potential [10].

5.2 Marine-Derived Microbial Enzymes

As industrial biocatalysts, industrial enzymes provide varied edges in terms of property and method potency over the standard natural process [11]. The marine setting is one of the world's most significant sources of bioactive compounds. Marine microbic communities secrete completely different enzymes from marine environments, counting on their environment and ecological functions. Because of their success in biogeochemical

Table 5.1 Enzymes from marine organisms.

S.n.	Source of microorganism	Enzyme	Application	Reference
Algae				
1.	*Porphyridium* sp.	α-1,3-Glucosidase glucosidase II glycoenzyme	Used for polysaccharide biosynthesis and other industrial, medical, and pharmaceutical chemicals are produced	[15]
2.	*Dunaliella salina Cylindrotheca, Closterium* and *Chaetoceros muelleri*	Extra - and intracellular *enzymes*	Diethyl phthalate (DEP) and di-n-butyl phthalate (DBP) as phthalates biodegrade and impede testosterone production, causing persistent toxicity	[16]
3.	*Saccharina japonica*	Mannuronan C5 epimerase	Alginate polysaccharide is polymerized, epimerized, and biosynthesized	[17]

(*Continued*)

Table 5.1 Enzymes from marine organisms. (*Continued*)

S.n.	Source of microorganism	Enzyme	Application	Reference
4.	*Botryococcus braunii* Race L	Squalene synthase	Squalene synthase (SS) produces C30 squalene, a precursor for sterols/steroids, lipoproteins, bile acids, and vitamin D biosynthesis in eukaryotes and hopanoids in some prokaryotes; geranylgeranyl pyrophosphate (GGPP), an intermediate in the process, is linked to lytic bone disease and other disorders such as neurodegenerative diseases, nonskeletal cancers and cardiovascular diseases	[18]
Bacteria				
5.	Marine shipworm bacterium	Protease	Cleansing additive	[19]
6.	*Vibrio costicola*	L-glutaminase	Anticancer properties	[20]
7.	*Bacillus* sp. KCTC 0377BP	Chitasonase	Production of COSs	[21]
8.	*Microbulbifer*-like JAMB-A94	Agarase	Production of NAS6	[22]

(*Continued*)

Table 5.1 Enzymes from marine organisms. (*Continued*)

S.n.	Source of microorganism	Enzyme	Application	Reference
9.	*Thalassomonas* sp. JAMB-A33	Agarase	Porphyran's antioxidant capacity is improved	[23]
10.	*Bacillus* sp. DAU101	Chitosanase	Transformation of soluble chitosan to (GlcN)24	[24]
11.	*Agarivorans* sp. JA-1	Agarase	Production of NAS2	[25]
12.	*Agrobacterium tumefaciens* strain C58 (Atu3025)	Alginate lyase	Synthesis of monosaccharides	[26]
13.	*Bacillus circulans* BM15	Protease	Laundry detergent formulations	[27]
14.	*Bacillus* sp. DAU101	Chitinase	Transformation of chitin to GlcNAc (GlcNAc)	[28]
15.	*Pseudomonas elongata* MTCC5261	Carrageenase	Isolation of protoplast	[29]
16.	*Bacillus mojavensis* A21	Protease	Detergent agent	[30]
17.	*Paenibacillus* sp. BME-14	Endoglucanase	Cellulase activities and uses in industry	[31]
18.	*Glaciecola mesophila* KMM 241	1,4-xylanase	Xylo-oligosaccharides production	[32]

(*Continued*)

Table 5.1 Enzymes from marine organisms. (*Continued*)

S.n.	Source of microorganism	Enzyme	Application	Reference
19.	*Bacillus cereus* TKU006	Protease and chitinase	Marine garbage is converted into a high-value commodity	[33]
20.	*Serratia* sp. TKU020	Chitinase	Chitin to GlcNAc conversion and (GlcNAc)2	[34]
21.	*Martelella mediterranea*	β-glucosidase	Industrial application	[35]
22.	*Bacillus licheniformis* DSM8785	Chitinase	Bio conversion chitin waste	[36]
23.	*Geobacillus thermoleovorans*	Xylanase	Bleaching of paper pulps before application and prebiotics	[37]
24.	*Streptomyces roseolus*	Chitosanase	Production of (GlcN)3	[38]
25.	*Bacillus subtilis* HD145	Chitosanase	Production of (GlcN)2-5	[39]
26.	*Zunongwangia profunda*	α-amylase	Applied biology	[40]
27.	*Catenovulum* sp. X3	Agarase	Production of NAS6, NAS8, NAS10, and NAS12	[41]
28.	*Saccharophagus* sp. AG21	Agarase	Production of NAS4 and NAS6	[42]
29.	*Zunongwangia profunda*	Xylanase	Food industry and bioethanol production	[43]

(*Continued*)

Table 5.1 Enzymes from marine organisms. (*Continued*)

S.n.	Source of microorganism	Enzyme	Application	Reference
30.	*Agarivorans gilvus* WH0801	Agarase	Production of NAS4	[44]
31.	*Bacillus cereus* GU-02	Chitosanase	Production of (GlcN)2-5	[45]
32.	*Rhodococcus* sp. strain YM12	Phosphinothricin N-acetyltransferase	Detoxification of herbicides	[46]
33.	*Alcanivorax* sp.	Serine hydroxymethyltransferase	Enzymatic synthesis of L-serine	[47]
34.	*Arthrobacter nicotianae*	Serine hydroxymethyltransferase	Used for cloning a novel glycogen	[48]
35.	*Aeromonas schubertii*	Chitinase	Conversion of colloidal chitin to (GlcNAc)6	[49]
36.	*Sphingomonas paucimobilis* PF-1	Fucoidanase	Causes fucoidan hydrolysis, making it a useful tool for structural investigation of fucoidans as well as the creation of beneficial fuco-oligosaccharides	[50]
37.	*Pseudoalteromonas hodoensis* H7	Agarase	Production of NAS4	[51]
38.	*Pseudoalteromonas sp.* DC14	Chitinase	Industrial application	[52]

(*Continued*)

Table 5.1 Enzymes from marine organisms. (*Continued*)

S.n.	Source of microorganism	Enzyme	Application	Reference
39.	*Vibrio campbellii*	beta-glucosidase	Use in industry	[53]
40.	*Pseudoalteromonas* sp. 129-1	Protease	Non-toxic anti-biofilm agent and laundry detergent	[54]
41.	*Bacillus* sp. NPST-AK15	serine protease	Laundry and pharmaceutical industries	[55]
42.	*Sphingopyxis alaskensis*	Ectoine synthase	Used for Ectoine biosynthesis	[56]
43.	*γ-Proteobacteria*	Diacylglycerol acyltransferase enzymes (DGAT) and Wax ester synthase (WS)	Neutral lipid synthesis and accumulation	[57]
44.	*Streptomyces* sp. CS147	Chitinase	Colloidal chitin to GlcNAc and (GlcNAc)2 conversion	[58]
45.	*Pseudomonas sp., Pseudomonas bolearic, Pseudomonas stutzeri, Bacillus firmus and Bacillus licheniformis species*	Xylanase	Biotechnological application	[59]
46.	*Bacillus* sp. TS	Chitasonase	Production of COSs	[60]

(*Continued*)

Table 5.1 Enzymes from marine organisms. (*Continued*)

S.n.	Source of microorganism	Enzyme	Application	Reference
47.	*Marinobacter hydrocarbonoclasticus* 584	β glucosidase and laminarinase	Bioremediation	[61]
48.	*M hydrocarbonoclasticus* MMRF-578 and 581	Phosphatase and aminopeptidase	Bioremediation	[61]
49.	*Paenibacillus barengoltzii*	Chitinase	The transformation of colloidal chitin into (GlcNAc)	[62]
50.	*Streptomyces anulatus* CS242	Chitinase	Bio conversion	[63]
51.	*Cobetia* sp. WG-007	Alginate lyase	Synthesis of alginate oligosaccharides with a low molecular weight	[64]
52.	*Cobetia* sp. NAP1	Alginate lyase	Production of monosaccharides	[65]
53.	*Streptomyces coelicolor* A3(2)	Chitinase	GlcNAc formation from crystalline chitin	[66]
54.	*Bacillus subtilis* 168	Chitasonase	Production of (GlcN)2-4	[67]
55.	*Paenibacillus* sp. S29	Alginate lyase	Production of monosaccharide	[68]
56.	*Gayadomonas joobiniege* G7 (AgaJ9)	Agarase	NAS2 and NAS4 production	[69]

(*Continued*)

Table 5.1 Enzymes from marine organisms. (*Continued*)

S.n.	Source of microorganism	Enzyme	Application	Reference
57.	*Streptomyces albolongus* ATCC 27414	Chitinase	Colloidal chitin to GlcNAc conversion and (GlcNAc)2	[70]
58.	*Gynuella sunshinyii*	Chitasonase	COSs can be prepared in a controlled manner using a specified DP	[71]
59.	*Cellulophaga* sp. SY116 (OalC17)	Alginate lyase	Alginate monomer production	[72]
60.	*Vibrio* sp. W13	Alginate lyase	Production of alginate monomer	[73]
61.	*Echinicola rosea* sp. Nov. JL3085T	Endoxylanase	Food, drug, and bioremediation application	[74]
Fungi				
62.	*Aspergillus* sp. Y2K	Chitosanase	Production of (GlcN)3-5	[75]
63.	*Marine beauveria* sp.	L-Glutaminase	Anticancer properties	[76]
64.	*Aspergillus niger*	Alkaline xylanases	Biobleaching	[77]
65.	*Aspergillus* sp. CJ22-326 (ChiA)	Chitosanase	Production of GlcN	[78]
66.	*Aspergillus awamori* BTMFW032	Tannase	Food industry	[79]

(*Continued*)

Table 5.1 Enzymes from marine organisms. (*Continued*)

S.n.	Source of microorganism	Enzyme	Application	Reference
67.	*Alternaria alternate, Alternaria solani, Epicoccum purpurascens, Gonatorrhodiella parasitica, Monascus ruber, Trematosphaeria mangrovei, Ulocladium Chartarum*	Laccase	Environmental application	[80]
68.	*Penicillium monoverticillium* CFR 2	Chitinase	Conversion of chitin to (GlcNAc)3	[81]
69.	*Aspergillus sclerotiorum* CBMAI 849	Lignin degrading enzymes	Environmental application	[82]
70.	*Penicillium* sp. D-1	Chitosanase	Production of COSs	[83]
71.	*Scopulariopsis* sp.	Protease	Detergent formulation	[84]
72.	*Pyrenochaeta* sp. CF-2008	Laccase and lipase	Industrial application	[85]
73.	*Dothideomycetes* sp. OY307, *Curvularia* sp. B34	Laccase	Industrial application	[85]
74.	*Trichoderma piluliferum* strain wxm37	Cellulase	Industrial application	[85]
75.	*Phoma* sp. ZH2.1, *Pleosporales* sp. LH241	Lipase	Industrial application	[85]

(*Continued*)

Table 5.1 Enzymes from marine organisms. (*Continued*)

S.n.	Source of microorganism	Enzyme	Application	Reference
76.	*Arthrinium phaeospermum* isolate T57, *Diaporthe* sp. H4236, *Trichoderma asperellum* isolate T29, *Arthrinium phaeospermum* isolate DFFSCS004, *Cladosporium* sp. JS1043, *Cladosporium sphaerospermum* isolate IJL07, *Ascomycota* sp. AR-2010 isolate TR063, *Cladosporium cladosporioides* isolate 2728	Lipase, cellulase and laccase	Industrial application	[85]
77.	*Hypocrea koningii* strain JZ-25	Laccase and cellulase	Industrial application	[85]
78.	*Mucor* sp. FJ09, *Aspergillus nomius* strain NRRL 26885, *Aspergillus flavipes* isolate BCT2-3, *Trichoderma hamatum* strain LXM1, *Hypocrea lixii isolate* SZMC 20858, *Arthrinium* sp. LH11, *Aspergillus terreus* strain 1B61	Lipase and cellulose	Industrial application	[85]

(*Continued*)

Table 5.1 Enzymes from marine organisms. (*Continued*)

S.n.	Source of microorganism	Enzyme	Application	Reference
79.	*Penicillium,* *Humicola,* *Gymnoascus, Endomysis,* *Zygorhynchus,* *Aspergillus, Scopulariopsis,* *Cephalosporium, Trichoderma,* *Zalerion, Pleospora,* *Chaetomium, Phoma,* *Botryphialophora*	Amylase, cellulase, chitinase, gelatinase, lipase	Remineralization and industrial application	[86]
80.	*Aspergillus* sp. W-2	Chitosanase	Production of (GlcN)2-6	[87]
81.	*Arthopyrenia* sp. CBMAI 1330 and *Nigrospora* sp. CBMAI 1328	Laccase	Saline environments/technological processes.	[88]
82.	*Alternaria alternata, Cochliobolus australiensis, Aspergillus terreus,* *Emericella nidulans, Cladosporium cladosporioides, Fusarium solani,*	Cellulase, xylanase, and pectinase	The food industry, water treatment, sugar fermentation, ethanol production, detergent manufacturing, and paper and textile industry	[89]
83.	*Geomyces pannorum*	α-amylase	Syrup industry	[90]

(*Continued*)

Table 5.1 Enzymes from marine organisms. (*Continued*)

S.n.	Source of microorganism	Enzyme	Application	Reference
84.	*Penicillium chrysogenum* MH745129	Chitinase	Control of citrus molds	[91]
Yeast				
85.	*Leucosporidium antarcticum*	Invertase and glucosidases	Pharmaceutical industries	[92]
86.	*Candida intermedia* YA01a, *Lodderomyces elongisporus* YF12c, *Candida quercitrusa* JHSb, *Pichia guilliermondii* N12c, *Candida parapsilosis* 3eA2, *Candia rugosa* wl8, *Yarrowia lipolytica* N9a, *Aureobasidium pullulans* HN2.3, *Rhodotorula mucilaginosa* L10-2	Lipase	Industrial application	[93]
87.	*Aureobasidium pullulan*	Lipase	Chemical industry	[94]
88.	*Yarrowia lipolytica*	Alginate lyase	Pharmaceutical application	[95]
89.	*Pichia guilliermondii*	Inulinase	Fuel and food industries	[96]

(*Continued*)

Table 5.1 Enzymes from marine organisms. (*Continued*)

S.n.	Source of microorganism	Enzyme	Application	Reference
90.	*Kodamea ohmeri*	Phytase	Industries that produce fedd	[97]
91.	*Glaciozyma antarctica* PI12	Chitinase	Biotechnological application	[98]
92.	*Rhodotorula mucilaginosa*	Protease	Feed industries	[99]
93.	*Candida davisiana, Guehomyces pullulans, Cryptococcus Adeliensis*	Endoxylanase	Chemical industries	[99]
94.	*Aureobasidium pullulans* 98	Cellulose	Cloning	[94]
95.	*Mrakia blollopis*	Cellulose	Industrial application	[100]
96.	*Tetracladium* sp.	Amylase	Industrial application	[100]

processes, vital ecological parts are considered vital within the marine setting. Because of their properties, marine microbic enzymes (e.g. proteolytic enzyme, lipase, agarase, polysaccharide, and other enzymes) are vital [12]. The marine setting is a vast resource, starting from water, sediments, and marine species, giving marine microorganisms a singular setting. It experiences completely different physical variations like high salinity, air mass, acidic pH, heat, or any combination thence, which might produce a singular setting for microorganisms to get new carbon skeleton secondary metabolites, high halogenation levels, and novel bioactivities [13]. The vital benefits of microbes over different species are that they are cheap to supply, their protein chemical action activities are a lot of stable and governable, rise among a brief time that produces a high protein yield, and because of the absence of differences due to the season, they are ready to give a relentless offer of culture. Moreover, relative to microbes, the tissues of plants and animals are thought of as a lot of hepatotoxic [14]. As per that, the enzymes derived from marine microbes are listed in Table 5.1.

5.3 Enzymes for Industrial Applications: The Potential of Marine Microbes

Enzymes are unit ideal metabolic catalysts that, through a well-defined pathway, permit completely different endogenous organic chemistry reactions to occur. By lowering the reaction energy of activation while not experiencing any permanent amendment, enzymes accelerate completely different biological reactions that are crucial in maintaining human life. Enzymes are present in nearly all present organisms, starting from a minute organism to plants and animals [11]. On the contrary, there are several benefits to the event of enzymes from microorganisms, like simple handling, speedy multiplication beneath regulated conditions, simple genetic manipulation, high yield of development, etc. Moreover, because of their chemical action activity, specificity, stability, non-toxicity, eco-friendly nature, cost-effectiveness, industrial use of microbic enzymes is gaining much attention. Beneath regulated conditions for manufacturing merchandise from good substrates, varied microbic enzymes are standard biocatalysts [101].

Additionally, several microbic enzymes convert dangerous compounds (nitriles, amines, carboxyl teams, and synthetic resin groups) into helpful things, either by biodegradation or bioconversion processes. Because of the wide variety of applications of microbic enzymes, their production is incredibly vital on an industrial scale beneath variable physical and chemical conditions. Beneath severe conditions of temperature and pH, microbic enzymes ought to perform well. The thermostable enzymes minimize the chance of microbic contamination on an advertisement scale throughout the prolonged fermentation method [102]. With the input of rDNA technology, supermolecule engineering, and metagenomics, varied new microbic enzymes are designed for many bioprocesses. Varied molecular techniques increase the yield and potency of the many microbic enzymes in food, prescription drugs, textiles, paper, leather, biotechnological industries [103]. Recombinant enzymes from completely different microorganisms are used to manufacture a good product on an industrial scale. Nowadays, the world demand for industrial microbic enzymes is incredibly competitive. In developing completely different bioprocesses on an industrial scale, microbic enzymes are of tremendous importance. Microbic enzymes are usually employed in varied industries, together with pulp, paper, leather, detergents, textiles, prescription drugs, cosmetics, beverages, chemicals, animal feed, and biofuels [104]. The economic applications of protein are depicted in Figure 5.1.

Since precedent days, microorganisms have been employed in food fermentation, and fermentation processes are still used to prepare

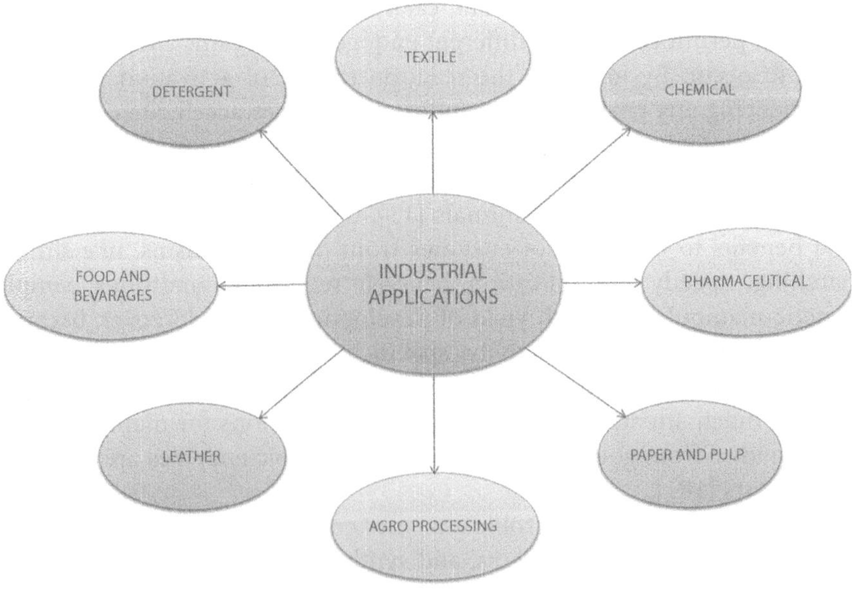

Figure 5.1 Applications of enzymes in industries.

many food things [105]. Within the food trade, microbic enzymes play a significant role since they are a lot of stable than plant and animal enzymes. They will be developed during a cost-efficient means through fermentation techniques with less time and house needs, and method modification and optimization are achieved simply because of their prime quality. Several enzymes notice varied applications in several industrial sectors, e.g., chemical action enzymes, within the food, detergent, paper, and textile industries. They are used to manufacture aldohexose syrups, high levulose corn syrups, crystalline aldohexose, disaccharide syrups, etc. [106]. They are used as additives within the detergent trade to eliminate starch-based stains. Appropriate paper coatings are wont to minimize starch body within the paper trade. Amylases are used for the warp filler of textile fibers within the textile trade. Similarly, enzymes like proteases, lipases, or xylanases within the food trade have broad applications.

Through entirely different molecular and organic chemistry strategies, microbic enzymes are simply changed. Overexpression of their genes can hyperproduce microbic enzymes with high specific activity [12]. Marine microorganisms are acknowledged to be concerned, especially pathways, within the degradation of organic matter and infrequently have faith inanimate thing enzymes to supply merchandise of low relative molecular mass which will be used as substrates. Lipases are crucial contributors for the institution of biocatalysis as a sound approach, resulting in the assembly of bulk chemicals for price intercalary merchandise during an extensive variety of areas and industries [107]. New/improved chemical action characteristics of those biocatalysts are pursued to address new issues and boost already applied processes to create a lot of environmentally friendly and competitive properties. With its enormous variety of physical and chemical environments, the marine setting should be a valuable supply of enzymes capable of operating beneath intense conditions, extending the variation of their helpful applications [108]. However, since the bizarre needs of marine organisms for growth as well as the requirements of recombinant organisms are typically necessary, the screening method typically proves to be tough, requiring optimization of fermentation media and culture conditions in either case. Bioprospecting for enzyme activity in marine organisms has allowed the identification of a good variety of sources comprising each microbic setting, often the foremost asked for, and better organisms [109]. As a result, new lipases are known; recovery methods and organic chemistry characterization ways are developed. Additionally, insight into the metabolic mechanisms of lipids in marine microbes has been obtained. The marine enzyme setting can increase the understanding of its functions within the biology of marine microbes and can contribute to more straightforward and more versatile enzyme isolation, thereby resulting in the more dissemination of business biocatalysis [110].

One of the initial hostile conditions on earth is the sea hypersaline hypoxia basins (DHABs). They host incredible microorganism diversity, even if DHABs have high saline environments, hypoxia, and significant hydrostatic pressure. Recent research has shown that fungi represent a significant cluster of eukaryotes inhabiting these systems. Here, the build-up of vast quantities of organic material will profit fungi. Marine fungi are also identified to come up with bioactive molecules, halophilic and halotolerant fungi, specifically, catalyst reservoirs and secondary metabolites with helpful applications in business, meditative and environmental biotechnology. Enzymes are created in giant quantities with varied industrial applications, which will be generated in DHABs (i.e., Aspergillus, Candida, Rhodotorula) by sure flora genera [111]. Lipases are necessary compounds for the process of food, biodiesel, and unsaturated fatty acids. Flora enzyme was isolated for the primary time in 1935 from fungus genus *Oxalicum* and *Aspergillus flavus* [112]. In many DHABs, several genera happiness to Aspergillus and fungus genus, similarly as fungus and Rhodotorula, are encountered and might actively change numerous oils, gaining extended interest as an owing to its end-use potential market for a wide range of products, including animal feed, biosensors, diagnostic instruments, detergents, oleochemicals, and bioremediation agents. Marine fungi have offered new incentives for investigating marine natural products, which are a repository of bioactive metabolites, in the area of Red Biotechnology. Since the invention of antibiotics, fungi are known nearly as good makers of antibiotics, which are low-molecular organic products with antibiotic activity and have an occasional concentration impact against alternative microorganism species [113]. Fungi are identified not solely as sources of antibiotics but conjointly producers of vast quantities of medication inhibitors and antitumor medicine. They create antitumor compounds with a number of the most promising compounds derived from a marine fungus, particularly those found in sea sediments. Asperolide E, and unique novel tetranorlabdane diterpenoid discovered from *Aspergillus wentii* SD-310 marine sediment-derived flora, is one example showing toxicity against cervical, breast, and carcinoma. Taxol® (generic name Paclitaxel) is one of the foremost vital associated-cancer compounds used to manufacture medicines with an annual price of various bucks [114]. Several fungi have shown the flexibility to compound this molecule, increasing the economic scale of synthesis. Curiously, the genus

Aspergillus aculeatinus Tax-6 was discovered in the Discovery, L'Atalante, and Urania basins have recently been documented to increase the assembly of taxol [115]. DHAB fungus may create bioactive compounds that will be employed as new treatment or prescription pharmaceuticals with enhanced options (safer, stronger, and with a broader variety). The growing diversity of extremophilic genomes and metagenomes will provide crucial data from DHAB specimens to identify novel flora enzymes and molecules relevant to biotechnology. Fungi's metabolic variations embody the flexibility to interrupt down a good form of novel enzymes and compounds, similarly to the synthesis of antibiotics, anticancer medicine, and alternative metabolites that will be accustomed to producing new Blue Biotechnologies [116].

Several marine organisms, together with phylum, invertebrate, Protozoa, Fungi, and Crustacean, turn out polysaccharides within the marine atmosphere because of the most plenteous advanced sugar compound. The ocean provides an excellent resource for chemicals with good applied potential as enzymes and a possible current value [117]. Due to the ease with which industrial enzymes may be grown and produced, the need for industrial enzymes, particularly those derived from marine microorganisms, has risen dramatically in recent years for their use in a wide range of processes [118]. Since most industrial processes are unit meted out beneath extreme chemistry conditions that cannot be definitively changed to the best conditions required for the activation of the enzymes obtainable it might be of nice profit to own enzymes that exhibit the best activity over a good vary of pH, temperature and salinity values [119]. Antifungal, insecticidal, antiprotozoal drug, medicine, hypocholesterolemic, and food quality enhancers, as well as antifungal, insecticidal, antiprotozoal drug, medicine, hypocholesterolemic, and food quality enhancers, are potential enzymes for applications in several fields of drugs, nutrition, biotechnology, industrial applications, and waste management. For their role within the biocontrol of flora phytopathogens, chitinases have received augmented attention. Additionally, these polysaccharides applications are notably valuable in agriculture as biocontrol agents for their ability to change the chitin parts within the flora cytomembrane in distinction to flora phytopathogens [120].

The use of marine protoctista has increased substantially in recent decades since diversity allows researchers to look for various bioactive chemicals in brown, red, and inexperienced marine protoctista. Marine protoctista is abundant source of dietary fiber [121]. They are conjointly valuable sources of structurally various bioactive chemicals with medicinal property, inhibitor capability, medication properties, anti-coagulant, anti-thrombotic, anti-viral activity. Moreover, marine protoctista contains vital quantities of advanced carbohydrates, called prebiotics [122]. Polysaccharides and oligosaccharides from marine organisms are of particular interest because they may be quickly cultivated and allow the production of certain bioactive substances to be controlled under optimal growing circumstances or via biotechnology. Additionally, the employment of microorganism enzymes throughout the extraction, purification, or alteration of that prebiotics is a vital instrument for novel structures during which they are biological operate is elucidated. Presentation as purposeful food ingredients of red, inexperienced, and algae for potential use, in conjunction with the likelihood of enhancing the event of polysaccharides and oligosaccharides from these sources through the employment of microorganism enzymes and its health benefits, as well as other applications [121].

An internationally recognized industrial catalyst cellulase is ready to change polysaccharide and polysaccharide spinoff organic compound bonds into soluble oligosaccharides and aldohexose while not cyanogenic by-products. Versatile cellulase applications embody the search for brand spanking new and novel catalyst sources. Each microorganism and fungi are deeply exploited for their ability to form a good varies of polysaccharide enzymes. A comprehensive analysis of cellulase derived from a marine fungus has already shown some promising results [123]. Amongst fourteen fungi derived from the marine sponge *Stelletta normani*, three halotolerant strains, *Emericellopsis* sp. TS11, *Cadophora* sp. TS2, and *Pseudogymnoascus* sp. TS12, showed high CMCase (Catalyzes spinoff compound, cellulose (CMC)) activity (obtained from a depth of 751 m). Marine cellulases have lately emerged as promising agents in the degradation of algae and hence the breakdown of wastes into low molecular fragments, indicating their outstanding promise in green chemistry and environmentally friendly processes. Moreover, they are employed in the transformation of bioethanol from lignocellulosic materials. Marine organism cellulases even have a high specific activity, thermostability, and alternative vital organic chemistry properties [124].

In recent years, researchers have extracted a variety of enzymes with specific activity from microbial sources, actinobacteria, fungi, and alternate marine microorganisms, and several other products have already been used in commercial applications. Specifically, an oversized variety of drug candidates have yielded some marine microorganism enzymes. Marine that types the large genetic and organic chemistry diversity of that remain in infancy are unit of current attention as a new promising source

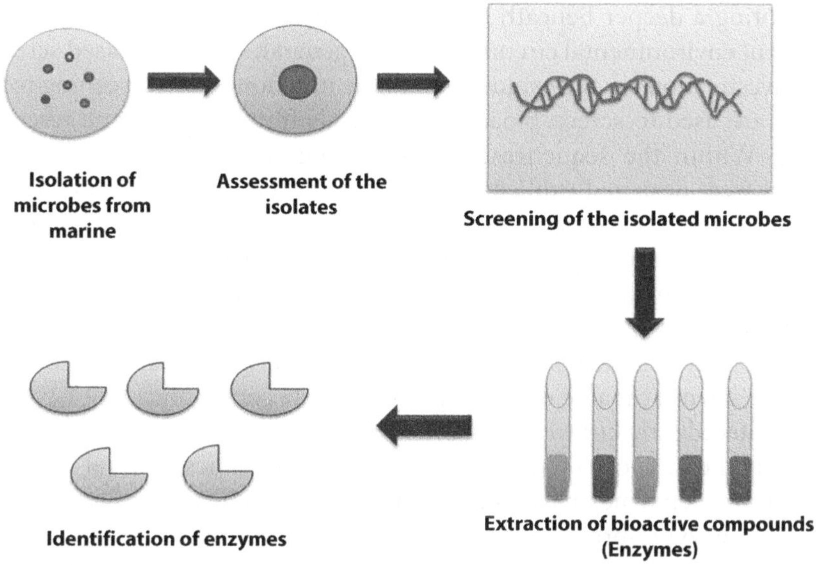

Figure 5.2 The isolation of microbes from marine for identification of enzymes.

of enzymes with uncertain application potential. The general procedure for the invention of biologically active natural compounds is shown in Figure 5.2.

5.4 Novel Enzyme Identification from Marine Microbes Through Metagenomics

Metagenomics may be a technology that involves researching polymer extracted from environmental samples while not the microorganism communities being extracted and civilized. This method was primarily accustomed discover novel microorganism products, similarly on samples, the microorganism diversity from completely different niches of the setting. The pooled polymer reservoir collected from every environmental sample has been mentioned because the metagenome acts as a supply of genetic info which will be processed for the detection of the latest organisms and the aim of characteristic antecedent unidentified organic chemistry pathways and functions [125]. Microorganism metagenomes have played a significant role in developing many novel enzymes, antimicrobials, therapeutic compounds, and various biochemically active compounds over the past 20 years. The microorganism metagenome analysis was conjointly promoting a deeper beneath standing of microbiomes under completely different environmental circumstances. In general, sequence-based screening investigations or the exploitation of the function-based screening technique are used to screen bioactive chemicals from metagenome research [126]. Within the sequence-based screening strategy, the primers and probes were designed explicitly on the idea of the sequence of previously known catalyst cryptography genes and alternative essential bioactive chemical cryptography genes. These genes were then known and later on sequenced by mistreatment PCR amplification. For the aim of cross-verification and over-expression, those genes would be cloned into entirely new expression systems. The purposeful screening approach detects and discovers genes capable of making entirely new compounds that have not been antecedently known [127].

With the support of microorganism metagenomics, several business significance enzymes have been known to date. Marine ecosystems host multiple enzyme-producing genes. Esterases, proteases, glycosyl hydrolases, dehalogenases, endo-1,4-β glucanases from the metagenomic analysis are several enzymes isolated from marine environments [128]. Sequence-based or functional-based screening experiments are accustomed to screening bioactive compounds in metagenome experiments. The primers and probes were constructed in sequence-based screening experiments based on the sequence of labeled catalyst cryptography genes and substitute important bioactive chemical cryptography genes. The purposeful screening will assist in detective work and finding genes that will turn out utterly new compounds that have not been antecedently known. The polymer is shredded up and cloned in functional-based screening to organize libraries that are then accustomed generate clones. Such clones could also be subject to composition screens, facilitating distinguishing compounds with a selected activity needed [11]. The steps meted out for marine microbes by metagenomics are delineated in Figure 5.3.

In many sectors, starting from the pharmaceutical business to the food business, most of the enzymes discovered by metagenome analysis have found entirely different applications. Amylases, an assortment of enzymes created by various microorganisms, archaea, and fungi with the flexibility to turn the breakdown of starch into aldohexose monomers or short aldohexose chains, are the number of the essential industrial enzymes used commercially. These enzymes are employed in the paper, fruit, clothing, detergent, and sugar industries, within the biodegradability of alkanes, and within the synthesis of gold nanoparticles. Amino acid hydrolase enzymes that change lipids are another class of enzymes referred to as lipases [129]. These enzymes were derived from entirely different microbic sources and

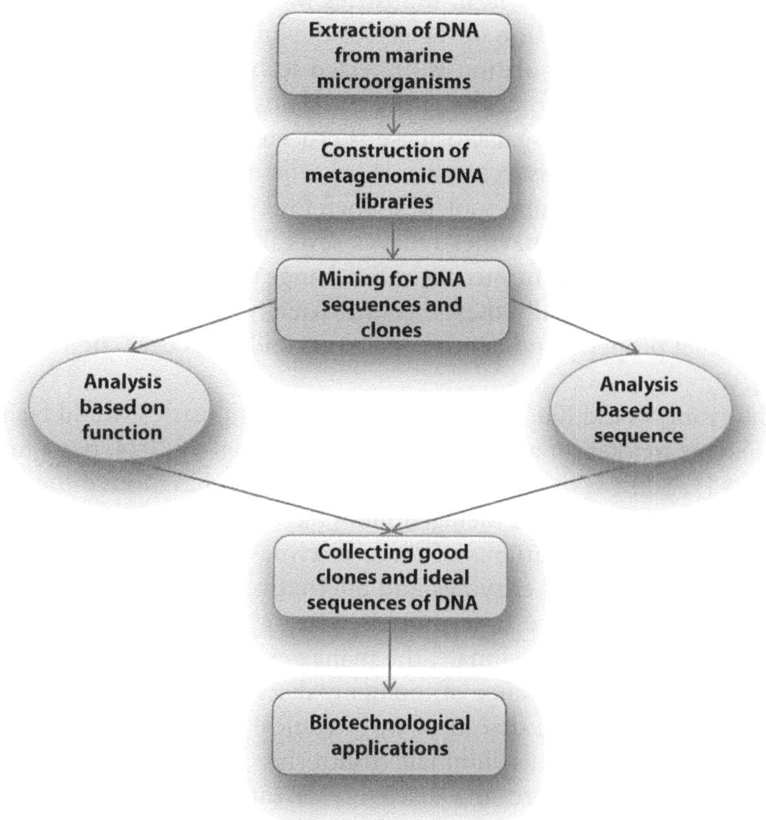

Figure 5.3 The steps involved in metagenomics.

bacterium and yeast. They were utilized in numerous industries, like the food trade to by selection hydrolyze fats and oils, the dairy farm trade to extract fat from milk merchandise, the detergent trade to extract onerous stains, the sewer water treatment for degrading macromolecule impurities, and therefore the cosmetic trade to come up with surface impurities [130]. By hydrolyzing the beta-1,4-linkage of N-acetyl glucosamine, the polyoseases catalyze the breakdown of the present chemical compound chitin. For this catalyst, microbic sources embody bacterium and fungi. Chitinases are used as associate degree antifungal medication to manage polysaccharide waste gift in sewer water, pests, biocontrol of assorted fungi and insects capable of ruin crops, and conjointly for the preparation of chitooligomers with numerous medical applications within the pharmaceutical trade. Another essential enzyme used principally within the food processor and dairy farm industries was beta galactosidases to catalyze the chemical reaction of the glycosidic linkages gift in disaccharide; 70% of the world's population is projected to be disaccharide intolerant, and it's so vital to reinforce and grow the lactose-free product. Therefore, it's been thought that β-galactosidase is crucial for assembling food merchandise freed from disaccharides [131]. For many industries, enzymes square important measure moieties. Therefore, the development of additional and additional enzymes relieves several industries from the pressure of exploitation venturous chemical change ways or making poisonous by-products. Additionally, these enzymes usually have relevant parameters like product property and delicate reaction conditions over chemical ways [132].

The rising would like for catalysts in several industries would facilitate the enzyme market hit a valuation of $17.50 billion by 2024, per a market forecast [133]. In several industries, these enzymes have found the proper smart connectedness, and metagenomics has assisted in the fast discovery of novel and helpful variants of those enzymes. Applications of metagenomics for identifying enzymes and alternative bioactive compounds of commercial significance have a light-emitting diode to property industrial growth. Within the climate, the management of microbes and toxins contributes to the hindrance of the many diseases also as epidemics. The microbic metagenomic design of early clinical identification of diseases helps to minimize also as eliminate health risks. Additionally, each bioremediation and, therefore, the discovery of innovative prescribed drugs square measure to enhance the standard of life globally [134].

5.5 Future Prospects

The 21st century is the associate degree ocean century, and therefore the ocean is an implausible treasure of human life. Most countries have faced similar issues recently, like high population, consumption of energy, and pollution. Marine biological advancement and growth provide persons with new sources and selections. Marine microbe enzymes, particularly extreme marine microbic enzymes, became progressively vital in their applications. Since enzymes have unexampled advantages, several corporations square measure keenly curious about adapting catalyst techniques to their method needs. The clinical application of enzymes is additionally an enormous increasing field. For instance, surgeons used chemical process enzymes for wound operation, and bound enzymes like enzyme, crystalline, and chymotrypsin are documented with promising clinical results.

Since the inflated therapeutic use of enzymes is presently unsure, fast progress may be expected during this space. Over the past few years, the utilization of enzymes has considerably inflated. The prospects for continued to expand the utilization of presently accessible enzymes square measure glorious in existing applications and new uses, also as within the use of recent enzymes for several functions. Additionally, some new areas and metagenomics give a robust lens for observant the microbic setting that will have the potential to revolutionize our understanding of the marine microbic enzymes. A technique for the invention of helpful bioactive agents from marine microorganisms has been developed to exploit several ecological variables of the marine setting. Many novel anti-malarial antibiotics, antitumor polysaccharides, glucan-degrading enzymes, and aminoglycoside antibiotics are known with relevance ocean water needs, development, and degradation of marine polymers and inclusion material. Together, due to marine biological diversity and therefore the specificity of biological metabolism, the analysis is merely beginning on a world scale; however, it is tremendous potential for industrial-beneficial growth and applications.

5.6 Conclusion

For many decades, microbic enzymes are utilized in completely different industries. With the arrival of biotechnology, the curiosity and demand for enzymes with novel properties have been increasing. Currently, microbes from terrestrial sources square measure utilized for the industrial development of enzymes, though the potential for synthesis of the many novel enzymes by marine microorganisms has been recognized. variety of catalyst detection studies are conducted in the Republic of India over the past twenty years, as well as phosphates, arylsulfatase, chitinase, l-asparaginase, l-glutaminase, amylase, protease, lipase, cellulase, urease, and lactamase isolated from a marine bacterium, blue-green algae, and fungi. Researchers started learning marine enzymes, and these days they are favored over the chemical catalysts for biotechnological applications and functions, encompassing numerous industries, namely, domestic, producing commercial, and aid. The sector of marine catalyst technology has recently assumed significance. It will, with success, replace alternative chemical sources of enzymes that square measure helpful for industrial and industrial functions and that convince be a valuable and eco-friendly different if it is any boosted. Marine catalyst technology can sustain the growing demand for catalyst output with spare directions and motivation. The marine setting has been progressively studied due to its colossal supply of potential biocatalysts. Marine metagenomics may be a speedily evolving and exciting field of genomic studies within which microbic species in marine environments may be investigated. The invention of biocatalysts has inflated with metagenomics, as several have shown novel characteristics. Even with improved practical screening capacities, the identification of biocatalysts remains a challenge. Recent developments within these studies have a light-emitting diode to a powerful interest in enlisting microbic enzymes for environmentally friendly trade growth.

References

1. Mouhid, L., Corzo-Martínez, M., Torres, C., Vázquez, L., Reglero, G., Fornari, T., Ramírez de Molina, A., Improving *in vivo* efficacy of bioactive molecules: An overview of potentially antitumor phytochemicals and currently available lipid-based delivery systems. *J. Oncol.*, 2017, 7351976, 2017.
2. Riera-Romo, M., Wilson-Savón, L., Hernandez-Balmaseda, I., Metabolites from marine microorganisms in cancer, immunity, and inflammation: A critical review. *J. Pharm. Pharmacogn. Res.*, 8, 368, 2020.
3. Mathur, S. and Hoskins, C., Drug development: Lessons from nature. *Biomed. Rep.*, 6, 612, 2017.
4. Stien, D., Marine microbial diversity as a source of bioactive natural products. *Mar. Drugs*, 18, 215, 2020.
5. Lauritano, C., Ferrante, M., II, Rogato, A., Marine natural products from microalgae: An-omics overview. *Mar. Drugs*, 17, 269, 2019.
6. Sun, Z.H., Gu, J., Ye, W., Wen, L.X., Lin, Q.B., Li, S.N., Zhang, W.M., Geospallins, A.–C., New thiodiketopiperazines with inhibitory activity against angiotensin-converting enzyme from a deep-sea-derived fungus *Geosmithia pallida* FS140. *Mar. Drugs*, 16, 464, 2018.
7. Heo, Y.M., Lee, H., Kim, K., Kwon, S.L., Park, M.Y., Kang, J.E., Kim, J.J., Fungal diversity in intertidal mudflats and abandoned solar salterns as a source for biological resources. *Mar. Drugs*, 17, 601, 2019.
8. Pye, C.R., Bertin, M.J., Lokey, R.S., Gerwick, W.H., Linington, R.G., Retrospective analysis of natural products provides insights for future discovery trends. *Proc. Natl. Acad. Sci.*, 114, 5601, 2017.
9. Jahromi, S.T. and Barzkar, N., Marine bacterial chitinase as sources of energy, eco-friendly agent, and industrial biocatalyst. *Int. J. Biol. Macromol.*, 120, 2147, 2018.
10. Gladfelter, A.S., James, T.Y., Amend, A.S., Marine fungi. *Curr. Biol.*, 29, R191, 2019.
11. Chapman, J., Ismail, A.E., Dinu, C.Z., Industrial applications of enzymes: Recent advances, techniques, and outlooks. *Catalysts*, 8, 238, 2018.

12. Sampath, K.G., Ghosh, T., Srinivas, T.N.R., Ecological perspective of marine microbial enzymes. *J. Emerg. Tech. Innov. Res.*, 6, 210, 2019.
13. Giddings, L.A. and Newman, D.J., Bioactive compounds from terrestrial extremophiles, in: *Bioactive Compounds from Terrestrial Extremophiles*, pp. 1–75, Springer, Cham, 2015.
14. Cheng, T.H., Ismail, N., Kamaruding, N., Saidin, J., Danish-Daniel, M., Industrial enzymes-producing marine bacteria from marine resources. *Biotechnol. Rep.*, 29, e00482, 2020.
15. Levy-Ontman, O., Fisher, M., Shotland, Y., Tekoah, Y., Arad, S.M., Insight into glucosidase II from the red marine microalga *Porphyridium sp.* (Rhodophyta). *J. Phycol.*, 51, 1075, 2015.
16. Gao, J. and Chi, J., Biodegradation of phthalate acid esters by different marine microalgal species. *Mar. Pollut. Bull.*, 99, 70, 2015.
17. Inoue, A., Satoh, A., Morishita, M., Tokunaga, Y., Miyakawa, T., Tanokura, M., Takao, O., Functional heterologous expression and characterization of mannuronan C5-epimerase from the brown alga *Saccharina japonica*. *Algal Res.*, 16, 282, 2016.
18. Thapa, H.R., Naik, M.T., Okada, S., Takada, K., Molnár, I., Xu, Y., Devarenne, T.P., A squalene synthase-like enzyme initiates production of tetraterpenoid hydrocarbons in *Botryococcus braunii* Race L. *Nat. Commun.*, 7, 11198, 2016.
19. Greene, R.V., Grifin, H.L., Cotta, M.A., Utility of alkaline protease from marine shipworm bacterium in industrial cleansing applications. *Biotechnol. Lett.*, 18, 759, 1996.
20. Prabhu, G.N. and Chandrasekaran, M., Impact of process parameters on l-glutaminase production by marine *Vibrio costicola* in solid state fermentation using polystyrene as an inert support. *Process Biochem.*, 32, 285, 1997.
21. Choi, Y.J., Kim, E.J., Piao, Z., Yun, Y.C., Shin, Y.C., Purification and characterization of chitosanase from *Bacillus* sp. strain KCTC 0377BP and its application for the production of chitosan oligosaccharides. *Appl. Environ. Microbiol.*, 70, 4522, 2004.
22. Ohta, Y., Hatada, Y., Nogi, Y., Li, Z., Ito, S., Horikoshi, K., Cloning, expression, and characterization of a glycoside hydrolase family 86 β-agarase from a deep-sea *Microbulbifer*-like isolate. *Appl. Microbiol. Biotechnol.*, 66, 266, 2004.
23. Hatada, Y., Ohta, Y., Horikoshi, K., Hyperproduction and application of alpha-agarase to enzymatic enhancement of antioxidant activity of porphyran. *J. Agric. Food Chem.*, 54, 9895, 2006.
24. Lee, Y.S., Yoo, J.S., Chung, S.Y., Lee, Y.C., Cho, Y.S., Choi, Y.L., Cloning, purification, and characterization of chitosanase from *Bacillus* sp. DAU101. *Appl. Microbiol. Biotechnol.*, 73, 113, 2006a.
25. Lee, D.G., Park, G.T., Kim, N.Y., Lee, E.J., Jang, M.K., Shin, Y.G., Lee, S.H., Cloning, expression, and characterization of a glycoside hydrolase family 50 beta-agarase from a marine Agarivorans isolate. *Biotechnol. Lett.*, 28, 1925, 2006b.

26. Ochiai, A., Hashimoto, W., Murata, K., A biosystem for alginate metabolism in Agrobacterium tumefaciens strain C58: Molecular identification of Atu3025 as an exotype family PL-15 alginate lyase. *Res. Microbiol.*, 157, 642, 2006.
27. Venugopal, M. and Saramma, A.V., An alkaline protease from *Bacillus circulans* BM15, newly isolated from a mangrove station: Characterization and application in laundry detergent formulations. *Indian J. Microbiol.*, 47, 298, 2007.
28. Lee, Y.S., Park, I.H., Yoo, J.S., Chung, S.Y., Lee, Y.C., Cho, Y.S., Choi, Y.L., Cloning, purification, and characterization of chitinase from *Bacillus* sp. DAU101. *Bioresour. Technol.*, 98, 2734, 2007.
29. Khambhaty, Y., Mody, K., Jha, B., Purification and characterization of κ-carrageenase from a novel γ-proteobacterium, *Pseudomonas elongata* (MTCC 5261) syn. *Microbulbifer elongatus* comb. Nov. *Biotechnol. Bioprocess Eng.*, 12, 668, 2007.
30. Haddar, A., Agrebi, R., Bougatef, A., Hmidet, N., Sellami-Kamoun, A., Nasri, M., Two detergent stable alkaline serine-proteases from *Bacillus mojavensis* A21: Purification, characterization and potential application as a laundry detergent additive. *Bioresour. Technol.*, 100, 3366, 2009.
31. Fu, X., Liu, P., Lin, L., Hong, Y., Huang, X., Meng, X., Liu, Z., A novel endoglucanase (Cel9P) from a marine bacterium *Paenibacillus* sp. BME-14. *Appl. Biochem. Biotechnol.*, 160, 1627, 2009.
32. Guo, B., Chen, X., Sun, C., Zhou, B., Zhang, Y., Gene cloning, expression and characterization of a new cold-active and salt-tolerant endo-β-1, 4-xylanase from marine *Glaciecola mesophila* KMM 241. *Appl. Microbiol. Biotechnol.*, 84, 1107, 2009.
33. Wang, S.L., Chao, C.H., Liang, T.W., Chen, C.C., Purification and characterization of protease and chitinase from *Bacillus cereus* TKU006 and conversion of marine wastes by these enzymes. *Mar. Biotechnol.*, 11, 334, 2009a.
34. Wang, S.L., Liou, J.Y., Liang, T.W., Liu, K.C., Conversion of squid pen by using *Serratia* sp. TKU020 fermentation for the production of enzymes, antioxidants, and N-acetyl chitooligosaccharides. *Process Biochem.*, 44, 854, 2009b.
35. Mao, X., Hong, Y., Shao, Z., Zhao, Y., Liu, Z., A novel cold-active and alkalistable β-glucosidase gene isolated from the marine bacterium *Martelella mediterranea*. *Appl. Biochem. Biotechnol.*, 162, 2136, 2010.
36. Songsiriritthigul, C., Lapboonrueng, S., Pechsrichuang, P., Pesatcha, P., Yamabhai, M., Expression and characterization of *Bacillus licheniformis* chitinase (ChiA), suitable for bioconversion of chitin waste. *Bioresour. Technol.*, 101, 4096, 2010.
37. Verma, D. and Satyanarayana, T., Cloning, expression and applicability of thermo-alkali-stable xylanase of *Geobacillus thermoleovorans* in generating xylooligosaccharides from agro-residues. *Bioresour. Technol.*, 107, 333, 2012.

38. Jiang, X., Chen, D., Chen, L., Yang, G., Zou, S., Purification, characterization and action mode of a chitosanase from *Streptomyces roseolus* induced by chitin. *Carbohydr. Res.*, 355, 40, 2012.
39. Kang, L.X., Chen, X.M., Fu, L., Ma, L.X., Recombinant expression of chitosanase from *Bacillus subtilis* HD145 in *Pichia pastoris*. *Carbohydr. Res.*, 352, 37, 2012.
40. Qin, Y., Huang, Z., Liu, Z., A novel cold-active and salt-tolerant α-amylase from marine bacterium *Zunongwangia profunda*: Molecular cloning, heterologous expression and biochemical characterization. *Extremophiles*, 18, 271, 2013.
41. Xie, W., Lin, B., Zhou, Z., Lu, G., Lun, J., Xia, C., Hu, Z., Characterization of a novel β-agarase from an agar-degrading bacterium *Catenovulum* sp. X3. *Appl. Biochem. Biotechnol.*, 97, 4907, 2013.
42. Lee, Y., Oh, C., De Zoysa, M., Kim, H., Wickramaarachchi, W.D., Whang, I., Lee, J., Molecular cloning, overexpression, and enzymatic characterization of glycosyl hydrolase family 16 β-Agarase from marine bacterium *Saccharophagus* sp. AG21 in *Escherichia coli*. *J. Microbiol. Biotechnol.*, 23, 913, 2013.
43. Liu, X., Huang, Z., Zhang, X., Shao, Z., Liu, Z., Cloning, expression and characterization of a novel cold-active and halophilic xylanase from *Zunongwangia profunda*. *Extremophiles*, 18, 441, 2014a.
44. Liu, N., Mao, X., Du, Z., Mu, B., Wei, D., Cloning and characterisation of a novel neoagarotetraose-forming-β-agarase, AgWH50A from *Agarivorans gilvus* WH0801. *Carbohydr. Res.*, 388, 147, 2014b.
45. Goo, B.G. and Park, J.K., Characterization of an alkalophilic extracellular chitosanase from *Bacillus cereus* GU-02. *J. Biosci. Bioeng.*, 117, 684, 2014.
46. Wu, G., Yuan, M., Wei, L., Zhang, Y., Lin, Y., Zhang, L., Liu, Z., Characterization of a novel cold-adapted phosphinothricin N-acetyltransferase from the marine bacterium *Rhodococcus* sp. strain YM12. *J. Mol. Catal. B Enzym.*, 104, 23, 2014.
47. Yuan, S., Jiang, W., Chen, L., Guo, Y., Liu, Z., A novel serine hydroxymethyltransferase from marine bacterium *Alcanivorax* sp. and its application on enzymatic synthesis of L-serine. *J. Mol. Catal. B Enzym.*, 109, 17, 2014.
48. Jiang, W., Chen, L., Hu, N., Yuan, S., Li, B., Liu, Z., A novel serine hydroxymethyltransferase from *Arthrobacter nicotianae*: Characterization and improving catalytic efficiency by rational design. *BMC Biotechnol.*, 14, 93, 2014.
49. Liu, C.L., Lan, C.Y., Fu, C.C., Juang, R.S., Production of hexaoligochitin from colloidal chitin using a chitinase from *Aeromonas schubertii*. *Int. J. Biol. Macromol.*, 69, 59, 2014c.
50. Kim, W., Park, J., Park, J., Choi, D., Park, Y., Purification and characterization of a fucoidanase (FNase S) from a marine bacterium *Sphingomonas paucimobilis* PF-1. *Mar. Drugs*, 13, 4398, 2015.

51. Chi, W.J., Park, J.S., Chang, Y.K., Hong, S.K., Cloning, Expression and Biochemical Characterization of a GH 16 agarase AgaH71 from *Pseudoalteromonas hodoensis* H7. *Appl. Biochem. Biotechnol.*, 175, 733, 2015.
52. Makhdoumi, A., Dehghani-Joybari, Z., Mashreghi, M., Jamialahmadi, K., Asoodeh, A., A novel halo-alkali-tolerant and thermo-tolerant chitinase from *Pseudoalteromonas* sp. DC14 isolated from the Caspian Sea. *Int. J. Environ. Sci. Technol.*, 12, 3895, 2015.
53. Wang, Z., Robertson, K.L., Liu, C., Liu, J.L., Johnson, B.J., Leary, D.H., Compton, J.R., Vuddhakul, V., Legler, P.M., Vora., G.J., A novel vibrio beta-glucosidase (LamN) that hydrolyzes the algal storage polysaccharide laminarin. *FEMS Microbiol. Ecol.*, 91, fiv087, 2015.
54. Wu, S., Liu, G., Zhang, D., Li, C., Sun, C., Purification and biochemical characterization of an alkaline protease from marine bacteria *Pseudoalteromonas* sp. 1291. *J. Basic Microbiol.*, 55, 1427, 2015.
55. Ibrahim, A.S., Al-Salamah, A.A., El-Badawi, Y.B., El-Tayeb, M.A., Antranikian, G., Detergent-, solvent- and salt-compatible thermoactive alkaline serine protease from halotolerant alkaliphilic Bacillus sp. NPST-AK15: Purification and characterization. *Extremophiles,* 19, 961, 2015.
56. Kobus, S., Widderich, N., Hoeppner, A., Bremer, E., Smits, S.H., Overproduction, crystallization and X-ray diffraction data analysis of ectoine synthase from the cold-adapted marine bacterium *Sphingopyxis alaskensis. Acta Crystallogr. F.*, 71, 1027, 2015.
57. Lanfranconi, M.P., Alvarez, A.F., Alvarez, H.M., Identification of genes coding for putative wax ester synthase/diacylglycerol acyltransferase enzymes in terrestrial and marine environments. *AMB Express,* 5, 42, 2015.
58. Pradeep, G.C., Yoo, H.Y., Cho, S.S., Choi, Y.H., Yoo, J.C., An extracellular chitinase from *Streptomyces* sp. CS147 releases N-acetyl-D-glucosamine (GlcNAc) as principal product. *Appl. Biochem. Biotechnol.*, 175, 372, 2015.
59. Bozoglu, C., Hundur, S., Alaylar, B., Karadayi, M., Gulluce, M., Isolation and molecular characterization of thermophilic bacteria with xylanase activity from thermal springs in Erzurum. *J. Life Sci. Technol.*, 3, 32, 2015.
60. Zhou, Z., Zhao, S., Wang, S., Li, X., Su, L., Ma, Y., Song, J., Extracellular overexpression of chitosanase from *Bacillus* sp. TS in *Escherichia coli. Appl. Biochem. Biotechnol.*, 175, 3271, 2015.
61. Abdulaziz, A., Sageer, S., Chekidhenkuzhiyil, J., Vijayan, V., Pavanan, P., Athiyanathil, S., Nair, S., Unicellular cyanobacteria synechocystis accommodate heterotrophic bacteria with varied enzymatic and metal resistance properties. *J. Basic Microbiol.*, 56, 845, 2016.
62. Yang, S.Q., Fu, X., Yan, Q.J., Guo, Y., Liu, Z., Jiang, Z.Q., Cloning, expression, purification and application of a novel chitinase from a thermophilic marine bacterium *Paenibacillus barengoltzii. Food Chem.*, 192, 1041, 2016.
63. Mander, P., Cho, S.S., Yun, H.C., Panthi, S., Choi, Y.S., Kim, H.M., Yoo, J.C., Purification and characterization of chitinase showing antifungal and

biodegradation properties obtained from *Streptomyces anulatus* CS242. *Arch. Pharm. Res.*, 39, 878, 2016.
64. Gong, J.S., Liu, X.M., Zhang, M.J., Li, H., Geng, Y., Purification and characterization of a high salt-tolerant alginate lyase from Cobetia sp. WG-007. *Biotechnol. Appl. Biochem.*, 64, 519, 2017.
65. Yagi, H., Fujise, A., Itabashi, N., Ohshiro, T., Purification and characterization of a novel alginate lyase from the marine bacterium *Cobetia* sp. NAP1 isolated from brown algae. *Biosci. Biotechnol. Biochem.*, 80, 2338, 2016.
66. Nguyen-Thi, N. and Doucet, N., Combining chitinase C and N-acetylhexosaminidase from *Streptomyces* coelicolor A3(2) provides an efficient way to synthesize N-acetylglucosamine from crystalline chitin. *J. Biotechnol.*, 220, 25, 2016.
67. Su, P.C., Hsueh, W.C., Chang, W.S., Chen, P.T., Enhancement of chitosanase secretion by *Bacillus subtilis* for production of chitosan oligosaccharides. *J. Taiwan Inst. Chem. Eng.*, 79, 49, 2017.
68. Kurakake, M., Kitagawa, Y., Okazaki, A., Shimizu, K., Enzymatic properties of alginate lyase from *Paenibacillus* sp. S29. *Appl. Biochem. Biotechnol.*, 183, 1455, 2017.
69. Jung, S., Lee, C.R., Chi, W.J., Bae, C.H., Hong, S.K., Biochemical characterization of a novel cold-adapted GH39 β-agarase, AgaJ9, from an agar-degrading marine bacterium *Gayadomonas joobiniege* G7. *Appl. Microbiol. Biotechnol.*, 101, 1965, 2017.
70. Gao, L., Sun, J.A., Secundo, F., Gao, X., Xue, C.H., Mao, X.Z., Cloning, characterization and substrate degradation mode of a novel chitinase from *Streptomyces albolongus* ATCC 27414. *Food Chem.*, 261, 329, 2018.
71. Qin, Z., Chen, Q.M., Lin, S., Luo, S., Qiu, Y.J., Zhao, L.M., Expression and characterization of a novel cold-adapted chitosanase suitable for chitooligosaccharides controllable preparation. *Food Chem.*, 253, 139, 2018.
72. Li, J., Xie, Y., Wang, R., Fang, Z., Fang, W., Zhang, X., Xiao, Y., Mechanism of salt-induced activity enhancement of a marine-derived laccase, Lac15. *Eur. Biophys. J.*, 47, 225, 2018.
73. Yu, Z., Zhu, B., Wang, W., Tan, H.D., Yin, H., Characterization of a new oligioalginate lyase from marine bacterium *Vibrio* sp. *Int. J. Biol. Macromol.*, 112, 937, 2018.
74. He, J., Liu, L., Liu, X., Tang, K., Isolation and characterization of a novel cold-active, halotolerant endoxylanase from *Echinicola rosea* sp. Nov. JL3085T. *Mar. Drugs*, 18, 245, 2020.
75. Cheng, C.Y. and Li, Y.K., An *Aspergillus* chitosanase with potential for large-scale preparation of chitosan oligosaccharides. *Biotechnol. Appl. Biochem.*, 32, 197, 2000.
76. Sabu, A., Keerthi, T., Kumar, S.R., Chandrasekaran, M., L-Glutaminase production by marine Beauveria sp. under solid state fermentation. *Process Biochem.*, 35, 705, 2000.

77. Raghukumar, C., Muraleedharan, U., Gaud, V.R., Mishra, R., Xylanases of marine fungi of potential use for biobleaching of paper pulp. *J. Ind. Microbiol. Biotechnol.*, 31, 433, 2004.
78. Chen, X.E., Xia, W.S., Yu, X.B., Purification and characterization of two types of chitosanase from *Aspergillus* sp. CJ22-326. *Food Res. Int.*, 38, 315, 2005.
79. Beena, P.S., Acidophilic tannase from marine *Aspergillus awamori* BTMFW032. *J. Microbiol. Biotechnol.*, 20, 1403, 2010.
80. Atalla, M.M., Zeinab, H.K., Eman, R.H., Amani, A.Y., Abd, E., AtyAbeer, A., Screening of some marine-derived fungal isolates for lignin degrading enzymes (LDEs) production. *Agric. Biol. J. North America*, 1, 591, 2010.
81. Suresh, P.V., Kumar, P.K.A., Sachindra, N.M., Thermoactive β-N-acetyl-hexosaminidase production by a soil isolate of *Penicillium monoverticillium* CFR 2 under solid state fermentation: Parameter optimization and application for N-acetyl chitooligosaccharides preparation from chitin. *World J. Microbiol. Biotechnol.*, 27, 1435, 2011.
82. Passarini, M.R.Z., Rodrigues, M.V.N., Da Silva, M., Sette, L.D., Marine-derived filamentous fungi and their potential application for polycyclic aromatic hydrocarbon bioremediation. *Mar. Pollut. Bull.*, 62, 364, 2011.
83. Zhu, X.F., Tan, H.Q., Zhu, C., Liao, L., Zhang, X.Q., Wu, M., Cloning and overexpression of a new chitosanase gene from *Penicillium* sp. D-1. *AMB Express*, 2, 13, 2012.
84. Niyonzima, F.N. and More, S., Purification and properties of detergent-compatible extracellular alkaline protease from *Scopulariopsis* spp. *Prep. Biochem. Biotechnol.*, 44, 738, 2014.
85. Li, L., Singh, P., Liu, Y., Pan, S., Wang, G., Diversity and biochemical features of culturable fungi from the coastal waters of Southern China. *AMB Express*, 4, 60, 2014.
86. Smitha, S.L., Correya, N.S., Philip, R., Marine fungi as a potential source of enzymes and antibiotics. *Int. J. Res. Mar. Sci.*, 3, 5, 2014.
87. Zhang, J., Cao, H., Li, S., Zhao, Y., Wang, W., Xu, Q., Yin, H., Characterization of a new family 75 chitosanase from *Aspergillus* sp. W-2. *Int. J. Biol. Macromol.*, 81, 362, 2015.
88. Passarini, M.R., Ottoni, C.A., Santos, C., Lima, N., Sette, L.D., Induction, expression and characterization of laccase genes from the marine-derived fungal strains *Nigrospora* sp. CBMAI 1328 and *Arthopyrenia* sp. CBMAI 1330. *AMB Express*, 5, 19, 2015.
89. Moubasher, A.A.H., Ismail, M.A., Hussein, N.A., Gouda, H.A., Enzyme producing capabilities of some extremophilic fungal strains isolated from different habitats of Wadi El-Natrun, Egypt. Part 2: Cellulase, xylanase and pectinase. *Eur. J. Biol. Res.*, 6, 103, 2016.
90. Gao, B., Mao, Y., Zhang, L., He, L., Wei, D., A novel saccharifying a-amylase of Antarctic psychotolerant fungi *Geomyces pannorum*: Gene, cloning, functional expression, and characterization. *Starch-Starke*, 67, 20, 2016.

91. Atalla, S.M., Gamal, N.G., Awad, H.M., Chitinase of marine *Penicillium chrysogenum* MH745129: Isolation, identification, production and characterization as controller for citrus fruits postharvest pathogens. *Jordan J. Biol. Sci.*, 13, 19, 2020.
92. Turkiewicz, M., Pazgier, M., Donachie, S.P., Kalinowska, H., Invertase and a-glucosidase production by the endemic Antarctic marine yeast *Leucosporidium antarcticum. Pol. Polar Res.*, 26, 125, 2005.
93. Wang, L., Chi, Z.M., Wang, X.H., Liu, Z.Q., Li, J., Diversity of lipase-producing yeasts from marine environments and oil hydrolysis by their crude enzymes. *Ann. Microbiol.*, 57, 495, 2007.
94. Rong, Y., Zhang, L., Chi, Z., Wang, X., A carboxymethyl cellulase from marine yeast (Aureobasidium pullulans 98): Its purification, characterization, gene cloning and carboxymethyl cellulose digestion. *J. Ocean Univ. China*, 14, 913, 2015.
95. Liu, G., Yue, L., Chi, Z., Yu, W., Chi, Z., Madzak, C., The surface display of the alginate lyase on the cells of *Yarrowia lipolytica* for hydrolysis of alginate. *Mar. Biotechnol.*, 11, 619, 2009.
96. Yu, X.J., Guo, N., Chi, Z.M., Gong, F., Sheng, J., Chi, Z., Inulinase overproduction by a mutant of the marine yeast *Pichia guilliermondii* using surface response methodology and inulin hydrolysis. *Biochem. Eng. J.*, 43, 266, 2009.
97. Li, X.Y., Chi, Z.M., Liu, Z.Q., Yan, K.R., Li, H.J., Phytase production by a marine yeast *Kodamea ohmeri* BG3. *Appl. Biochem. Biotechnol.*, 149, 183, 2008.
98. Ramli, A.N.M., Mahadi, N.M., Rabu, A., Illias, R.M., Molecular cloning, expression and biochemical characterization of a cold-adapted novel recombinant chitinase from *Glaciozyma antarctica* PI12. *Microb. Cell Fact.*, 10, 94, 2011.
99. Duarte, A.W.F., Dayo-Owoyemi, I., Nobre, F.S., Sette, L.D., Taxonomic assessment and enzymes production by yeasts isolated from marine and terrestrial Antarctic samples. *Extremophiles*, 17, 1023, 2013.
100. Carrasco, M., Villarreal, P., Barahona, S., Alcaino, J., Cifuents, V., Baeza, M., Screening and characterization of amylase and cellulase activities in psychrotolerant yeasts. *BMC Microbiol.*, 16, 1, 2016.
101. Hoseinifar, S.H., Sun, Y.Z., Caipang, C.M., Short-chain fatty acids as feed supplements for sustainable aquaculture: An updated view. *Aquac. Res.*, 48, 1380, 2017.
102. Morgan, N.K., Wallace, A., Bedford, M.R., Choct, M., Efficiency of xylanases from families 10 and 11 in production of xylo-oligosaccharides from wheat arabinoxylans. *Carbohydr. Polym.*, 167, 290, 2017.
103. Corinaldesi, C., Barone, G., Marcellini, F., Dell'Anno, A., Danovaro, R., Marine microbial-derived molecules and their potential use in cosmeceutical and cosmetic products. *Mar. Drugs*, 15, 118, 2017.

104. Li, S., Wang, L., Chen, X., Zhao, W., Sun, M., Han, Y., Cloning, expression, and biochemical characterization of two new oligoalginate lyases with synergistic degradation capability. *Mar. Biotechnol.*, 20, 75, 2018.
105. Guerrand, D., Lipases industrial applications: Focus on food and agroindustries. *OCL Oilseeds Fats Crops Lipids*, 24, D403, 2017.
106. Sajith, S., Priji, P., Sreedevi, S., Benjamin, S., An overview on fungal cellulases with an industrial perspective. *J. Nutr. Food Sci.*, 6, 461, 2016.
107. Romano, S., Jackson, S.A., Patry, S., Dobson, A.D., Extending the "one strain many compounds" (OSMAC) principle to marine microorganisms. *Mar. Drugs*, 16, 244, 2018.
108. Jiménez, C., Marine natural products in medicinal chemistry. *ACS Med. Chem. Lett.*, 9, 959, 2018.
109. Nicoletti, R. and Andolfi, A., The marine-derived filamentous fungi in biotechnology, in: *Grand Challenges in Marine Biotechnology*, pp. 157–189, Springer, Cham, 2018.
110. Navvabi, A., Razzaghi, M., Fernandes, P., Karami, L., Homaei, A., Novel lipases discovery specifically from marine organisms for industrial production and practical applications. *Process Biochem.*, 70, 61, 2018.
111. Danovaro, R., Corinaldesi, C., Dell'Anno, A., Snelgrove, P.V., The deep-sea under global change. *Curr. Biol.*, 27, R461, 2017.
112. Merlino, G., Barozzi, A., Michoud, G., Ngugi, D.K., Daffonchio, D., Microbial ecology of deep-sea hypersaline anoxic basins. *FEMS Microbiol. Ecol.*, 94, fiy085, 2018.
113. Steinle, L., Knittel, K., Felber, N., Casalino, C., de Lange, G., Tessarolo, C., Treude, T., Life on the edge: Active microbial communities in the Kryos $MgCl_2$-brine basin at very low water activity. *ISME J.*, 12, 1414, 2018.
114. Grossart, H.P., Van den Wyngaert, S., Kagami, M., Wurzbacher, C., Cunliffe, M., Rojas-Jimenez, K., Fungi in aquatic ecosystems. *Nat. Rev. Microbiol.*, 17, 339, 2019.
115. Deshmukh, S.K., Prakash, V., Ranjan, N., Marine fungi: A source of potential anticancer compounds. *Front. Microbiol.*, 8, 2536, 2018.
116. Barone, G., Varrella, S., Tangherlini, M., Rastelli, E., Dell'Anno, A., Danovaro, R., Corinaldesi, C., Marine fungi: Biotechnological perspectives from deep-hypersaline anoxic basins. *Diversity*, 11, 113, 2019.
117. Barzkar, N., Jahromi, S.T., Poorsaheli, H.B., Vianello, F., Metabolites from marine microorganisms, micro, and macroalgae: Immense scope for pharmacology. *Mar. Drugs*, 17, 464, 2019.
118. Mohanrasu, K., Rao, R.G.R., Raja, R., Arun, A., Bioremediation process by marine microorganisms. *Encycloped. Mar. Biotech.*, 4, 2211, 2020.
119. Cardozo, F.A., Facchinatto, W.M., Colnago, L.A., Campana-Filho, S.P., Pessoa, A., Bioproduction of N-acetyl-glucosamine from colloidal α-chitin using an enzyme cocktail produced by *Aeromonas caviae* CHZ306. *World J. Microbiol. Biotechnol.*, 35, 114, 2019.

120. Birolli, W.G., Lima, R.N., Porto, A.L.M., Applications of marine-derived microorganisms and their enzymes in biocatalysis and biotransformation, the underexplored potentials. *Front. Microbiol.*, 10, 1453, 2019.
121. de Borba Gurpilhares, D., Moreira, T.R., da Luz Bueno, J., Cinelli, L.P., Mazzola, P.G., Pessoa, A., Sette, L.D., Algae's sulfated polysaccharides modifications: Potential use of microbial enzymes. *Process Biochem.*, 51, 989, 2016.
122. Wells, M.L., Potin, P., Craigie, J.S., Raven, J.A., Merchant, S.S., Helliwell, K.E., Smith, A.G., Camire, M.E., Brawley, S.H., Algae as nutritional and functional food sources: Revisiting our understanding. *J. Appl. Phycol.*, 29, 949, 2017.
123. Jayasekara, S. and Ratnayake, R., Microbial cellulases: An overview and applications, in: *Cellulose*, p. 22, Intech Open, London, 2019.
124. Barzkar, N. and Sohail, M., An overview on marine cellulolytic enzymes and their potential applications. *Appl. Microbiol. Biotechnol.*, 104, 6873, 2020.
125. Alves, L.D.F., Westmann, C.A., Lovate, G.L., de Siqueira, G.M.V., Borelli, T.C., Guazzaroni, M.E., Metagenomic approaches for understanding new concepts in microbial science. *Int. J. Genomics*, 2018, 2312987, 2018.
126. Abia, A.L.K., Alisoltani, A., Ubomba-Jaswa, E., Dippenaar, M.A., Microbial life beyond the grave: 16S rRNA gene-based metagenomic analysis of bacteria diversity and their functional profiles in cemetery environments. *Sci. Total Environ.*, 655, 831, 2019.
127. Datta, S., Rajnish, K.N., Samuel, M.S., Pugazlendhi, A., Selvarajan, E., Metagenomic applications in microbial diversity, bioremediation, pollution monitoring, enzyme and drug discovery. A review. *Environ. Chem. Lett.*, 18, 1229, 2020.
128. Liu, P., Wang, W., Zhao, J., Wei, D., Screening novel β-galactosidases from a sequence-based metagenome and characterization of an alkaline β-galactosidase for the enzymatic synthesis of galactooligosaccharides. *Protein Expr. Purif.*, 155, 104, 2019.
129. Thornbury, M., Sicheri, J., Slaine, P., Getz, L.J., Finlayson-Trick, E., Cook, J., McCormick, C., Characterization of novel lignocellulose-degrading enzymes from the porcupine microbiome using synthetic metagenomics. *PloS One*, 14, e0209221, 2019.
130. Appna, N.R., Nagiri, R.K., Korupolu, R.B., Kanugala, S., Chityal, G.K., Thipparapu, G., Banda, N., Design and synthesis of novel 4-hydrazone functionalized/1, 2, 4-triazole fused pyrido [2, 3-d] pyrimidine derivatives, their evaluation for antifungal activity and docking studies. *Med. Chem. Res.*, 28, 1509, 2019.
131. Xavier, J.R., Ramana, K.V., Sharma, R.K., β-galactosidase: Biotechnological applications in food processing. *J. Food Biochem.*, 42, e12564, 2018.
132. Mahapatra, G.P., Raman, S., Nayak, S., Gouda, S., Das, G., Patra, J.K., Metagenomics approaches in discovery and development of new bioactive compounds from marine actinomycetes. *Curr. Microbiol.*, 77, 645, 2020.

133. Uygut, M.A. and Tanyildizi, M. S., Determination of effective parameters for alpha-amylase production in a modified rotating drum bioreactor. *Arab. J. Sci. Eng.*, 43, 3381, 2018.
134. Dubey, R.K., Tripathi, V., Prabha, R., Chaurasia, R., Singh, D.P., Rao, C.S., Abhilash, P.C., Metatranscriptomics and metaproteomics for microbial communities profiling, in: *Unraveling the Soil Microbiome*, pp. 51–60, Springer, Cham, 2020.

6

Cyanobacteria as a Source of Novel Bioactive Compounds

Pankaj Kumar Rai[1], Amit Joshi[1,2], Gerard Abraham[3], Richa Saxena[1], Subhomoi Borkotoky[1], Ravindra Kumar Yadav[4], Akhilesh Kumar Pandey[1] and Keshawanand Tripathi[1]*

[1]Department of Biotechnology, Invertis University, Bareilly, India
[2]Domain of Bioinformatics, Lovely Professional University, Phagwara, Punjab, India
[3]ICAR-Center for Conservation and Utilization of Blue-Green Algae, IARI, New Delhi, India
[4]Department of Environmental Studies, University of Delhi, Delhi, India

Abstract

The occurrence of cyanobacterial blooms in nutrients enriched aquatic ecosystems is a widespread phenomenon. Cyanobacterial blooms synthesized several bioactive compounds, and cyanotoxin is among one of them. The most commonly reported cyanotoxin are Nodularin, Microcystins Anatoxin-a, Saxitoxins, and Cylindrospermopsin. Biochemically, cyanotoxins are polyketides, alkaloids, nonribosomal peptides, potent toxins, and trypsin inhibitors. Pharmaceutically and neutraceutically these are significant, and exhibits, a diverse range of bioactivity, mainly antiviral, antibacterial, antitumor, anticancerous, antifungal, etc. Several oligopeptides are synthesized through nonribosomal peptide synthetase (NRPSs) in prokaryotes and lower eukaryotes. Cyanobacterial bioactive compounds are not utilized in growth and developments but are a promising source for the novel compounds. Nutrient composition especially nitrogen and phosphorus significantly influences types of bioactive compounds. The common cyanotoxin frequently occurred in cyanobacterial blooms is Microcystin, and is encoded by *mcy* gene cluster in freshwater and brackish water ecosystems. There are several methods used for purification and characterizations of the bioactive compounds from blooms forming cyanobacterial such as HPLC, UPLC, LC-MS-MS,

**Corresponding author*: tripathikn009@gmail.com

Sanjay Kumar, Narendra Kumar and Shahid-ul-Islam (eds.) Role of Microbes in Industrial Products and Processes, (145–170) © 2023 Scrivener Publishing LLC

and MALDI-TOF-MS. This review mainly focused on novel bioactive compounds obtained from cyanobacteria.

***Keywords*:** Cyanobacteria, cyanotoxins, secondary metabolites, bioactive compounds, non-ribosomal peptide

6.1 Introduction

In aquatic ecosystems, cyanobacterial bloom formation is a global phenomenon and poses a severe threat to aquatic plants and human health. Moreover, some species of cyanobacteria grow excessively in freshwater, lakes, ponds, and rivers due to eutrophication, which leads to harmful bloom formation all over the world [1]. The consequence of the harmful cyanobacterial bloom proliferation leads to the accumulation of several bioactive compounds such as cyanotoxins in water bodies causing high risk to global public health and aquatic organisms [1]. Cyanotoxin production is increasing consistently due to the enormous growth of various cyanobacterial sp. and their level of toxins production [2]. Light intensity, pH, temperature, macronutrients such as nitrogen (N) and phosphorus (P) significantly influenced cyanobacteria growth and their toxin biosynthesis [1, 3]. Due to greenhouse gases, the earth's temperature rises, causing dramatic changes in species composition and promoting harmful phytoplankton blooms [4].

Cyanobacteria (the term given by Roger Stanier 1970) are Gram-negative, photoautotrophic microbes [5, 6]. They are considered pioneer organisms of the Earth, resemble prokaryotic cellular organizations, and exhibit similarities of oxygen-evolving algae and plants. Cyanobacteria structures are similar to bacterial cellular organizations and thus placed into photosynthetic prokaryotes [5]. Indeed, the plastids of photosynthetic plants are transferred from cyanobacteria and proved the origin of plastids from endosymbiotic cyanobacteria [7]. The fossil record shows that cyanobacteria originated in the Precambrian era, and it convert an anaerobic to the aerobic environment on the Earth [8]. It globally exists in different habitats ranging from terrestrial to aquatic ecosystems such as deserts, salt pans, hypersaline water, ice caps, and thermal vents [8, 9].

Cyanobacteria show similarity with plants due to the presence of both photosystems (PSI and PSII). They are photosynthesis microbe; the morphology of the cyanobacteria is unicellular, filamentous, and colonial or mat form [10, 11]. Cyanobacteria classified in to five order *Chroococcales*,

Pleurocapsales, Oscillatoriales, Nostocales, and Stigonematales. Cyanobacteria showed three different types of specialized cells; vegetative cells (responsible for photosynthesis due to presence of photosystems PSI and PS II), akinetes (spore formation), and heterocyst (responsible for nitrogen fixation and PSII deficient) [11]. In nitrogen-deficient paddy soil, cyanobacterial forms often present more than 50% of the species, and play a significant role and potentially contribute towards biological nitrogen fixation, phosphate solubilization, siderophore, phytohormone synthesis, improve soil fertility and crop productivity [12–14]. Cyanobacteria show unique symbiosis with other organisms such as *Azolla* and provide 25–40 kg N/ha per season, and also enhance crop yields and soil quality of paddy fields [3]. The genome size of the cyanobacteria varies 1.6–8.6 $\times 10^9$ Da and is closely related to eubacteria [15].

Cyanotoxins played a significant role in the survival of cyanobacteria, and provide chemical defense over the other microbes and phytoplanktons [16]. The role of chemical signaling is also involved, and their specific role in the ecological aspect and biotechnological implication is still unexplored. However, in recent years screening isolation and identification of cyanotoxin from cyanobacteria are less explored [17].

6.2 Factors Affecting Toxicity

6.2.1 Growth Stage

The growth stage played significantly on toxin production of cyanobacteria [16]. *Microcystis aeruginosa* releases toxins at the mid-exponential growth phase; and several species of cyanobacterial toxin biosynthesis and secretion at the mid to late exponential stage [59, 60]. Similarly, algaecides compounds secreted through efflux mechanism in the medium at the mid-exponential stage and positively correlated with the biomass of *Oscillatoria* spp. [18]. *Aphanizomenon* sp. LMECYA 31 releases toxins in the medium through cell lysis [19]. Cyanobacterial age plays significantly role in the secretion of toxins. *Microcystis* sp. synthesized Microcystin and Anatoxin-Awas retained inside the cell in favorable situations, whereas secreted in the medium during stress conditions. Microcystin was released in the medium during the log phase, and the maximum was found at the late log stage. Similarly, under Cu and Fe stress conditions, the maximum amount of Anatoxin-a was noticed at the late log stage in *Phormidium autumnale* [20].

6.2.2 Nutritional Factors

6.2.2.1 Nitrogen Sources

The concentration of N and P in the nutrient medium plays determining factor in the toxin production of cyanobacteria [21]. N concentration affects the production of cyanotoxin in nitrogen and non-nitrogen-fixing cyanobacteria [22]. *Microcystis aeruginosa* synthesized 10 times more toxin as compared with N deficient strains [23]. Watanabe and Odishi [23] noticed a slight reduction of toxin production at a lower nitrate concentration of *M. aeruginosa.* Nitrate concentration and toxin production are directly related to *Oscillatoria agardhii* [24]. A non-nitrogen fixer genus *Oscillatoria, Microcystis, Aphanocapsa, Raphidiopsis, Planktothrix,* and *Woronochinia* needs nitrogen for the stimulation of toxin production and biosynthesis of peptide nature of the cyanotoxin [24].

6.2.2.2 Phosphorus

Phosphorus is the second limiting nutrient after Nitrogen and is essential for all forms of life [25, 26]. Like fossil fuels, P is a non-renewable resource, and its availability in the future is limited P is available on the earth in the form of orthophosphate (PO_4^{3-}) and is often immobilized with salts, therefore its bioavailability to phytoplankton and cyanobacteria is limited. Under natural conditions, P availability is typically limited in aquatic water bodies. Nutrient enrichment is a major factor for the cyanobacterial blooms formation in water bodies [26a].

In the freshwater ecosystem, P is the main limiting factor for algal bloom formation [27], whereas N is the limiting factor in marine ecosystems [25]. The immense use of Pi-containing chemical-based fertilizers in the agriculture sector and detergents released in households increased the P concentration into water bodies and thus promoted cyanobacterial blooms. Thus, P represents a pollutant for aquatic living systems. Alkaline phosphatase is a phytoplankton-produced enzyme that hydrolyzes complex organic phosphorus molecules to release orthophosphate, a form of phosphorus that is easily accessible to phytoplanktons [25, 26a]. In phytoplanktons, alkaline phosphatase activity is recognized as the most common sign of P-stress. Chemical pollution in freshwater ecosystems could be monitored using the alkaline phosphatase assay [26a].

Alkaline phosphatases are considered as an initial indicator of eutrophication in wetlands [28]. In limnological studies, determining the phosphatase activity of an organism is considered a significant approach. It can be used in conjunction with other indicators to determine the degree of P restriction in aquatic ecosystems or single species. A low level of P concentration is essential in the external medium for toxin production and a higher concentration showed an inhibitory effect [25]. A direct correlation was not observed in anatoxin-a production and external Pi concentrations in *Anabaena* sp. and *Aphainzomemon* sp. [29]. Several reports on Pi-limited conditions stimulated the toxin production of *M. aeruginosa* [24]. Physiological parameters also influence the toxin production of cyanobacteria. Temperature, light, enhanced algicidal and fungicidal activity in extracellular spent media of two freshwater cyanobacteria (*Anabaena* sp. and *Calothrix* spp.) [30–32]. The highest cyanotoxins productions of *Anabaena* sp. strain 90 were noticed when cyanobacterium grown 2.6 μg/L Pi concentrations in the nutrient medium [32]. On the other hand, algicide production by filamentous cyanobacterium *Oscillatoria lativirens* and Pi concentrations were inversely related [32b]. Types of toxin production by *Aphanizomenon* sp. significantly affect Pi concentrations [32a]. Moreover, there is no general relationship between Pi concentration and toxin production in terms of nutritional requirements, especially Pi.

6.3 Biosynthesis of Bioactive Compounds

6.3.1 Non-Ribosomal Peptides Synthesis

Cyanobacteria synthesized a large number of small peptides termed as non-ribosomal peptides [33]. The non-ribosomal polypeptides are approximately 50 amino acids long, and catalyzed by non-ribosomal peptide synthetases (NRPSs). Anabaenopeptins, aeruginosins, microcystins, cyanopeptolins, microginins, and microviridins are common non-ribosomal polypeptides [34, 35]. Structural framework biosynthetic enzymes are modules that are responsible for the addition of a single unit of natural product. The modules are synthesized by a single polypeptide chain (Type I), as reported in most fungal strains, or several interacting proteins (Type II) in bacterial and cyanobacterial systems. The modular structure of NRPSs consists of different domains are known as activation (A), thiolation (T), and condensation (C) domains that are easily recognized by their signature sequences [36]. Some other conserved domains also exist like M (methyltransferases), E-domains (epimerases), TE domains (thioesterase), halogenases, and ABC transporters [37].

6.3.2 Novel Bioactive Compounds from Cyanobacteria

Cyanobacteria contain wide range of biological products, and synthesized of bioactive compounds that is important to pharmaceutical and nutraceutical industry [38, 39]. These bioactive compounds can be classified as peptides, alkaloids, polyketides, and peptidic amides; and synthesized through non-ribosomal biosynthetic pathways [34]. In this pathway, peptides are different families of natural compounds and a diverse range of biological activities [40]. Non-ribosomal peptide synthesis takes place with the help of multifunctional enzymes that act as a catalyst for the assembly monomer of amino acids [40]. Total 38 novel compounds have been reported from cyanobacteria. Among the 38 compounds, 7 are antibacterial, 8 are anti-protozoal, 2 are antiviral, 7 are protease inhibitor, 6 are cytotoxic, and 1 is Ca^{2+} channel inhibitor in nature.

Due to the high degree of diversity, cyanobacteria synthesized a broad spectrum of bioactive compounds. Biochemically, they are biochemically different in nature such as amino acids, fatty acids, lipopeptides, and macrolides [41]. Cyanobacterial bioactive compounds synthesis mainly belongs to the genera of cyanobacteria *Oscillatoriales*, *Nostocales*, *Chroococcales*, *Pleurocapsales*, and *Stigonematales*. Out of that, 50% of secondary metabolites have been reported from *Microcystis* sp., *Anabaena* sp., *Spirulina* sp., *Phormidium* sp., *Nostoc* sp., and *Lyngbya* sp. [39]. Besides this, they are also a rich source of protein, phycocyanin, carotenoid, astaxanthin, fatty acids, lipopeptides, and other bioactive compounds. Genome sequencing analysis revealed that bioactive compounds of cyanobacteria are mainly synthesized non-ribosomal peptides, ribosomal, and polyketides molecules. In near future, cyanobacteria sources will be potential candidates for novel bioactive compounds (Table 6.1).

6.3.3 Cyanobacterial Drug for Anticancerous Compounds

Currently available drugs in the market show resistance against tumor cells, therefore, there is an immediate requirement for new anti-cancerous drugs. In addition, new types of cancer cells (Glioblastoma) are rapidly increasing. As a result, cancer is still a high mortality rate suffering worldwide. In the year 2007, 187.6 million people died in the world, reported by the American Cancer Society. During the years of 1990s, screening and characterization of cyanobacterial extracts for novel anti-tumorigenic compounds were started at Oregon State University and the University of Hawaii by Moore and Gerwick. Cyanobacteria are potential photosynthetic microbes that contain a vast range of anti-cancerous compounds. The main

Table 6.1 Bioactive compounds from cyanobacteria.

S.n.	Bioactive compounds	Cyanobacteria	Bioactivity	References
1	Lagunamide	*Lyngbya majuscula*	Anticancer	[5]
2	Symplocamide A	*Symploca* sp.	Anticancer	[78]
3	Symplostatin	*Symploca* sp.	Anticancer	[79]
4	Apratoxins A	*Lyngbya majuscula*	Anticancer	[80]
5	Aplysiatoxin	*Lyngbya majuscule*	Anticancer	[81]
6	Acutiphycin	*Oscillatoria acutissima*	Anticancer	[82]
7	Dragonamide C, D	*Lyngbya polychroa*	Anticancer	[83]
8	Cryptophycins	*Nostoc* sp.	Anticancer	[84]
9	Borophycin	*Nostoc linckia*	Anticancer	[85]
10	Curacin A	*Lyngbya majuscula*	Anticancer	[86]
11	Synthadoti	*Lyngbya* sp.	Anticancer	[87]
12	Malyngamide C, J, and K	*Lyngbya majuscula*	Anticancer	[88]
13	Tolyporphin	*Tolypohrix nodosa Bharadwaja*	Anticancer	[89]
14	Largazole	*Symploca* sp.	Anticancer	[90]

(Continued)

Table 6.1 Bioactive compounds from cyanobacteria. (*Continued*)

S.n.	Bioactive compounds	Cyanobacteria	Bioactivity	References
15	C-phycocyanin	*Aphanizomenon flos-aquae*	Anticancer	[91]
16	Caylobolide	*Phormidium* sp.	Anticancer	[92]
17	Obyanamide	*Lyngbya confervoides*	Anticancer	[93]
18	Ulongapeptin	*Lyngbya* sp.	Anticancer	[94]
19	Ambiguines	*Fischerella* sp.	Antibacterial	[95]
20	Bis-(v-butyrolactones)	*Anabena variabilis*	Antibacterial	[96]
21	Hapalindole	*Nostoc* CCC537	Antibacterial	[97]
22	Abietane diterpenes	*Microcoleous lacustris*	Antibacterial	[98]
23	Nostocine A	*Nostoc spongiaeforme*	Antibacterial	[99]
24	Noscomin	*Nostoc commune*	Antibacterial	[100]
25	Didehydromirabazole	*Scytonema mirabile*	Antibacterial	[101]
26	*Tolyporphin*	*Tolypothrix nodosa*	Antibacterial	[102]

(*Continued*)

Table 6.1 Bioactive compounds from cyanobacteria. (*Continued*)

S.n.	Bioactive compounds	Cyanobacteria	Bioactivity	References
27	Muscoride	*Nostoc muscorum*	Antibacterial	[103]
28	Ambiguine	*Fischerella ambigua*	Antibacterial	[104]
29	Cyanopeptolin	*Microcystis* PCC 9631	Protease inhibitors	[105]
30	Lyngbyastatin 5–7	*Lyngbya*. sp.	Protease inhibitors	[106]
31	Symplocamide A	*Symploca* sp.	Protease inhibitors	[107]
32	Planktocyclin	*Planktothrix rubescens*	Protease inhibitors	[108]
33	Carbamidocyclophanes	*Nostoc*.sp.CAVN 10	Antibacterial	[109]
34	Pahayokolide A	*Lyngbya* sp.	Antibacterial	[110]
35	Nostoflan	*Nostoc flagelliforme*	Antiviral	[111]
36	Ichthyopeptins	*Microcystis ichthyoblabe*	Antiviral	[112]
37	β-carbolines bauerines A–C	*Dichothrixbauerian*	Antiviral	[113]
38	Indolocarbazoles	*Nostoc sphaericum*	Antiviral	[114]

(*Continued*)

Table 6.1 Bioactive compounds from cyanobacteria. (*Continued*)

S.n.	Bioactive compounds	Cyanobacteria	Bioactivity	References
39	Dolastatin 10	*Symploca* sp.	Anticancer	[115]
40	Cryptophycin-8	*Nostoc spongiaeforme*	Anticancer	[116]
41	Lyngbyastatin 5	*Lyngbya* sp.	Protease inhibitors	[117]
42	Gallinamide A	*Schizothrix* sp..	Protease inhibitors	[118]
43	Microginins 299-C/D	*Microcystis aeruginosa*	Protease inhibitors	[119]
44	Banyasides A	*Nostoc* sp.	Protease inhibitors	[120]
45	Scyptolin A	*Scytonema hofmanni*	Protease inhibitors	[121]
46	Kempopeptins A	*Lyngbya* sp..	Protease inhibitors	[122]

targets of cyanobacterial anti-cancerous drugs on actin and tubulin molecules on eukaryotes cells [42]. Two anticancerous compounds, Dolastatin 10 and 12 have been reported in *Leptolyngbya* spp. and *Symploca* spp. [42]. *Lyngbya majuscule* showed antiproliferative activity due to bioactive compounds, Curacin A [43]. Cryptophytes inhibit leukemia cell lines and compounds were isolated from *Nostoc* sp. and *Calothrix* sp. The two bioactive compounds Calothrixin (A and B) were extracted [44] from organic solvents (Table 6.1) and reduced the multiplication of the human cancer HeLa cell line [45]. *Lyngbya* spp. synthesized Cyclodepsipeptide, which is Apratoxin A effective *in vivo* condition, on early-stage of human adenocarcinoma cell lines and apoptosis. It also inhibits on cell cycle, mainly the G-1 phase. Photosynthetic pigments of cyanobacteria are the source of c-phycocyanin, which exhibits different pharmacological activities such as anti-cancer and anti-inflammatory. These properties could be an effective source of cancer therapy. Glycosylated macrocycles tolyporphin was isolated from a lipophilic extract of *Tolypothrix nodosa* that was active against tumor cells [46, 46a]. There are several other anticancer compounds which not covered in the section their mechanisms of action have not been determined.

6.3.4 Cyanobacterial Drug for Antiviral Compounds

Different types of viruses such as avian influenza (H5N1), acquired immune deficiency syndrome (HIV), and Corona-virus spread over the world and showed catastrophic effects. The Food and Drug Administration has approved highly active antiretroviral therapy (HAART) used for HIV treatment. According to Luescher-Mattli (2003), HAART is an efficient therapy for reducing HIV progression, but it is also hazardous. Antiviral action is exhibited by cyanobacteria at different phases of viral propagation by limiting viral absorption or penetration and reducing viral proliferation in cells. The extract of *Lyngbyala gerheimeii* protects human lymphoid blastoid T-cells against the cytopathic impact of HIV infection [47]. A novel class of HIV inhibitors observed in the cyanobacterial extract is sulfonic acid-containing glycolipid. CVN (cyanovirin-N), a 101-amino-acid long peptide with an 11-kDa molecular weight and inhibitors of cell-to-cell and cell-to-virus fusion, was identified from cyanobacteria [48, 77]. Scytovirin is produced by *Scytonema varium* and is a 9.7 kDa polypeptide with a 95-aa long peptide, which is effective against HIV envelope glycoprotein. *Microcystis ichthyoblabe* inhibitors of antiviral action against influenza A virus, and identified two bioactive compounds Ichthyopeptins (A and B).

Sulfoglycolipids isolated from lipophilic extracts of cyanobacteria inhibit DNA polymerases enzyme activity of reverse transcriptase of HIV-1 [48].

6.3.5 Cyanobacterial Drug for Antibacterial Compounds

Some diseases become untreatable if the rate of bacterial resistance keeps going up. Vancomycin-resistant Enterococci, AmpC beta-lactamase synthesizing *Enterobacteriaceae*, and *S. aureus* (methicillin-resistant) are common examples of multidrug-resistant bacteria. The treatments hurdles are a major source of worry across the world. As a result, the search for novel antibiotics to treat bacterial infections is essential. Some diseases may become untreatable if the bacterial resistance increases per day. As a result, new antibiotics for treating bacterial infections are desperately needed. Screening, isolation, and characterization of cyanobacterial extract of antibacterial activity in the quest to develop new antibiotics, and they have discovered that they are potentially active against a variety of bacteria. Unfortunately, characterization and structural identification of antibacterial substances from cyanobacterial are not much more explored (Table 6.1). Noscomin were extracted from *N. commune* exhibited antibacterial activity against bacteria such as *S. epidermidis*, *E. coli*, and *B. cereus*. Cyanobacterial extracts are a source of antibacterial activity developing new antibiotics that are potentially active against a variety of bacteria. *Anabaena* sp. extract was found antibacterial activity against vancomycin-resistant microorganisms [49]. The two antibacterial substances (carbamidocyclophanes and paracyclophanes) were extracted from *Nostoc* sp. CAVN 10,and exhibited considerable activity against *Staphylococcus aureus*. Fifty bioactive compounds were reported from *Fischerella* spp. and showed antimicrobial activity. Another antibacterial activity of Ambiguine-I-isonitrile isolated from *Fischerella* spp. showed activity against *B. subtilis* and *S. albus* much more effectively than streptomycin [50].

6.4 Methods for Detection of Cyanotoxin

Extraction of cyanotoxin can be carried out by several methods. Two methods (simultaneously and sequential intracellular metabolites) are generally used for the extraction of metabolites simultaneously. First, the metabolic activity is quenched using liquid nitrogen in both modes of extraction. Simultaneous extraction involves quenching followed by extraction with a combination of organic and water solvent (hexane/butanol/chloroform/water/methanol/dichloromethane), whereas sequential extraction involves biomass separation followed by extraction according to the sequentially decreasing polarity of solvent (water, aqueous methanol, methanol, and hexane). Cyanotoxin extraction from cyanobacteria solid-liquid extractions (SLE) and liquid-liquid extraction (known as solvent extraction, LLE) are also used [51, 52]. Extraction of solute commonly SLE methods are used, in which matrix or solid (fixed phase) and liquid solvent (mobile phase) is involved. LLE methods involved separating secondary metabolites, and separation based on relative solubility of non-miscible compounds, mainly polar and non-polar solvents. Both phases are in the liquid phase, and separation of the solute is determined by the distribution coefficient of solute in the mobile phase. Both types of extraction methods need huge quantities of solvents, prolonged time for extraction, and limited chance of consistency of results; another method used for extraction is known as supercritical fluid extraction (SFE) [51]. It is also known as pressurized liquid extraction (PLE), which is low-cost chemical-free green extraction (GE) method that has recently been successfully used (PLE). SFE separates dissolved solutes from raw materials by employing particular gases (as solvents) above critical thresholds. For the extraction of biomolecules, carbon dioxide (CO_2) gas is a common solvent. Along with CO_2, some researchers are now using 10% ethanol as a co-solvent. SFE–CO_2 (220 bars, 26.7°C) co-solvent) 10% ethanol was used for the extraction of antibacterial components from *S. platensis*. Bioactive phenols were extracted from freshwater algae by the application of solid-phase and liquid extraction methods. Detection and purification of biomolecules from cyanobacterial extract, several analytical techniques are available (see Figure 6.1). The simplest method for identifying and separating bioactive molecules is thin layer chromatography (TLC) followed by spectrophotometric analysis. Small cyanobacterial peptides are separated by using high-performance TLC (HTPLC) plates are used [53]. On the other hand, separation by application of TLC/HPTLC is un-specific and less sensitive. Identification and characterization through high-resolution liquid chromatography (HPLC) has risen significantly. The more advanced methods of liquid chromatography are ultra-performance liquid chromatography (UPLC) as compared with HPLC nowadays. Mass spectrometry (LC-MS) is another technique after liquid chromatography is used accurate identification of osmolytes or bioactive compounds [54, 55]. Rapid atom bombardment (FAB-LC–MS), electrospray ionization (ESI-LC–MS), and electrospray ionization (LC-MS-MS) advanced techniques are also used for exact identification and characterization of secondary metabolites [57, 57a]. A combination of liquid chromatography and LC-MS are frequently utilized cyclopeptide identification of cyanobacteria [56, 57b]. However, enzyme-linked immunosorbent assays (ELISA),

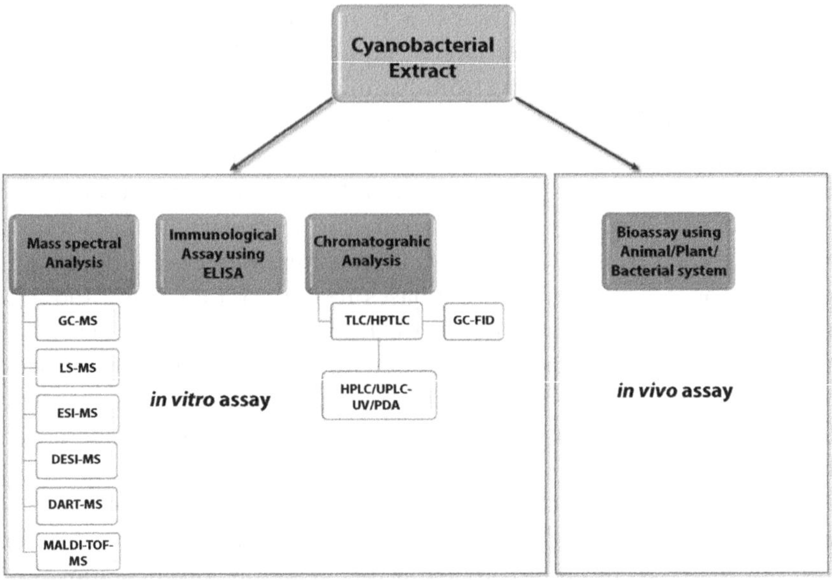

Figure 6.1 Different characterization methods of secondary metabolites or bioactive compounds extracted from cyanobacteria [58].

which are highly sensitive and used for cyclopeptide quantification and detection of cyanobacteria reported. Immunoassay methods are frequently used for bio-molecule characterization because of their easy operation, more sensitivity, and high specificity. MALDI-TOF-MS is utilized for the detection of secondary metabolites. In this technique the small number of samples are required does not need separation or purification [58]. To identify secondary metabolites of cyanobacterial origin, these techniques are frequently used for the characterization of metabolites. MALDI-TOF MS is commonly used to identify peptides, but it may also be used to detect alkaloids. The mass fragmentation pattern of a given analytes in MALDI-TOF MS is different, and it can change depending on the mass spectrometry ionization method and the charge state of the molecule [34].

6.5 Genetic Basis of Cyanotoxin Production

Heptatotoxins, neurotoxins, and saxitoxin are common examples of cyanotoxins [61]. The most common transmission routes to humans of cyanotoxin are drinking water and recreational activities as well as fish, mussels, and larger crustaceans [62]. World Health Organization (WHO) defines permissible concentrations of microcystin-LR as reported in drinking water in the years 1997 and analyzed impacts on human health and environmental risk [63]. According to WHO, the majority of the chemicals are inorganic natures. Microcystin-LR is the only organic compound [64]. The structure of Anatoxin-a was reported by different worker [65], and later on, hepatotoxin microcystin-LA was characterized [66]. Since 1996, cyanobacterial toxin synthesis was characterized and identified a group of genes involved in toxin synthesis [67]. Different types of cyanobacteria such as toxic and non-toxic strains coexist in water over a long time [68], they are unable to be differentiated in a simple microscope. However, the genetic basis could be one of the possible approaches to understanding these phenomena. Our aim is to the genetic basis of toxin production from available information, as well as how genetic approaches have been used to explore toxin biosynthesis. The different types of cyanotoxins including Nodularin, Microcystin, and Cylindrospermopsin as well as the neurotoxic Saxitoxin are discussed in this review.

6.5.1 Genetic Basis of Microcystin and Nodularin Production

Microcystis, Anabaena, Anabaenopsis, Planktothrix, and Nostoc are planktonic freshwater genera that synthesize Microcystins [69]. However, some genera of cyanobacteria including filamentous (*Spirulina, Oscillatoria, Phormidium,* and *Plectonema*) and unicellular *Synechococcus* and *Synechocystis* also synthesize microcystin in both fresh and brackish water. The thio-template mechanism that is found in polyketide synthesis (PKS), fatty acid synthesis, non-ribosomal peptide synthesis (NRPS) is often used to produce Microcystins and Nodularin [70]. Peptide synthetases (encoded by a cluster of genes *mcyABC*), polyketides synthases (encoded by genes of *mcyD*), and tailoring enzymes (synthesized by groups of genes) of cyanobacterial genera *Microcystis* (Unicellular) and *Anabaena* (filamentous) [71] (see Figure 6.2). Multifunctional peptide synthetase enzymes are made-up of different functional domains known as aminoacyl adenylation (for activation), thioesterification (peptide carrier-domain), and condensation domain responsible for elongation of the amino acid [71]. The mcy gene cluster of cyanobacterial genera *Planktothrix Microcystis* and *Anabaena* are closely related to the nodularin synthetase gene cluster (*nda*) of Nodularia [72]. In *Microcystis aeruginosa* PCC7806, the mcy genes in microcystin synthesis were acquired through insertional inactivation [67] that involves inserting alien DNA into a specific gene and becoming inactive. Wild-type strains, mcy genes are completely blocked by the knock-out mutant, and microcystin production is inhibited [67]. Few cyanobacteria, in general,

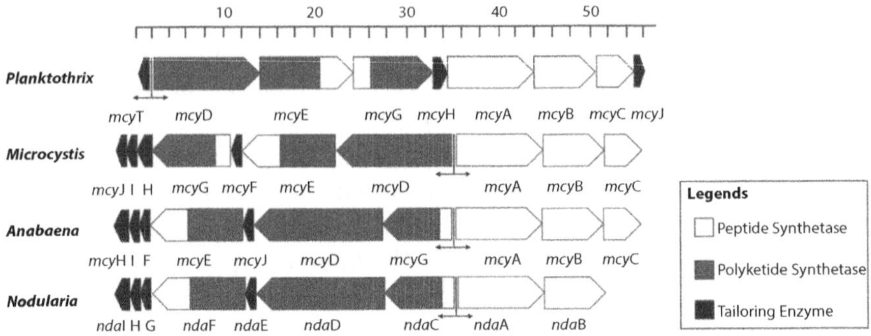

Figure 6.2 Schematic representation of gene clusters of microcystin (mcy) from *Planktothrix* [75], *Microcystis* [71], *Anabaena* [76], and *Nodularia*. Arrows show that direction of transcription, and scale indicates the size of the gene cluster (kbp).

are amenable to genetic manipulation [73], with toxin-producing cyanobacteria *Planktothrixagardhii* that has been genetically manipulated many times [74]. The red-pigmented *P. rubescens* is yet another species amenable to genetic manipulation by the process of insertional inactivation and mcyD genes inhibited microcystin production in *P. rubescens* strain No21/2 as compared to the wild type.

6.6 Conclusions

Among the natural sources, photosynthetic microbe's cyanobacteria synthesized a group of secondary metabolites. In aquatic environments, cyanobacteria are dominant as compared with other ecosystems. Among the groups of microbes, photosynthetic Gram-negative cyanobacteria are frequently utilized as a microbial cell factory to obtained cyanotoxin nowadays. Under favorable growth conditions, cyanobacteria proliferate unduly in aquatic ecosystems. Nutrient enrichment in water bodies is mainly responsible for bloom formation. Today is the era of natural products, which can be easily obtained through photosynthetic microbes at cheaper rate and shorter time frame. Moreover, a limited number of cyanobacterial origin secondary metabolites/bioactive compounds/cyanotoxin have been explored. Cyanobacteria exhibit a wide range of activities such as antiviral, antifungal, anticancerous, and microbial. Limited numbers of bioactive compounds have been reported from cyanobacteria, and some of them are in clinical trials. Screening, isolation and characterization of pharmaceutical products of cyanobacterial origin is more important for interdisciplinary research in the scientific community. Further research in the field of bioactive molecules is still awaited.

Acknowledgments

The authors are grateful to HOD, Department of Biotechnology, Invertis University Bareilly, India for constant encouragement.

References

1. Rastogi, R.P., Madamwar, D., Incharoensakdi, A., Bloom dynamics of cyanobacteria and their toxins: Environmental health impacts and mitigation strategies. *Front. Microbiol.*, 6, 1254, 2015.
2. Plaas, H.E. and Paerl, H.W., Toxic cyanobacteria: A growing threat to water and air quality. *Environ. Sci. Technol.*, 55, 1, 44–64, 2020.
3. Yadav, S., Anam, G.B., Ahn, Y.H., Influence of abiotic factors on the growth of cyanobacteria isolated from Nakdong River, South Korea. *J. Phycol.*, 57, 3, 874–885, 2021.
4. Basu, S. and Mackey, K.R., Phytoplankton as key mediators of the biological carbon pump: Their responses to a changing climate. *Sustainability*, 10, 3, 869, 2018.
5. Sharma, N.K., Tiwari, S.P., Tripathi, K., Rai, A.K., Sustainability and cyanobacteria (blue-green algae): Facts and challenges. *J. Appl. Phycol.*, 23, 6, 1059–1081, 2011.
6. Tripathi, K., Kumar, N., Abraham, G. (Eds.), *The Role of Photosynthetic Microbes in Agriculture and Industry*, Nova Science Publisher's, Incorporated, NY, US, 2018.
7. McFadden, G., II, Origin and evolution of plastids and photosynthesis in eukaryotes. *Cold Spring Harb. Perspect. Biol.*, 6, 4, a016105, 2014.
8. Zhou, P., Wang, L., Liu, H., Li, C., Li, Z., Wang, J., Tan, X., CyanoOmicsDB: An integrated omics database for functional genomic analysis of cyanobacteria. *Nucleic Acids Res.*, 50, D1, D758–D764, 2022.
9. Cirés, S., Casero, M.C., Quesada, A., Toxicity at the edge of life: A review on cyanobacterial toxins from extreme environments. *Mar. Drugs*, 15, 7, 233, 2017.
10. Van Hassel, W.H.R., Andjelkovic, M., Durieu, B., Marroquin, V.A., Masquelier, J., Huybrechts, B., Wilmotte, A., A summer of cyanobacterial blooms in Belgian waterbodies: Microcystin quantification and molecular characterizations. *Toxins*, 14, 1, 61, 2022.
11. Thajuddin, N. and Subramanian, G., Cyanobacterial biodiversity and potential applications in biotechnology. *Curr. Sci.*, 35, 4, 47–57, 2005.

12. Kageyama, H., Tripathi, K., Rai, A.K., Cha-Um, S., Waditee-Sirisattha, R., Takabe, T., An alkaline phosphatase/phosphodiesterase, PhoD, induced by salt stress and secreted out of the cells of Aphanothecehalophytica, a halotolerant cyanobacterium. *Appl. Environ. Microbiol.*, 77, 15, 5178–5183, 2011.
13. Tripathi, K., Kumar, N., Singh, M., Singh, R.K., Fungal siderophore: Biosynthesis, transport, regulation, and potential applications, in: *Rhizosphere Microbes*, pp. 387–408, Springer, Singapore, 2020.
14. Gr, S., Yadav, R.K., Chatrath, A., Gerard, M., Tripathi, K., Govindsamy, V., Abraham, G., Perspectives on the potential application of cyanobacteria in the alleviation of drought and salinity stress in crop plants. *J. Appl. Phycol.*, 33, 6, 3761–3778, 2021.
15. Herdman, M., Janvier, M., Rippka, R., Stanier, R.Y., Genome size of cyanobacteria. *Microbiology*, 111, 1, 73–85, 1979.
16. Zanchett, G. and Oliveira-Filho, E.C., Cyanobacteria and cyanotoxins: From impacts on aquatic ecosystems and human health to anticarcinogenic effects. *Toxins*, 5, 10, 1896–1917, 2013.
17. Singh, R., Parihar, P., Singh, M., Bajguz, A., Kumar, J., Singh, S., Prasad, S.M., Uncovering potential applications of cyanobacteria and algal metabolites in biology, agriculture and medicine: Current status and future prospects. *Front. Microbiol.*, 8, 515, 2017.
18. Ray, S. and Bagchi, S.N., Nutrients and pH regulate algicide accumulation in cultures of the cyanobacterium *Oscillatoria laetevirens*. *New Phytol.*, 149, 3, 455–460, 2001.
19. Dias, E., Pereira, P., Franca, S., Production of paralytic shellfish toxins by Aphanizomenon sp. LMECYA 31 (cyanobacteria). *J. Phycol.*, 38, 4, 705–712, 2002.
20. Harland, F.M., Wood, S.A., Moltchanova, E., Williamson, W.M., Gaw, S., Phormidiumautumnale growth and anatoxin-a production under iron and copper stress. *Toxins*, 5, 12, 2504–2521, 2013.
21. Kaebernick, M. and Neilan, B.A., Ecological and molecular investigations of cyanotoxin production. *FEMS Microbiol. Ecol.*, 35, 1, 1–9, 2001.
22. Codd, G., Cyanobacterial toxins, in: *Biochemistry of the Algae and Cyanobacteria*, p. 283, 1988.
23. Watanabe, M.F. and Oishi, S., Effects of environmental factors on toxicity of a cyanobacterium (Microcystis aeruginosa) under culture conditions. *Appl. Environ. Microbiol.*, 49, 5, 1342–1344, 1985.
24. Sivonen, K., Effects of light, temperature, nitrate, orthophosphate, and bacteria on growth of and hepatotoxin production by Oscillatoria agardhii strains. *Appl. Environ. Microbiol.*, 56, 9, 2658–2666, 1990.
25. Tripathi, K., Sharma, N.K., Kageyama, H., Takabe, T., Rai, A.K., Physiological, biochemical and molecular responses of the halophilic cyanobacterium Aphanothecehalophytica to pi-deficiency. *Eur. J. Phycol.*, 48, 4, 461–473, 2013.

26. Tripathi, K., Sharma, N.K., Rai, V., Rai, A.K., Low cellular p-quota and poor metabolic adaptations of the freshwater cyanobacterium *Anabaena fertilissima* Rao during Pi-limitation. *Antonie van Leeuwenhoek*, 103, 2, 277–291, 2013.
26a. Singh, S., Rai, P.K., Chau, R., Ravi, A.K., Neilan, B.A., Asthana, R.K., Temporal variations in microcystin-producing cells and microcystin concentrations in two fresh water ponds. *Water Res.*, 69, 131–142, 2015.
27. Bláha, L., Babica, P., Maršálek, B., Toxins produced in cyanobacterial water blooms-toxicity and risks. *Interdiscip. Toxicol.*, 2, 2, 36–41, 2009.
28. Newman, S., McCormick, P.V., Backus, J.G., Phosphatase activity as an early warning indicator of wetland eutrophication: Problems and prospects. *J. Appl. Phycol.*, 15, 1, 45–59, 2003.
29. Rapala, J., Sivonen, K., Luukkainen, R., Niemela, S., II, Anatoxin-a concentration in Anabaena and Aphanizomenon under different environmental conditions and comparison of growth by toxic and non-toxic Anabaena-strains—A laboratory study. *J. Appl. Phycol.*, 5, 6, 581–591, 1993.
30. Oh, H.M., Lee, S.J., Jang, M.H., Yoon, B.D., Microcystin production by Microcystis aeruginosa in a phosphorus-limited chemostat. *Appl. Environ. Microbiol.*, 66, 1, 176–179, 2000.
31. Jaiswal, P., Singh, P.K., Prasanna, R., Cyanobacterial bioactive molecules—An overview of their toxic properties. *Can. J. Microbiol.*, 54, 9, 701–717, 2008.
32. Repka, S., Meyerhöfer, M., von Bröckel, K., Sivonen, K., Associations of cyanobacterial toxin, nodularin, with environmental factors and zooplankton in the Baltic Sea. *Microb. Ecol.*, 47, 4, 350–358, 2004.
32a. Ballot, A., Fastner, J., Wiedner, C., Paralytic shellfish poisoning toxin-producing cyanobacterium *Aphanizomenon gracile* in northeast Germany. *Appl. Environ. Microbiol.*, 76, 4, 1173–1180, 2010.
32b. Ray, S. and Bagchi, S.N., Nutrients and pH regulate algicide accumulation in cultures of the cyanobacterium *Oscillatoria laetevirens*. *New Phytol.*, 149, 3, 455–460, 2001.
33. Namikoshi, M. and Rinehart, K.L., Bioactive compounds produced by cyanobacteria. *J. Ind. Microbiol.*, 17, 5, 373–384, 1996.
34. Welker, M., & Von Döhren, H., 2006. Cyanobacterial peptides—Nature's own combinatorial biosynthesis. *FEMS Microbiol. Rev.*, 30, 4, 530–563.
35. Menzella, H.G. and Reeves, C.D., Combinatorial biosynthesis for drug development. *Curr. Opin. Microbiol.*, 10, 3, 238–245, 2007.
36. Mootz, H.D., Schwarzer, D., Marahiel, M.A., Construction of hybrid peptide synthetases by module and domain fusions. *Proc. Natl. Acad. Sci.*, 97, 11, 5848–5853, 2000.
37. Rounge, T.B., Rohrlack, T., Nederbragt, A.J., Kristensen, T., Jakobsen, K.S., A genome-wide analysis of nonribosomal peptide synthetase gene clusters and their peptides in a Planktothrix rubescens strain. *BMC Genom.*, 10, 1, 1–11, 2009.

38. Teuscher, E., Lindequist, U., Mundt, S., Cyanobakterien, quellenbiogener Wirkstoffe. *Pharm. Ztg. Wiss*, 137, 57–69, 1992.
39. Tan, L.T., Bioactive natural products from marine cyanobacteria for drug discovery. *Phytochemistry*, 68, 7, 954–979, 2007.
40. Sieber, S.A. and Marahiel, M.A., Molecular mechanisms underlying nonribosomal peptide synthesis: Approaches to new antibiotics. *Chem. Rev.*, 105, 2, 715–738, 2005.
41. Bhadury, P. and Wright, P.C., Exploitation of marine algae: Biogenic compounds for potential antifouling applications. *Planta*, 219, 4, 561–578, 2004.
42. Catassi, A., Cesario, A., Arzani, D., Menichini, P., Alama, A., Bruzzo, C., Russo, P., Characterization of apoptosis induced by marine natural products in non-small cell lung cancer A549 cells. *Cell. Mol. Life Sci.*, 63, 19, 2377–2386, 2006.
43. Nagle, D.G., Geralds, R.S., Yoo, H.D., Gerwick, W.H., Kim, T.S., Nambu, M., White, J.D., Absolute configuration of curacin A, a novel antimitotic agent from the tropical marine cyanobacterium Lyngbyamajuscula. *Tetrahedron Lett.*, 36, 8, 1189–1192, 1995.
44. Smith, C.D., Zhang, X., Mooberry, S.L., Patterson, G.M., Moore, R.E., Cryptophycin: A new antimicrotubule agent active against drug-resistant cells. *Cancer Res.*, 54, 14, 3779–3784, 1994.
45. Rickards, R.W., Rothschild, J.M., Willis, A.C., de Chazal, N.M., Kirk, J., Kirk, K., Smith, G.D., Calothrixins A and B, novel pentacyclic metabolites from Calothrix cyanobacteria with potent activity against malaria parasites and human cancer cells. *Tetrahedron*, 55, 47, 13513–13520, 1999.
46. Luescher-Mattli, M., Algae, a possible source for new drugs in the treatment of HIV and other viral diseases. *Curr. Med. Chem. Anti-Infect. Agents*, 2, 3, 219–225, 2003.

46a. Safaei, M., Maleki, H., Soleimanpour, H., Norouzy, A., Zaheri, S.H., Vali, H., Nogabi, K.A., Development of novel methods for the purification of C-phycocyanin pigment from a local cyanobacterial strain Limnothrixsp. NS01 and evaluation of its anticancerous properties. *Sci. Rep.*, 9, 1, 1–16, 2019.

47. Gustafson, K.R., Cardellina, J.H., Fuller, R.W., Weislow, O.S., Kiser, R.F., Snader, K.M., Boyd, M.R., AIDS-antiviral sulfolipids from cyanobacteria (blue-green algae). *J. Natl. Cancer Inst.*, 81, 16, 1254–1258, 1989.
48. Yang, H.N., Lee, E.H., Kim, H.M., Spirulina platensis inhibits anaphylactic reaction. *Life Sci.*, 61, 13, 1237–1244, 1997.
49. Bhateja, P., Mathur, T., Pandya, M., Fatma, T., Rattan, A., Activity of blue green microalgae extracts against *in vitro* generated Staphylococcus aureus with reduced susceptibility to vancomycin. *Fitoterapia*, 77, 3, 233–235, 2006.
50. Raveh, A. and Carmeli, S., Antimicrobial ambiguines from the cyanobacterium Fischerella sp. collected in Israel. *J. Nat. Prod.*, 70, 2, 196–201, 2007.

51. Mendiola, J.A., Jaime, L., Santoyo, S., Reglero, G., Cifuentes, A., Ibañez, E., Señoráns, F.J., Screening of functional compounds in supercritical fluid extracts from Spirulina platensis. *Food Chem.*, 102, 4, 1357–1367, 2007.
52. Onofrejová, L., Vašíčková, J., Klejdus, B., Stratil, P., Mišurcová, L., Kráčmar, S., Vacek, J., Bioactive phenols in algae: The application of pressurized-liquid and solid-phase extraction techniques. *J. Pharm. Biomed. Anal.*, 51, 2, 464–470, 2010.
53. Pelander, A., Ojanperä, I., Lahti, K., Niinivaara, K., Vuori, E., Visual detection of cyanobacterial hepatotoxins by thin-layer chromatography and application to water analysis. *Water Res.*, 34, 10, 2643–2652, 2000.
54. Harada, K., II, Nakano, T., Fujii, K., Shirai, M., Comprehensive analysis system using liquid chromatography–mass spectrometry for the biosynthetic study of peptides produced by cyanobacteria. *J. Chromatogr. A*, 1033, 1, 107–113, 2004.
55. Spoof, L., Vesterkvist, P., Lindholm, T., Meriluoto, J., Screening for cyanobacterial hepatotoxins, microcystins and nodularin in environmental water samples by reversed-phase liquid chromatography–electrospray ionisation mass spectrometry. *J. Chromatogr. A*, 1020, 1, 105–119, 2003.
56. Ferranti, P., Fabbrocino, S., Nasi, A., Caira, S., Bruno, M., Serpe, L., Gallo, P., Liquid chromatography coupled to quadruple time-of-flight tandem mass spectrometry for microcystin analysis in freshwaters: method performances and characterisation of a novel variant of microcystin-RR. *Rapid Commun. Mass Spectrom.*, 23, 9, 1328–1336, 2009.
57. Lindner, P., Molz, R., Yacoub-George, E., Dürkop, A., Wolf, H., Development of a highly sensitive inhibition immunoassay for microcystin-LR. *Anal. Chim. Acta*, 521, 1, 37–44, 2004.
57a. Welker, M., Fastner, J., Erhard, M., von Döhren, H., Applications of MALDI-TOF MS analysis in cyanotoxin research. *Environ. Toxicol. An Int. J.*, 17, 4, 367–374, 2002.
57b. Erhard, M., von Döhren, H., Jungblut, P., Rapid typing and elucidation of new secondary metabolites of intact cyanobacteria using MALDI-TOF mass spectrometry. *Nat. Biotechnol.*, 15, 9, 906–909, 1997.
58. Dixit, R.B. and Suseela, M.R., Cyanobacteria: Potential candidates for drug discovery. *Antonie Van Leeuwenhoek*, 103, 5, 947–961, 2013.
59. Carmichael, W.W., Health effects of toxin-producing cyanobacteria: "TheCyanoHABs". *Hum. Ecol. Risk Assess. An Int. J.*, 7, 5, 1393–1407, 2001.
60. Ibelings, B.W. and Chorus, I., Accumulation of cyanobacterial toxins in freshwater "seafood" and its consequences for public health: A review. *Environ. Pollut.*, 150, 1, 177–192, 2007.
61. Chorus, I. and Welker, M., *Toxic Cyanobacteria in Water: A Guide to Their Public Health Consequences, Monitoring and Management*, p. 858, Taylor & Francis, London, UK, 2021.

62. World Health Organization, WHO, & World Health Organization Staff, *Guidelines for Drinking-Water Quality, Vol. 1*, World Health Organization, Geneva, Switzerland, 2004.
63. Devlin, J.P., Edwards, O.E., Gorham, P.R., Hunter, N.R., Pike, R.K., & Stavric, B., Anatoxin-a, a toxic alkaloid from Anabaena flos-aquae NRC-44h. *Can. J. Chem.*, 55, 8, 1367–1371, 1977.
64. Botes, D.P., Tuinman, A.A., Wessels, P.L., Viljoen, C.C., Kruger, H., Williams, D.H., Hammond, S.J., The structure of cyanoginosin-LA, a cyclic heptapeptide toxin from the cyanobacterium Microcystis aeruginosa. *J. Chem. Soc. Perkin Trans.* 1, 1, 2311–2318, 1984.
65. Dittmann, E., Neilan, B.A., Erhard, M., Von Döhren, H., Börner, T., Insertional mutagenesis of a peptide synthetase gene that is responsible for hepatotoxin production in the cyanobacterium Microcystis aeruginosa PCC 7806. *Mol. Microbiol.*, 26, 4, 779–787, 1997.
66. Carmichael, W.W. and Gorham, P.R., The mosaic nature of toxic blooms of cyanobacteria, in: *The Water Environment*, pp. 161–172, Springer, Boston, MA, 1981.
67. Sivonen, K. and Jones, G., Cyanobacterial toxins, in: *Encyclopedia of Microbiology*, pp. 290–307, 2009.
68. Tillett, D. and Neilan, B.A., Xanthogenate nucleic acid isolation from cultured and environmental cyanobacteria. *J. Phycol.*, 36, 1, 251–258, 2000.
69. Jaiswal, D., Sahasrabuddhe, D., Wangikar, P.P., Cyanobacteria as cell factories: The roles of host and pathway engineering and translational research. *Curr. Opin. Biotechnol.*, 73, 314–322, 2022.
70. Moffitt, M.C. and Neilan, B.A., Characterization of the nodularin synthetase gene cluster and proposed theory of the evolution of cyanobacterial hepatotoxins. *Appl. Environ. Microbiol.*, 70, 11, 6353–6362, 2004.
71. Flores, E., Muro-Pastor, A.M., Meeks, J.C., Gene transfer to cyanobacteria in the laboratory and in nature, in: *The Cyanobacteria: Molecular Biology, Genomics and Evolution*, pp. 45–57, 2008.
72. Philmus, B., Christiansen, G., Yoshida, W.Y., Hemscheidt, T.K., Post-translational modification in microviridin biosynthesis. *ChemBioChem*, 9, 18, 3066–3073, 2008.
73. Christiansen, G., Fastner, J., Erhard, M., Borner, T., Dittmann, E., Microcystin biosynthesis in Planktothrix: Genes, evolution, and manipulation. *J. Bacteriol.*, 185, 2, 564–572, 2003.
74. Rouhiainen, L., Vakkilainen, T., Siemer, B.L., Buikema, W., Haselkorn, R., Sivonen, K., Genes coding for hepatotoxic heptapeptides (microcystins) in the cyanobacterium Anabaena strain 90. *Appl. Environ. Microbiol.*, 70, 2, 686–692, 2004.
75. Loya, S., Reshef, V., Mizrachi, E., Silberstein, C., Rachamim, Y., Carmeli, S., Hizi, A., The inhibition of the reverse transcriptase of HIV-1 by the natural sulfoglycolipids from cyanobacteria: Contribution of different moieties to their high potency. *J. Nat. Prod.*, 61, 7, 891–895, 1998.

76. Linington, R.G., Edwards, D.J., Shuman, C.F., McPhail, K.L., Matainaho, T., Gerwick, W.H., Symplocamide A, a potent cytotoxin and chymotrypsin inhibitor from the marine Cyanobacterium Symploca sp. *J. Nat. Prod.*, 71, 1, 22–27, 2008.
77. Joshi, A., Pathak, D.C., Mannan, M., Kaushik, V., *In-silico* designing of epitope-based vaccine against the seven banded grouper nervous necrosis virus affecting fish species. *Netw. Model. Anal. Health Inform. Bioinform.*, 10, 1, 1–12, 2021.
78. Luesch, H., Yoshida, W.Y., Moore, R.E., Paul, V.J., New apratoxins of marine cyanobacterial origin from Guam and Palau. *Bioorg. Med. Chem.*, 10, 1973–1978, 2002a.
79. Mynderse, J.S., Moore, R.E., Kashiwagi, M., Norton, T.R., Antileukemia activity in the Oscillatoriaceae: Isolation of debromoaplysiatoxin from Lyngbya. *Science*, 196, 538–540, 1977.
80. Barchi, J.J., Moore, R.E., Patterson, G.M.L., Acutiphycin and 20, 21-didehydroacutiphycin, new antineoplastic agents from the cyanophyte *Oscillatoria acutissima*. *J. Am. Chem. Soc.*, 106, 8193–8197, 1984.
81. Gunasekera, S.P., Ross, C., Paul, V.J., Matthew, S., Luesch, H., Dragonamides C and D, linear lipopeptides from the marine cyanobacterium brown *Lyngbyapolychroa*. *J. Nat. Prod.*, 71, 887–890, 2008.
82. Moore, R.E., Corbett, T.H., Patterson, G.M.L., Valeriote, F.A., The search for new antitumor drugs from blue green algae. *Curr. Pharma. Des.*, 2, 317–330, 1996.
83. Hemscheidt, T., Puglisi, M.P., Larsen, L.K., Patterson, G.M.L., Moore, R.E., Rios, J.L., Clardy, J., Structure and biosynthesis of borophycin, a new boeseken complex of boric acid from a marine strain of the blue-green alga *Nostoc linckia*. *J. Org. Chem.*, 59, 3467–3471, 1994.
84. Simmons, T.L., Andrianasolo, E., McPhail, K., Flatt, P., Gerwick, W.H., Marine natural products as anticancer drugs. *Mol. Cancer Ther.*, 4, 333–342, 2005.
85. Kwan, J.C., Rocca, J.R., Abboud, K.A., Paul, V.J., Luesch, H., Total structure determination of grassypeptolide, a new marine cyanobacterial cytotoxin. *Org. Lett.*, 10, 789–792, 2008.
86. Bhatnagar, I. and Kim, S.K., Immense essence of excellence: Marine microbial bioactive compounds. *Mar. Drugs*, 8, 2673–2701, 2010.
87. Prinsep, M.R., Caplan, F.R., Moore, R.E., Patterson, G.M.L., Smith, C.D., Tolyphorin, a novel multidrug resistance reversing agent from the blue green algae Tolypothrix nodosa. *J. Am. Chem. Soc.*, 114, 385–387, 1992.
88. Taori, K., Paul, V.J., Luesch, H., Structure and activity of largazole, a potent antiproliferative agent from the Floridian marine cyanobacterium Symploca sp. *J. Am. Chem. Soc.*, 130, 1806–1807, 2008.
89. Tokuda, H., Nishino, H., Shirahashi, H., Murakami, N., Nagatsu, A., Sakakibara, J., Inhibition of 12-O-tetradecanoylphorbol- 13-acetate

promoted mouse skin papilloma by digalactosyl diacylglycerols from the fresh water cyanobacterium *Phormidium tenue. Cancer Lett.*, 104, 91–95, 1996.

90. Salvador, L.A., Biggs, J.S., Paul, V.J., Luesch, H., Veraguamides, A.G., Cyclic hexadepsipeptides from a dolastatin 16-producing cyanobacterium Symploca of hydnoides from Guam. *J. Nat. Prod.*, 74, 917–927, 2011.
91. Williams, P.G., Yoshida, W.Y., Moore, R.E., Paul, V.J., Isolation and structure determination of obyanamide, a novel cytotoxic cyclic depsipeptide from the marine cyanobacterium Lyngbyaconfervoides. *J. Nat. Prod.*, 65, 29–31, 2002.
92. Raveh, A. and Carmeli, S., Antimicrobial ambiguines from the cyanobacterium Fischerella sp. collected in Israel. *J. Nat. Prod.*, 70, 196–201, 2007.
93. Ma, L.X. and Led, J.J., Determination by high field NMR spectroscopy of the longitudinal electron relaxation rate in Cu (II) plastocyanin form *Anabaena variabilis. Am. Chem. Soc.*, 122, 7823–7824, 2000.
94. Gutierrez, M., Suyama, T.L., Engene, N., Wingerd, J.S., Matainaho, T., Gerwick, W.H., Apratoxin D, a potent cytotoxic cyclodepsipeptide from Papua New Guinea collections of the marine cyanobacteria Lyngbyamajuscula and Lyngbyasordida. *J. Nat. Prod.*, 71, 1099–1103, 2008.
95. Hirata, K., Yoshitomi, S., Dwi, S., Iwabe, O., Mahakhant, A., Polchai, J., Miyamoto, K., Bioactivities of nostocine A produced by a freshwater cyanobacterium Nostoc spongiaeforme TISTR 8169. *J. Biosci. Bioeng.*, 95, 512–517, 2003.
96. Jaki, B., Orjala, J., Heilmann, J., Linden, A., Vogler, B., Sticher, O., Novel extracellular diterpenoids with biological activity from the cyanobacterium *Nostoc commune. J. Nat. Prod.*, 63, 339–343, 2000.
97. Stewart, J.B., Bomemann, V., Chen, J.L., Moore, R.E., Caplan, F.R., Karuso, H., Larsen, L.K., Patterson, G.M., Cytotoxic, fungicidal nucleosides from blue-green algae belonging to the Scytonemataceae. *J. Antibiot.*, 41, 1048–105, 1988.
98. Prinsep, M.R., Caplan, F.R., Moore, R.E., Patterson, G.M.L., Smith, C.D., Tolyphorin, a novel multidrug resistance reversing agent from the blue green algae *Tolypothrix nodosa. J. Am. Chem. Soc.*, 114, 385–387, 1992.
99. Nagatsu, A., Kajitani, H., Sakakibara, J., Muscoride a: A new oxazole peptide alkaloid from freshwater cyanobacterium *Nostoc muscorum. Tetrahedron Lett.*, 36, 4097–4100, 1995.
100. Raveh, A. and Carmeli, S., Antimicrobial ambiguines from the cyanobacterium Fischerella sp. collected in Israel. *J. Nat. Prod.*, 70, 196–201, 2007.
101. Bister, B., Keller, S., Baumann, H., II, Nicholson, G., Weist, S., Jung, G., Jüttner, F., Cyanopeptolin 963A, a chymotrypsin inhibitor of Microcystis PCC 7806. *J. Nat. Prod.*, 67, 10, 1755–1757, 2004.
102. Taori, K., Paul, V.J., Luesch, H., Structure and activity of largazole, a potent antiproliferative agent from the Floridian marine cyanobacterium Symploca sp. *J. Am. Chem. Soc.*, 130, 6, 1806–1807, 2008.

103. Kodani, S., Komaki, H., Hemmi, H., Miyake, Y., Kaweewan, I., Dohra, H., Streptopeptolin, a cyanopeptolin-type peptide from *Streptomyces olivochromogenes. Am. Chem. Soc.*, 3, 7, 8104–8110, 2018.
104. Taori, K., Matthew, S., Rocca, J.R., Paul, V.J., Luesch, H., Lyngbyastatins 5–7, potent elastase inhibitors from Floridian marine cyanobacteria, Lyngbya spp. *J. Nat. Prod.*, 70, 10, 1593–1600, 2007.
105. Linnington, R.G., Edward, D.J., Shuman, F.C., Macphail, K.L., Metainaho, T., Gerwick, W.H., Symplocamide A potent cytototoxin and chymotrypsin inhibitor from marine cyanobacterium. Symploca sp. *J. Nat. Prod.*, 71, 22–27, 2008.
106. Baumann, H.I., Keller, S., Wolter, F.E., Nicholson, G.J., Gunj, G., Sussmuth, R.D., Juttner, F., Planktocyclin, a cyclooctapeptide protease inhibitor produced by the freshwater cyanobacterium Planktothrixrubescens. *J. Nat. Prod.*, 70, 1611–1615, 2007.
107. Bui, H.T., Jansen, R., Pham, H.T., Mundt, S., Carbamidocyclophanes A– E, chlorinated paracyclophanes with cytotoxic and antibiotic activity from the Vietnamese cyanobacterium nostoc sp. *J. Nat. Prod.*, 70, 4, 499–503, 2007.
108. Berry, J., Gantar, M., Gawley, R.E., Wang, M., Rein, K.S., Pharmacology and toxicology of phayokolide A, a bioactive metabolite from a fresh water species of Lyngbya isolated from the Florida everglades. *Comp. Biochem. Physiol.*, 139, 231–238, 2004.
109. Kanekiyo, K., Lee, J.B., Hayashi, K., Takenaka, H., Hayakawa, Y., Endo, S., Hayashi, T., Isolation of an antiviral polysaccharide, nostoflan, from a terrestrial cyanobacterium, Nostoc flagelliforme. *J. Nat. Prod.*, 68, 7, 1037–1041, 2005.
110. Zainuddin, E.N., Mentel, R., Wray, V., Jansen, R., Nimtz, M., Lalk, M., Mundt, S., Cyclic depsipeptides, ichthyopeptins A and B, from Microcystis ichthyoblabe. *J. Nat. Prod.*, 70, 7, 1084–1088, 2007.
111. Larsen, L.K., Moore, R.E., Patterson, G.M., β-Carbolines from the blue-green alga Dichothrixbaueriana. *J. Nat. Prod.*, 57, 3, 419–421, 1994.
112. Knübel, G., larsen, L.K., Moore, R.E., Levine, I.A., Patterson, G.M., Cytotoxic, antiviral indolocarbazoles from a blue-green alga belonging to the Nostocaceae. *J. Antibiot.*, 43, 10, 1236–1239, 1990.
113. McPhail, K.L., Correa, J., Linington, R.G., Gonzalez., J., Ortega-Barrıa., E., Capson, T.L., Gerwick, W.H., Ant malarial linear lipopeptides from a Panamanian strain of the marine cyanobacterium Lyngbya majuscule. *J. Nat. Prod.*, 70, 984–988, 2007.
114. Urones, J.G., Basabe, P., Marcos, I.S., Pineda, J., Lithgow, A.M., Moro, R.F., Gravalos, M.D.G., Meroterpenes from Cystoseirausneoides. *Phytochemistry*, 31, 179–182, 1992.
115. Taori, K., Paul, V.J., Luesch, H., Structure and activity of largazole, a potent antiproliferative agent from the Floridian marine cyanobacterium Symploca sp. *J. Am. Chem. Soc.*, 130, 1806–1807, 2008.

116. Miller, B., Friedman, A.J., Choi, H., Hogan, J., McCammon, J.A., Hook, V., Gerwick, W.H., The marine cyanobacterial metabolite gallinamide a is a potent and selective inhibitor of human Cathepsin L. *J. Nat. Prod.*, 77, 92–99, 2014.
117. Ishida, K., Matsuda, H., Murakami, M., Four new microginins, linear peptides from the cyanobacterium Microcystis aeruginosa. *Tetrahedron*, 54, 44, 13475–13484, 1998.
118. Pluotno, A. and Carmeli, S., Banyasin A and banyasides A and B, three novel modified peptides from a water bloom of the cyanobacterium Nostoc sp. *Tetrahedron*, 61, 3, 575–583, 2005.
119. Von, E., Berer, L., Merkel, P., Huhn, T., Blom, J.F., Cyanopeptolins954, a chlorine-containing chymotrypsin inhibitor of *Microcystis aeruginosa* NIVACYA 43. *J. Nat. Prod.*, 68, 1324–1327, 2005.
120. Taori, K., Paul, V.J., Luesch, H., Kempopeptins A and B, serine protease inhibitors with different selectivity profiles from a marine cyanobacterium, Lyngbya sp. *J. Nat. Prod.*, 71, 1625–1629, 2008.
121. Matern, U., Oberer, L., Falchetto, R.A., Erhard, M., Konig, W.A., Herdman, M., Weckesser, J., Scyptolin A and B, cyclic depsipeptides from axenicultures of Scytonemahofmanni PCC 7110. *Phytochemistry*, 58, 1087–1095, 2001.
122. Murakami, M., Sun, Q., Ishida, K., Matsuda, H., Okino, T., Yamaguchi, K., Microviridins, elastase inhibitors from the cyanobacterium *Nostoc minutum* (NIES-26). *Phytochemistry*, 45, 1197–1202, 1997.

7

Actinobacteria in Natural Product Research: Avenues and Challenges

Santhoshini Hazari[1†], Uzma Tabassum[1†], Anum Jehan Siddiqui[1†], Shivani Hazari[1†], Addagatla Ravindar[2], Sanjeeb Kumar Mandal[1], Sanjay Kumar[3] and Bishwambhar Mishra[1*]

[1]Department of Biotechnology, Chaitanya Bharathi Institute of Technology, Hyderabad, India
[2]Department of Chemical Engineering, Osmania University, Hyderabad, India
[3]Department of Chemical Engineering, IIT Delhi, New Delhi, India

Abstract

Actinobacteria are well-known biosynthetic factories that generate a wide range of secondary metabolites. In numerous ways, recent genetic discoveries appear to impact the exploitation of these metabolically diverse microorganisms. Specifically, different methodologies progress continuously, from the isolation of new species through the findings of new chemicals. Over the past decades, researchers have developed industrially competitive strains of actinomycetes with excellent performances. This chapter summarizes the occurrence, habitat, and diversity of actinobacteria in the natural environment. Novel bioactive compounds and major enzymes synthesized by the groups of actinomycetes have been summarized. Furthermore, the current challenges and future perspectives for industrial strains more systematically through metabolic engineering strategies are also discussed.

***Keywords*:** Actinobacteria, bioactive compound, antibiotic, metabolic engineering, enzymes, streptomyces

7.1 Introduction

Actinobacteria are nothing but Gram-positive bacteria which contain a GC-rich genome [23]. These groups of bacteria possess the biosynthetic

**Corresponding author*: mishra.bishwambhar@gmail.com
†These authors contributed equally to this work

Sanjay Kumar, Narendra Kumar and Shahid-ul-Islam (eds.) Role of Microbes in Industrial Products and Processes, (171–194) © 2023 Scrivener Publishing LLC

potential to produce secondary metabolites, have a wide structural variety, and are important commercially [1]. The majority of these species have a complicated mycelial lifecycle, forming aerial and substrate mycelia with distinct coloration with a high degree of sporulation. As a result of increasing research into their biotechnological value, actinobacteria have been discovered from various sources throughout aquatic and terrestrial ecosystems. Actinobacteria can be found in various freshwater, marine soil environments, animals, and plants [1]. They play critical defense and nutrient recycling functions in their environment or linked host [3]. Actinobacteria have evolved varied metabolic potentials with the outcome of their converting to such diverse habitats and interactions, allowing them to create a vast range of secondary metabolites [2]. Various natural products synthesized by actinobacteria have been illustrated in Figure 7.1.

Two-thirds of all consistently generated antibiotics and a variety of immunosuppressive medications, antifungal, anticancer, and anthelmintic, come from Actinobacteria [29]. Tetracyclines, streptomycin, neomycin, chloramphenicol, vancomycin, erythromycin, rifamycin, kanamycin, gentamicin, platensimycin, and daptomycin are among the clinically

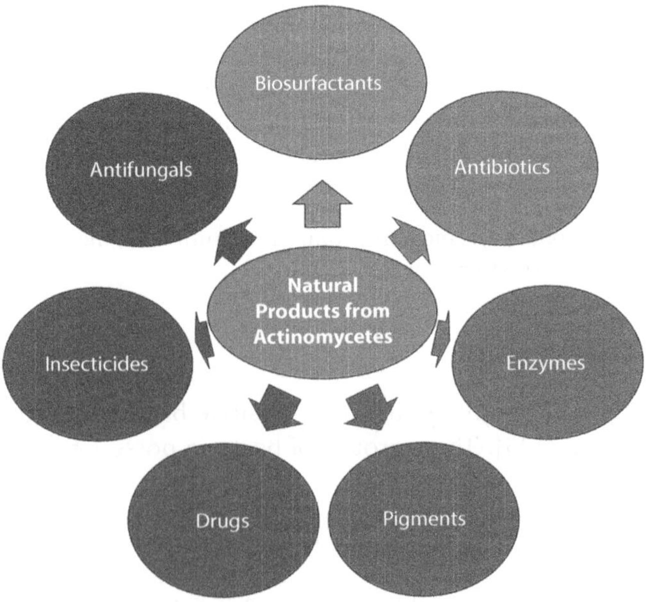

Figure 7.1 Natural products from Actinobacteria.

relevant antiinfective drugs [4]. Anthracyclines, bleomycin, mitosanes, and enediynes are examples of successful anticancer drugs generated from Actinobacteria [38]. Furthermore, various newly developed antimicrobials (such as actinonin, GE2270A, and deoxyactagardine B), as well as anticancer substances (such as marthiapeptide A, sungsanpin, salinosporamide A, and diketopiperazines), are being tested [5]. Tacrolimus and rapamycin (immunosuppressives), milbemycins and avermectins (antiparasitics), and spinosyns (spinosyns) are among the other notable bioactive substances identified in Actinobacteria. Enzyme inhibitors produced by actinobacteria have many industrial and biotechnology uses [4]. They have an important role in creating nanoparticles with crucial bioactivities [5].

Actinobacteria research has led to the invention of numerous vital drugs that continue to save many human lives worldwide after nearly 80 years. However, the appearance of (antibiotic) resistance frequently furnishes existing medications unproductive, necessitating the development of advanced drugs [4]. Opportunely, the advanced discovery of *Streptomyces coelicolor* A3(2) prospective has resulted in the sequencing and mining of actinobacterial genomes. The genetic capability disclosed impacts how chemical diversity is harnessed and opens up new paths for natural product discovery [4]. The occurrence, habitat, and diversity of actinobacteria in the natural environment are summarized in this chapter. The groups of actinomycetes have generated novel bioactive chemicals and powerful enzymes, which have been summarized. Furthermore, the current obstacles and future prospects for industrial strains are examined more comprehensively using metabolic engineering methodologies.

7.2 Occurrence, Habitat, and Diversity of Actinobacteria

Actinobacteria can be found in various ecosystems, including soil, plants, and water. The leading cause of actinobacteria population density is habitat and environmental conditions.

7.2.1 Soil Habitat

The soil is perhaps the most crucial home for actinobacteria. They are established in concentrations of 10^6–10^9 cells per gram of soil [2]. Other environmental parameters, such as soil moisture, temperature, and pH, significantly impact inhabitants' density and actinobacteria kinds [33].

The ideal growth temperatures for several mesophilic actinobacteria are in the range of 25°C to 30°C. Thermophilic actinobacteria, on the other hand, grow at temperatures between 50°F and 60°F. The preferred pH range for actinobacteria growth seems to be between 6.0 and 9.0. At pH 7.0, they developed to their full potential, demonstrating that they act similarly as neutrophils. Actinobacteria, particularly those of the genus *Streptomycetes*, do not grow well at pH levels below 3.5 [3]. Actinobacteria might develop vegetative propagation in low-humidity conditions. However, their growth is impeded in parched soils with low moisture content [2].

7.2.2 Plant Habitat

Endophytic actinomycetes are indeed plentiful and extensive in tropical regions. The plant which has a reaction towards the host is a crucial factor throughout this connection. Understanding whether endophytic populations or rhizospheric bacteria are more valuable to plants has long been an interesting area of research. On the other hand, plant endospheres contain a diverse range of microbial endophytes and form a multiplex micro-ecosystem [4]. Endophytes can gain particular genetic information and facilitate metabolic pathways. These can manufacture certain bioactive substances comparable to host plants using genetic exchange as a result of their symbiotic relationship with plants. As a result of this host-endophyte relationship, the host feeds nutrients to the microbial endophytes, while the microbial population aids in resistance mechanisms. Despite the fact that many studies have been done on the molecular relationship between bacteria and plants, there is little knowledge of the molecular scale operation of endophyte-host partnerships. The far more ecologically diverse and life forms ecosystems on the earth are temperate and tropical rainforests, which are also acknowledged as the richest storehouse for the spread and development of endophytes [4].

7.2.3 Marine Habitat

Many experts have now looked into the spread of actinobacteria in the marine ecosystem. Marine bacteria can also be a valuable source of natural compounds, making them a promising research topic [5]. There have also been discoveries of unusual marine actinobacteria with high cellulolytic activity. (NRPS) Nonribosomal polyketide synthase and (PKS) Polyketide synthase pathways are found in most marine actinobacteria, showing their potential to create bioactive compounds.

Table 7.1 Classification of Actinobacteria.

Class	Sub-class	Order	Sub-Order	Family
Actinobacteria	*Antinobacteridae*	*Actinomycetales* *Nitriliruptorales* *Bifidobacteriales*	*Frankineae* *Streptomycineae* *Actinopolysporineae* *Micromonosporineae* *Propionibacteriaceae* *Pseudonocardineae* *Glycomycineae* *Corynebacterineae* *Micrococcineae*	*Dietziaceae* *Mycobacteriaceae* *Frankiaceae* *Acidothermaceae* *Microbacteriaccae* *Micrococcaceae* *Streptomycetaceae* *Actinopolysporaceae* *Micromonosporaceae* *Nocardioidaceae* *Pseudonocardiaceae* *Glycomycetaceae* *Nitriliruptoraceae*
	Acidimicrobidae	*Acidimicrobiales*	*Acidimicrobineae*	*Acidimicrobiaceae*
	Coriobacteridae	*Coriobacteriales*		*Coriobacteriaceae*
	Rubrobacteridae	*Rubrobacteriales* *Thermoleophilales* *Solirubrobacterales*	*Rubrobacterineae*	*Rubrobacteracaea* *Thermoleophilaceae* *Conexibacteraceae* *Patulibacteraceae* *Solirubrobacteraceae*
	Sphaerobacteridae	*Sphaerobacterales*		*Sphaerobacteraceae*

7.2.4 Diversity

In a biodiversity-rich habitat, microbial members with great diversity can be found. Endophytic microorganisms predominated in rainforests, and many actinobacterial representatives belonged to them [5]. Endophytic microbe isolation competencies are also being studied for medicinal plants with a lengthy indigenous medicinal history. According to the researchers, around 560 bioactive rhizosphere actinomycetes were discovered in Chinese medicinal plants. These endophytic actinomycetes have a broad-spectrum antibacterial activity [6]. A total of 37 medicinal plants' endophytic heterogeneity has just been found, resulting in 600 actinobacteria belonging to 34 genera and some seven unknown genera. Several papers have identified plant-linked actinobacteria; nevertheless, weather patterns may alter the richness and species allocation amongst some of the host plants.

Streptomyces and *Micromonospora* are the most common actinomycetes found in soil. They produce a variety of hydrolytic exoenzymes and play a vital part in the carbon cycle. *Streptomyces* has a phylogenetic diversity that is both broad and rational. Furthermore, *Streptomyces* is known as nature's most proficient chemist, producing a vast array of naturally occurring substances. According to Bergey's handbook, soil *Micromonospora* has 32 unique species, and an additional 50 genera of *Micromonospora* are currently being reported [6]. A multitude of marine actinomycetes capable of developing bioactive compounds is identified in the South Sea of China and the Yellow Sea. *Actinobacillus* is a bacterial domain with five subclasses, six suborders, and fourteen sub-subspecies. As a result, every species in this phylum displayed a diverse array of metabolic capabilities, morphology, and physiology [6]. The classification of actinobacteria has been summarized in Table 7.1.

7.3 Natural Products from Actinobacteria

7.3.1 Discovery of Novel Bioactive Compound

Bioactive substances have activities in the body that may be beneficial to one's health. They're being researched for cancer, heart disease, and other disorders. Among soil bacteria, the genus *Streptomyces* abounds, prevention [27]. Lycopene, resveratrol, lignan, tannins, and indoles are examples of bioactive chemicals. Actinomycetes have unique chemical structures that could develop novel medications to tackle diseases resistant to antibiotics [9].

7.3.1.1 Antibiotics

Actinobacterial secondary metabolites have been at the forefront of the hunt for new chemicals with a variety of biological activities, most notably antibacterial characteristics. As a result, significant effort has gone into isolating actinobacteria capable of producing new and potent antibacterial compounds (antibiotics) to combat microbial diseases. So far, actinobacteria have had the most significant impact on human health worldwide. The fact that actinobacteria have evolved with a significant number of biosynthetic gene clusters indicates that it has a lot of potential for creating biologically active natural product scaffolds [12]. Antibiotics, enzyme inhibitors, immunomodulators, plant growth promoters, and a range of other compounds helpful to humans are produced by this category of natural bioactive secondary metabolite producers. Antibiotics such as -lactams, lactones, aminoglycosides, anthracyclines, peptides, anthraquinones, macrolides, and a range of other compounds are produced by Actinobacteria. Extensive variances in the spectrum of antibacterial action occur across the diverse aminoglycosides, just as they do with some other antibiotic categories [16].

7.3.1.1.1 β-Lactam

β-lactam antibiotics impede terminal transpeptidation via obstructing with cross-linking of peptidoglycans in the bacterial cell wall by binding to the penicillin-binding proteins. Bacteria are cytolyzed as a result of osmotic pressure. The first antibiotic was found, and it was penicillin. This class of antibiotics was discovered as a naturally occurring molecule. Carbapenems, penicillins and their derivatives, cephalosporins, monobactams, and -lactam inhibitors are all lactam antibiotic family members. Cephalosporins are classed as first, second, third, or fourth-generation based on their activity spectrum and the year of their release. The first-generation cephalosporins are effective. It is resistant to gram-positive bacteria, whereas gram-negative bacteria are only moderately impacted. Second-generation cephalosporins are antibiotics that help to fight infections caused by microbes. They are more effective against gram-negative bacteria than gram-positive bacteria [12]. Third-generation cephalosporins› gram-negative activity has improved, while their gram-positive activity has been inconsistent. However, Fourth-generation -lactams show antibacterial activity against gram-negative and gram-positive microorganisms. Because they readily disseminate among bacteria, carbapenems are broad-spectrum -lactam antibiotics.

7.3.1.1.2 Aminoglycosides

Aminoglycosides work by compromising the integrity of the bacterial cell membrane and reducing protein synthesis. Their antimicrobial spectrum is quite extensive. Streptomyces Griseus discovered the first aminoglycoside antibiotic, streptomycin, in 1940. Kanamycin was discovered in 1949 and 1957 due to the antibiotics neomycin and other *Streptomyces* species. *Micromonospora purpurea* was discovered in the early 1960s. Sisomicin is the source of amikacin, a kanamycin derivative. Because aminoglycosides typically function in conjunction with other antibiotics, they are effective anti-infectives. *Streptomyces* sp. (*Streptomyces kanamyceticus, Streptomyces spectabilis*) make aminoglycosides [13].

7.3.1.1.3 Chloramphenicol

Chloramphenicol inhibits protein synthesis and peptide chain elongation by binding to the peptidyl transferase of the 50S ribosomal subunit of 70S ribosomes. Chloramphenicol was originally synthesized in 1947 by *Streptomyces venezuelae* [17]. Chloramphenicol has a straightforward chemical structure. Only a few synthetic variants with no antibacterial activity have been developed. Chloramphenicol kills both aerobic and anaerobic bacteria and a variety of bacteria that are both positive and negative [13].

7.3.1.1.4 Tetracyclines

Tetracyclines inhibit microorganism growth by interfering with protein synthesis. It interacts with ribosomes inside the bacterial cell. Tetracycline derivatives also have an effect on bacterial membranes. Tetracycline molecules are made up of four tetracyclic rings that have been fused together. The nucleus is related to various functional groups. Tetracyclines are antibiotics that have a broad spectrum of activity against gram-positive bacteria [14]. The first tetracycline antibiotic, chlortetracycline, was found in *Streptomyces aureofaciens*.

7.3.1.1.5 Quinolone

Quinolones obstruct DNA replication by inhibiting the activities of topoisomerase IV and DNA gyrase in bacteria. Nalidixic acid, which has a bactericidal effect against gram-negative bacteria, was discovered in 1962. Various fluoroquinolones, such as norfloxacin, were developed in the 1980s and 1990s. Ciprofloxacin, ofloxacin, and sparfloxacin are examples of broad-spectrum antimicrobials that have been created. *Helicobacter pylori* is resistant to quinolones. Quinolones were discovered in an actinomycete [14].

7.3.1.1.6 Sulfonamides

Organic sulfur compounds with the radical -SO_2NH_2 are known as sulfonamides. This class of chemicals is the oldest synthetic medication on the market, having been launched in 1932. Sulfonamides are antimicrobial drugs that work by decreasing bacteria's development and activity. P-aminobenzoic acid (p-aminobenzoic acid), the target of these chemicals, is acidic, interfering with bacteria's metabolic operations. Sulfonamides were the first antibiotics to work against bacteria septicemias. These medications are effective in treating Nocardia-related fungal infections as well as tissue infections. *Streptococci* are to blame. Some *Streptomyces* species produce sulfonamide antibiotics. The overwhelming bulk of these secondary metabolite gene clusters is either underexpressed or absent. As a result, different cryptic gene cluster expression strategies must be devised. After that, strain improvement will help with product development and commercialization, which would be a vital phase for the sector [14]. Unidentified "biosynthetic substances" harvesting in the industrial setting, mutagenesis and protoplast fusion procedures for recombination are available for actinobacteria strain improvement.

7.3.1.2 Insecticidal Compound

Spinosad is a type of insecticide derived from chemical components found in *Saccharopolyspora spinosa* bacterium species. In 1985, *Saccharopolyspora* was isolated from crushed sugarcane. Aerial hyphae produced by bacteria are yellowish-pink in color. In the Virgin Islands, *Saccharopolysporaspinosa* was isolated from dirt collected within a non-operational sugar mill. Spinosad is a combination of the spinosyn family with a generalized network that contains a distinctive tetracyclic ring system linked to a neutral sugar and an amino sugar (D-forosamine) [53]. Spinosad is a nonpolar compound that is difficult to dissolve in water. Spinosad is a novel configuration pesticide developed from a group of natural compounds produced by the fermentation of *S. spinosa* [53]. Spinosyns have been discovered in over 20 natural forms, more than 200 synthesized forms (spinosoids) have also been developed in the lab. Spinosad is a highly efficient insecticide with low toxicity to mammals and the environment. However, the original manufacturer, *Saccharopolyspora spinosa*, is unable to supply the tremendous global demand for Spinosad.

7.3.1.3 Antifungal and Antibacterial Agents

7.3.1.3.1 Antifungal Agent

Actinomycetes create a novel antifungal intermediate product called sceliphrolactam. *Streptomyces* sp. with the yellow and black mud dauber, *Sceliphron caementarium*, was isolated in 2011 from *Streptomyces* species. The structure, which is made up of 26-membered macrocyclic lactam that is polyunsaturated and polyoxygenated, is very sensitive to high temperatures and light. Despite its shortcomings, *sceliphrolactam* has been investigated to show antifungal activity against amphotericin B-resistant *Candida albicans*, and the MIC value was found to be 4 g/m [28]. The three other novel polyene macrolide members (16-17-epoxyfilipin V;15-glycidylfilipin III 16-17-epoxyfilipin V) were recovered from cultures of *S. lavendulaegriseus*, a soil actinomycete. Only 15-glycidylfilipin III had a strong inhibitory effect on *C. albicans* mycelial growth (MIC = 6.25 g/ml compared to 3.13 g/ml for the control). The other compounds had significantly lower fungicidal activity (MIC = 200 g/ml). Secondary metabolites from bacteria with an epoxide activity are more poisonous or have more cytostatic characteristics than their non-epoxide counterparts. A Streptomyces sp. strain was found in a soil sample collected from a coal mine at a thickness of 5 cm in Nanchang, Jiangxi, People's Republic of China, in 2017. Enduspeptides A-F is six new cyclic octa depsipeptides produced by this strain. With IC50 values of 5.33, 1.72, and 8.13 g/ml, enduspeptides were the most effective antifungal agents against *C. glabrata* ATCC 90030 [51].

7.3.1.3.2 Antibacterial Agent

Antibacterial activity of actinomycetes isolates was tested against eight harmful microorganisms, including (*Streptococcus pyogenes, Proteus vulgaris, Staphylococcus aureus, Pseudomonas aeruginosa, Bacillus subtilis, Salmonella typhimurium, Klebsiella pneumoniae*) [52]. On nutritional agar media, bacteria were cultivated. The inhibitory zone diameter was used to calculate the antibacterial activity [11]. Various bioactive compounds isolated from actinobacteria have been summarized in Table 7.2.

Table 7.2 Bioactive compounds isolated from actinobacteria.

Name of actinobacteria	Compound isolated	Activity shown	Reference
Streptomyces sp.	Dentigerumycin	Antitumoral	[1]
Streptomyces sp.S2A	Pyrrolo[1–a]pyrazine-1,4-dione,hexahydro-3-(2-methyl propyl)	Antioxidant, Antimicrobial, Enzyme inhibitory, and Cytotoxic	[3]
Streptomyces sp. ZZ745	Bagremycins	Antimicrobial	[2]
Streptomyces sp. Al Dhabi-90	3-methylpyridazine, n-hexadecanoic acid, indazol-4- one, octadecanoic acid, and 3a-methyl-6-((4- methyl phenyl	Antibacterial	[5]
Streptomyces chumphonensis	Not identified	Antimicrobial	[7]
Streptomyces sp.	2-Alkyl-4-hydroxyquinolines	Antifungal	[4]
Streptomyces sp. 120,454	Mayamycin B	Antibacterial	[4]
Streptomyces sp. 166	Sekgranaticin	Anticancer	[8]
Streptomyces sp. NA06554	Borrelidin J	Antibacterial	[11]
Streptomyces sp. NA4286	Murayaquinone D	Anticancer	[9]
Streptomyces sp. YO15-A001	YO-001A	Antifungal	[10]
Streptomyces sp. SBT345	Ageloline A	Antioxidant	[8]

7.3.2 Discovery of Prominent Enzymes

Actinobacteria have been used to produce a variety of industrially important enzymes. Apart from pharmaceutically important metabolites, proteases, amylase, lipase, cellulase, xylanase, dextranase, inulinase, and keratinase were some reactions catalyzed by actinobacteria. The enzymes that break down complex polysaccharides are frequently overlooked. Screening has indeed been aided mostly by the availability of actinobacteria genome and protein sequencing data. These bacteria produce cellulases, proteases, amylases, chitinases, xylanases, and pectinases [18].

7.3.2.1 Amylases

Amylases are hydrolytic enzymes that break down starch molecules into smaller glucose polymers and dextrins. Endo and exo starch digesting enzymes are the two starch digesting enzymes [20]. In the food, textile, fermentation, and amylases, paper sectors assume a key contribution. Amylases can be found across both plants and animals, although microorganisms are one of the finest suppliers for starch garbling in the business world. Amylases within the Streptomyces genus of actinobacteria are frequently thought to be a source of amylolytic enzymes [22]. Amylolytic enzymes that are heat stable are now being studied to improve industrial starch breakdown processes. Alpha-amylases, which belong to the glycoside hydrolysis family 13, are enzymes that break down carbohydrates, are starch-degrading metalloenzymes for consistency and interaction. Enzymatic hydrolysis of starch has several advantages over acid hydrolysis, including product stability, specificity, lower energy, and the elimination of neutralization stages. Researchers are always looking for novel enzymes with improved characteristics and cost-effective manufacturing methods to address the growing demand for these enzymes. It has been discovered that endophytic actinobacteria *Nocardiopsis* sp. produce-amylase in the presence of yam bean [20], as has been identified from shrimp ponds, *Actinobacteria* sp. AE-19 produces -amylase. *Streptomyces* sp. D1 was isolated from the west coast of India and demonstrated a generation of amylase in another investigation. *Streptomyces gulbargensis* DAS, an alkali-thermotolerant strain, produced optimized extracellular -amylase [19]. The marine ecology of India has been investigated for haloalkaliphilic Saccharopolyspora sp. strain A9 that produces surfactants and oxidant stable, Ca21 ion independent, and detergent stable amylase for detergent applications [20].

7.3.2.2 Pectinases

Another notable industrially employed enzyme is pectinase, which is used extensively in the food industry, for example, to clarify fruit juices and the production of wine. The genus *Streptomycetes* produced a lot of pectinases, but there are just a few reports from other genera [24].

7.3.2.3 Xylanases

Arabinosyl, acetyl, and glucuronosyl units and D-xylosyl residues are substituted to make xylan [26]. The decomposition of xylan is necessary for its optimal use in nature. There are two types of xylanases: Internal hydrolysis of 1,4-glycosidic bonds by GH family 43,11,10,8,7 and GH families 54, 52, 43, and 39 cleave xylose monomer from non-reducing ends of oligosaccharides. Over-expression of xylanases genes with altered biochemical properties has been modified and cloned in heterologous and homologous hosts for profit-oriented over-expression. In the catalytic process, the substrate-binding site amino acids like tyrosine and tryptophan and aspartate residues and glutamate have been discovered. Numerous investigations on the generation of xylanases with actinomycetes have been reported. Alpha-L-arabinofuranosidase, and Alpha-N-arabinofuranosidase have all been found to be produced by *T. bifida*. *Cellulomonas fimi* has been found to produce mannosidase, -mannanase, endo-, and exo-xylanase [24]. *Cellulomonas flavigena* ATCC 482 produces xylanases from the GH81, GH51, GH43, GH30, GH26, GH16, GH13, GH11, and GH10 families [25]. Xylanases have received little study, despite the fact that actinobacteria represent a significant origin of bioactive compounds. The lack of co-secretion of cellulases affiliated with xylan metabolism by actinobacteria over fungi is an advantage. Actinobacterial with xylanases have several benefits, including high thermostability, non-cellulolytic activity, and action throughout a wide pH range, all of which are important in various industrial uses. Endotoxin xylanases with high alkaliphilic *Thermoactinomyces thalophilus*, *Actinomadura* sp., and other species [26] have been reported to have temperature-active features. In India, xylanases from actinomycetes have been found, which have been characterized from soil samples.

7.3.2.4 Proteases

Dehairing, silver recovery, detergents, baking, food, and brewing, from photographic film, and bioremediation are just a few of the applications for proteases, which account for B60 percent of the overall enzyme industry.

Table 7.3 Different types of prominent enzymes.

Name of the enzymes	Name of the actinobacteria	Application	Reference
Amylase	*Streptomyces erumpens*, *Nocardiopsis* sp., *Streptomyces* sp., *Thermobifidafusca*	• The softness of bread and volume in baking. • Removal of stains in Detergent • Drinking, drainage improvement in paper and pulp. • Clarification, mashing in Beverage. • Production of glucose and fructose syrups in the starch industry.	[24]
Pectinases	Streptomyces lydicus	• Scouring in textile.	[15, 26]
Xylanase	*Actinomadura* sp., *Streptomyces* sp.	• Conditioning of dough in baking. • Bleach boosting paper and pulp. • Digestibility in Animal feed.	[24]
Protease	*Thermoactinomyces* sp., *Nocardiopsis* sp., *Streptomyces pactum*, *Streptomycesthermoviolaceus*, *Streptomyces* sp.	• Cheese making in food. • Clarification – low-calorie beer in brewing. • Detergent.	[30]

(*Continued*)

Table 7.3 Different types of prominent enzymes. (*Continued*)

Name of the enzymes	Name of the actinobacteria	Application	Reference
Cellulase	*Streptomyces* sp., *Thermobifida halotolerant*, *Streptomyces* sp., *Thermomonospora* sp., *Streptomyces ruber*.	• Denim finishing • Softening of cotton in textile. • Removal of stains in Detergent.	[32]
Glucose Oxidases	*Streptomyces coelicolor*	• Strengthening of dough in baking.	[34]
Lipase	Streptomyces griseus	• Cheese flavoring in Dairy	[34]
Cholesterol Oxidase	*Rhodococcus* sp.	• Pharmaceutical formulation, • agricultural and waste management	[35]
L-Asparaginase	*Streptomycesgulbargensis*, *Streptomyces Griseus*	• Anti-cancerous in medical, • Acrylamide formation in the food industry.	[36]
Laccase	*Streptomyces lavendulae*, *Streptomyces griseus*	• Attenuation of dyes in textile, Cellulose bleaching in paper and pulp industry, bioremediation.	[15, 37]
Prenyl Transferase	*Streptomyces* sp. CNQ-509	• Aromatic compounds (flavonoids and stilbenoids)	[37]
Halogenase	Salinispora tropica	• Anti-tumor in medicine	[39]

Alkaline proteases have been found in many types of taxa, including actinobacteria, and have been found in organic synthesis, wastewater treatment, textiles, and leather products. Thermophilic proteases from Streptomyces sp. from Brazilian with soil, extracellular proteases from *Streptomyces gulbargensis* DAS 131, *Streptomyces nogalator* strain Ac 80 [26] alkaliphilic actinomycetes, the flesh captivate of the Mud crab, marine actinobacteria, *Scylla serrata* soil samples from the Northern Himalayas were all investigated for protease production. *Streptomyces lactamdurans*) and *Streptomyces peucetius* have been shown to coordinate cellular protein recycling-related with secondary metabolism and morphogenesis. Actinomycetes from both (mesophilic and thermophilic) environments seemed deliberate for their proteases. Another Nocardiopsis sp. NCIM 5124 alkaline proteases existing previously identified.

Using prospective microorganisms to hydrolyze keratin-containing wastes provides an appealing, efficient, pharmaceutical, and alternative technique for reprocessing keratin wastes from the poultry processing industries, leather, textile, feed, cosmetic, and wastewater treatment, as well as the leather industry. *Streptomyces albus* AZA from Turkey, *Streptomyces* sp. from Egypt, *Streptomyces pactum* DSM 40530, and Streptomyces strain BA7 from Turkey have all been examined for their ingenuity to produce keratinase. Furthermore, numerous large arrays of *Streptomyces* species, *Streptomyces fradiae*, *Streptomycesalbus*, *Streptomyces thermoviolaceus*, primarily *Streptomyces pactum* and *Streptomyces sclerotialus*, were discovered to be an accomplishment for keratinase synthesis, according to various researchers. The Egyptian herbal plant's endophytic Micromonospora strains were likewise found to have keratinolytic activity.

Different groups of enzymes produced by actinomycetes groups of microorganisms have been summarized in Table 7.3.

7.4 Metabolic Engineering of Natural Product Biosynthesis

Actinomycetes create more than two-thirds of all known secondary metabolites [41]. These secondary metabolites have been successfully used in medicine and agriculture and various biological activities, including immunosuppressive (rapamycin), anticancer (mithramycin), antibiotics (teicoplanin), and many more. However, identifying and developing natural products into medications is a time-consuming and expensive procedure [41]. Furthermore, due to a high rediscovery rate for known molecules, several pharmaceutical companies have shifted their focus away from NP discovery and toward target-based screening of synthetic compound libraries. As a result, microbes were regarded as a source of discovery that had previously been exhausted. Actinobacteria, specifically the *Streptomyces* genus, is recognized as a reliable starting point for discovering secondary metabolites or natural products. Techniques like metabolic engineering were effectively utilized to a number of useful natural products in recent years, resulting in larger concentrations of the target chemical substances and improvement of the desired activities [2].

7.4.1 Tools for Metabolic Engineering

The availability of genetic tools has a significant impact on the efficiency of strain optimization. Several advanced genetic engineering techniques have been developed to improve actinomycete's host properties. Some examples of these are SSRs (Site-Specific Recombinases), CRISPR/Cas9 system, controlling genetic elements, and endonucleases [46].

7.4.1.1 SSR (Site-Specific Recombinases)

They were due to the ability to catalyze exact inversions, integrations, and DNA excisions between their different acceptance sites. SSR systems are considered strong and effective tools for high-throughput manipulation of genomes. Manufacturing genes encoding Dre, Cre, and FLP recombinases with optimum GC content were created for use in Actinomycetes [42]. These recombinases could effectively abolish antibiotic resistance markers having outstanding leaving, efficiency, and only one recognition site [31]. The nucleocidin nucleoside antibiotic is a priority for organic compounds for generations; however, owing to reduced yield and poor repeatability, microbiological humongous manufacturing has yet to be accomplished [21].

But, the successive deletions of the added genetic code in a similar hereditary framework conceivably restricted ascribed to the undesirable transposition of DNA caused by numerous target sites left after each deletion. The challenge was solved by deploying an incremental marker excision system (IMES) predicated upon that phage phiC31 intron and its mutant at various places [44]. Because the gene encoding does not recognize the aberrant location that remains there in the chromosome following recombination, it can be employed for a series of complex genomic engineering. Fine-tuning of gene expression is identified as crucial expression vectors and varied perspectives on biotechnological processes. The sequence upstream or downstream of 10 and 35 consensus regions were randomized using deformed primers, giving libraries of promoters with strengths ranging from two to 319% compared to ermEp1 and ermEp2 [45]. Strong promoters have recently been used to provide a connector strategy for remodeling quiet biosynthetic gene clusters. Identification and characterization of around forty strong actinomycetes promoters with firmness up to 75% of the power of ermEp were made [43]. The renewing of silent spectinabilin and polycyclic tetramate macrolactam (PTM) pathways from heterologous *Streptomyces* sp. was achieved using these promoters.

7.4.1.2 CRISPR/Cas9 (Clustered Regularly Interspaced Short Palindromic Repeats)

CRISPR/Cas9 is a precise genome-editing system used by many organisms [46]. The advantage of this approach is that it has unprecedented flexibility in introducing a double-strand break in a particular target genomic sequence, owing to a customized single guide RNA [50]. Though the system is new, its approach has already seen widespread applications in genome engineering projects [47]. The modified CRISPR/Cas9 system has generated targeted chromosomal deletions of various sizes in four Streptomyces species. Furthermore, Cas9 was able to govern the gene expression in Streptomyces coelicolor efficiently and reversibly [48].

Another method that can generate double-strand breaks into the DNA is the I-SceI endonuclease system, enhancing homologous recombination at a specific locus on the chromosome. I-SceI also has the added benefit of acting as a counter-selection tool for co-integrated [49].

7.5 Future Outlook

Actinobacteria produce novel substances with diverse biological potential, including antibacterial, anticancer, and antioxidant properties. These chemicals could be used as medicinal leads or as catalysts for the chemical production of novel molecules. Furthermore, various analogs and derivatives of existing chemicals with a broader range of bioactivities broaden the drug discovery possibilities. Bioprospecting actinobacteria provide a large reservoir of industrially and pharmaceutically relevant macromolecules for effective breakdown, including cellulose, lignin, starches, hemicelluloses, and chitin. The large spectrum of genes responsible for the polysaccharide hydrolysis in the genome of actinobacteria highlights the potential of these microorganisms for industrial applications [40]. As a result, ongoing research into the discovery and characterization of promising genes and enzymes for various industrial applications is crucial for developing long-lasting and cost-effective manufacturing for commercialization. Because of their complicated structure, the metabolic engineering of actinobacteria provides for a more concentrated approach to increasing strain productivity as well as the generation of analogs that would be difficult to manufacture by chemical methods. To be commercially viable, metabolic engineering approaches will need to improve present genetic tools and develop new simplified multiplex genome engineering procedures.

7.6 Conclusion

The Actinobacteria genomic data is massive, and the degree of traditionally unexplored physiological functions is tremendous. Thousands of actinobacterial individuals belonging to various taxonomic groups have been acquired throughout the last five years due to consistent isolation research methods in order at numerous ecological niches. New Actinobacteria's "enduring origins" appear to be even more marine ecosystems and plants. In the hunt for new Actinobacteria and related metabolites, terrestrial ecosystems serve as "renewed reservoirs."

Because of the significant assistance from next-generation modern biotechnology and metabolomic techniques, more attention should be dedicated to the continual investigation of all uninvestigated ecosystems, particularly co-evolved ecological niches. Typically expressed actinobacteria provide a large reservoir of industrially and pharmaceutically significant macromolecules for effective breakdown, including lignin, hemicellulose, chitin, cellulose, and starches. Antibiotic synthesis by rare Actinobacteria against multidrug-resistant infections has significant potential. The ability to identify rare Actinobacteria using both conventional and molecular methods is an important tool. However, more research is needed, such as DNA-DNA hybridization and additional chemotaxonomic and biochemical testing, to determine their species level. The antibacterial potential of rare Actinobacteria with unusual metabolic pathways can be detected using molecular techniques to identify biosynthetic gene clusters. This will aid the discovery of new bioactive metabolites. The current evaluation contributes to the control of antimicrobial drug resistance and will improve health care worldwide. Furthermore, it provides promising approaches to aid drug discovery research.

Metabolic engineering of Actinobacteria provides a more targeted method to improving strain productivity. It can also be used to develop analogs that would be difficult, if not impossible, to obtain using chemical procedures because of their complicated structures. To become industrially viable, metabolic engineering approaches will need to refine existing genetic tools and develop fresh simplified multiplex genome engineering procedures. To direct more metabolic potential toward the creation of heterologous macromolecules, cutting-edge approaches like systems and synthetic biology must be used to create a new generation of Streptomyces-based chassis with reduced and fine-tuned genomes.

References

1. Rahlwes, K.C., Sparks, I.L., Morita, Y.S., Cell walls and membranes of Actinobacteria, in: *Bacterial Cell Walls and Membranes*, pp. 417–69, 2019.
2. Weber, T., Charusanti, P., Musiol-Kroll, E.M., Jiang, X., Tong, Y., Kim, H.U., Lee, S.Y., Metabolic engineering of antibiotic factories: New tools for antibiotic production in actinomycetes. *Trends Biotechnol.*, 33, 15–26, 2015.
3. Hwang, K.S., Kim, H.U., Charusanti, P., Palsson, B.O., Lee, S.Y., Systems biology and biotechnology of Streptomyces species for the production of secondary metabolites. *Biotechnol. Adv.*, 32, 255–268, 2014.
4. Baltz, R.H., Genetic manipulation of secondary metabolite biosynthesis for improved production in Streptomyces and other actinomycetes. *J. Ind. Microbiol. Biotechnol.*, 43, 343–370, 2015.
5. Prakash, D., Nawani, N., Prakash, M., Bodas, M., Mandal, A., Khetmalas, M. *et al.*, Actinomycetes: A repertory of green catalysts with a potential revenue resource. *Biomed. Res. Int.*, 2013, 264020, 2013.
6. Bilyk, O. and Luzhetskyy, A., Metabolic engineering of natural product biosynthesis in actinobacteria. *Curr. Opin. Biotechnol.*, 42, 98–107, 2016.
7. Tao, H., Zhang, Y., Deng, Z., Liu, T., Strategies for enhancing the yield of the potent insecticide spinosad in actinomycetes, 14, 1700769, 2018. https://doi.org/10.1002/biot.201700769.
8. Aggarwal, N., Thind, S.K., Sharma, S., Role of secondary metabolites of actinomycetes in crop protection, in: *Plant Growth Promoting Actinobacteria*, pp. 99–121, Springer, Singapore, 2016.
9. Jakubiec-Krześniak, K., Rajnisz-Mateusiak, A., Guspiel, A., Ziemska, J., Solecka, J., Secondary metabolites of actinomycetes and their antibacterial, antifungal and antiviral properties. *Pol. J. Microbiol.*, 67, 3, 259, 2018.
10. El-Khawaga, M.A. and Megahed, M.M.M., Antibacterial and insecticidal activity of actinomycetes isolated from sandy soil of (Cairo-Egypt). *Egypt. Acad. J. Biol. Sci.*, 4, 1, 53–67, 2012.

11. Sekurova, O.N., Schneider, O., Zotchev, S.B., Novel bioactive natural products from bacteria via bioprospecting, genome mining and metabolic engineering. *Microb. Biotechnol.*, 12, 5, 828–844, 2019.
12. Berdy, J., Thoughts and facts about antibiotics, Where we are now and where we are heading. *J. Antibiot.J. Antibiot.*, 5865, 1–26385–395, 2012.
13. Andes, D.R. and Craig, W.A., Cephalosporins, in: *Principles and Practice of Infectious Diseases*, G.L. Mandell, J.E. Bennett, R. Dolin, (Eds.), pp. 294–311, Churchill Livingstone, Philadelphia, PA, 2005.
14. Petri, W.A., Penicillins, cephalosporins, and other β-lactam antibiotics, in: *Goodman & Gilman's, The Pharmacologic Basis of Therapeutics*, L.L. Brunton, J.S. Lazo, K.L. Parker, (Eds.), p. 11271154, The McGraw-Hill Companies, New York, 2006.
15. Marshall, W.F., Vink, A., Orenstein, R., Wilson, J.W., Estes, L.L., Infectious diseases, in: *Mayo Clinic Internal Medicine Review*, T.M. Habermann, A.K. Ghosh, D.J. Rhodes, (Eds.), p. 589, Taylor & Francis Group, New York, NY, 2006.
16. Vakulenko, S.B. and Mobashery, S., Versatility of aminoglycosides and prospects for their future. *Clin. Microbiol. Rev.*, 16, 430–450, 2003.
17. Schwarz, S., Kehrenberg, C., Doublet, B., Cloeckaert, A., Molecular basis of bacterial resistance to chloramphenicol and florfenicol. *FEMS Microbiol. Rev.*, 28, 519–542, 2004.
18. Chopra, I. and Roberts, M., Tetracycline antibiotics: Mode of action, applications, molecular biology, and epidemiology of bacterial resistance. *Microbiol. Mol. Biol. Rev.*, 65, 2, 232–260, 2001.
19. King, D.E., Malone, R., Lilley, S.H., New classification and update on the quinolone antibiotics. *Am. Fam. Phys.*, 61, 2741–2748, 2000.
20. Siegmund, K., Maheshwary, S., Narayanan, S., Connors, W., Riedrich, M., Printz, M. *et al.*, Molecular details of quinolone DNA interactions, solution structure of an unusually stable DNA duplex with covalently linked nalidixic acid residues and non-covalent complexes derived from it. *Nucl. Acids Res.*, 33, 4838–4848, 2005.
21. Fukuda, K., Tamura, T., Segawa, Y., Mutaguchi, Y., Inagaki, K., Enhanced production of the fluorinated nucleoside antibiotic nucleocidin by a rif R-resistant mutant of Streptomyces calvus IFO13200. *Actinomycetologica*, 23, 51–55, 2019.
22. Charlop-Powers, Z., Owen, J.G., Reddy, B.V.B., Ternei, M.A., Guimaraes, D.O., Global biogeographic sampling of bacterial secondary metabolism. *eLife*, 4, 1–10, 2015.
23. Jose, P.A. and Jha, B., New dimensions of research on actinomycetes, quest for next generation antibiotics. *Front. Microbiol.*, 7, 1295, 2016.
24. Abdulla, H.M., El-Shatoury, S.A., Abdel Rahman, M.A., El-Shenawy, N.S., Isolation and antibiotic prospective of endolithic actinobacteria inhabiting diorite rocks of South Sinai, Egypt. *Egypt. J. Bot.*, 58, 3, 477–490, 2018. https://doi.org/10.21608/ ejbo.2018.1021.1090.

25. Guo, X., Liu, N., Li, X., Ding, Y., Shang, F., Gao, Y., Ruan, J., Huang, Y., Red soils harbor diverse culturable actinomycetes that are promising sources of novel secondary metabolites. *Appl. Environ. Microbiol.*, 81, 9, 3086–3103, 2015. https://doi. org/10.1128/aem.03859-14.
26. Rateb, M.E., Ebel, R., Jaspars, M., Natural product diversity of actinobacteria in the Atacama desert. *Antonie Van Leeuwenhoek*, 111, 8, 1467–1477, 2018. https://doi. org/10.1007/s10482-018-1030-z.
27. Takagi, M. and Shin-ya, K., Construction of a natural product library containing secondary metabolites produced by actinomycetes. *J. Antibiot.*, 65, 9, 443–447, 2012. https://doi.org/10.1038/ja.2012.52.
28. Goodfellow, M., Williams, S.T., Mordarski, M., *Actinomycetes in Biotechnology*, Academic Press, London, 1988.
29. Hussain, A.A., Mostafa, S.A., Ghazal, S.A., Ibrahim, S.Y., Studies on antifungal antibiotic and bio insecticidal activities of some actinomycete isolates. *Afr. J. Mycol. Biotechnol.*, 10, 63–80, 2002.
30. Johnstone, D.B. and Waskman, S.A., Streptomycin II, an antibiotic substance produced by a new species of Streptomyces. *Proc. Soc. Exp. Biol. Med.*, 65, 294–295, 2002.
31. Walsh, C., *Antibiotics: Actions, Origins and Resistance*, pp. 107–120, 222–226, ASM Press, Washington, DC, 2003.
32. Hwang, B.K., Ahn, S.J., Moon, S.S., Production, purification, and antifungal activity of the antibiotic nucleoside, tubercidin, produced by Streptomyces violaceoniger. *Can. J. Bot.*, 72, 4, 480–485, 2002.
33. Kumar, N., Singh, R.K., Mishra, S.K., Isolation and screening of soil actinomycetes as sources of antibiotics active against bacteria. *Int. J. Microbiol. Res.*, 2, 12–16, 2010.
34. Jeffrey, L.S., Isolation, characterization and identification of actinomycetes from agriculture soils at Semongok, Sarawak. *Afr. J. Biotechnol.*, 7, 3697–3702, 2008.
35. Arifuzzaman, M., Khatun, M.R., Rahman, H., Isolation and screening of actinomycetes from sundarbans soil for antibacterial activity. *Afr. J. Biotechnol.*, 9, 4615–4619, 2010.
36. Oskay, M., Usame, A., Azeri, C., Antibacterial activity of some actinomycetes isolated from farming soils of Turkey. *Afr. J. Biotechnol.*, 3, 441–446, 2004.
37. Torsvik, V. and Ovreas, L., Microbial diversity and function in soil: From genes to ecosystems. *Curr. Opin. Microbiol.*, 5, 240–245, 2020.
38. Silva, L.J., Crevelin, E.J., Souza, D.T. *et al.*, Actinobacteria from Antarctica as a source for anticancer discovery. *Sci. Rep.*, 10, 13870, 2020. https://doi. org/10.1038/s41598-020-69786-2.
39. Gohain, A., Actinobacteria: Diversity and biotechnological applications, in: *Recent Advancements in Microbial Diversity*, pp. 217–231, 2020.
40. Shivlata, L. and Satyanarayana, T., Thermophilic and alkaliphilic Actinobacteria: Biology and potential applications. *Front. Microbiol.*, 6, 1014, 2015.

41. Hwang, K.S., Systems biology and biotechnology of Streptomyces species for the production of secondary metabolites. *Biotechnol. Adv.*, 32, 255–268, 2014.
42. Gaj, T., ZFN, TALEN, and CRISPR/Cas-based methods for genome engineering. *Trends Biotechnol.*, 31, 397–405, 2013.
43. Demain, A.L. and Sanchez, S., Microbial drug discovery: 80 years of progress. *J. Antibiot.*, 62, 1, 5–16, 2009.
44. Barka, E.A., Vatsa, P., Sanchez, L., Gaveau-Vaillant, N., Jacquard, C., Meier-Kolthoff, J.P., Klenk, H.P., Clement, C., Ouhdouch, Y., van Wezel, G.P., Taxonomy, physiology, and natural products of actinobacteria. *Microbiol. Mol. Biol. Rev.*, 80, 4, 1–43, 2016.
45. Kieser, T., Bibb, M., Buttner, M., Chater, K., Hopwood, D., *Practical Streptomyces Genetics*, John Innes Foundation, Norwich, UK, 2000.
46. Weber, T., Charusanti, P., Musiol-Kroll, E.M., Jiang, X., Tong, Y., Kim, H.U., Lee, S.Y., Metabolic engineering of antibiotic factories: New tools for antibiotic production in actinomycetes. *Trends Biotechnol.*, 33, 15–26, 2015.
47. Jose, P.A. and Jha, B., Intertidal marine sediment harbours Actinobacteria with promising bioactive and biosynthetic potential. *Sci. Rep. Uk*, 7, 1, 10041–10041, 2017.
48. Haft, D.H., Selengut, J., Mongodin, E.F., Nelson, K.E., A guild of 45 CRISPR-associated (Cas) protein families and multiple CRISPR/Cas subtypes exist in prokaryotic genomes. *PloS Comput. Biol.*, 1, e60, 2005.
49. Barrangou, R., Fremaux, C., Deveau, H., Richards, M., Boyaval, P., Moineau, S., Romero, D.A., Horvath, P., CRISPR provides acquired resistance against viruses in prokaryotes. *Science*, 315, 1709–1712, 2007.
50. Horvath, P. and Barrangou, R., CRISPR/Cas, the immune system of bacteria and archaea. *Science*, 327, 167–170, 2010.
51. Um, S., Kim, Y.-J., Kwon, H., Wen, H., Kim, S.-H., Kwon, H.C., Park, S., Shin, J., Oh, D.-C., Sungsanpin, a lasso peptide from a deep-sea streptomycete. *J. Nat. Prod.*, 76, 5, 873–879, 2013.
52. Mohamed, H., Miloud, B., Zohra, F., Garcia-Arenzana, J.M., Veloso, A., Rodriguez Couto, S., Isolation and characterization of actinobacteria from algerian sahara soils with antimicrobial activities. *Int. J. Mol. Cell. Med.*, 6, 2, 109–120, 2017.
53. Kirst, H., The spinosyn family of insecticides: Realizing the potential of natural products research. *J. Antibiot.*, 63, 101–111, 2010.

8

Production of Bioethanol from Lignocellulosic Waste Parali

Navodit Kumar Singh[1], Arpita Vats[2], Akruti Singh[3], Anushka Tyagi[3], Santosh Kumar Mishra[3], Narendra Kumar[4] and Sanjay Kumar[1]*

[1]Department of Biochemical Engineering and Biotechnology, Indian Institute of Technology Delhi, New Delhi, India
[2]Department of Chemical and Environmental Engineering, University of Nottingham, Nottingham, United Kingdom
[3]Department of Biotechnology, IMS Engineering College, Ghaziabad, Uttar Pradesh, India
[4]Department of Biotechnology and Bioengineering, Institute of Advanced Research, Koba Institutional Area, Gandhinagar, Gujarat, India

Abstract
As a result of the rapid pace of urbanization, industrialization, and technology-driven industry, the world's energy demand is increasing on a daily basis. For many decades, it has been seen and reported that fossil-fuel-based energy sources are the primary energy source. However, due to the non-renewable nature of these traditional energy sources, they will be depleted in the near future. Furthermore, the continual decline in fossil-fuel-based oil production, the low number of finds of newer oil reserves, the instability of oil-producing countries, and the skyrocketing pricing of these fuels render these fuels unsustainable for use in the future. Earlier findings on bioethanol production were made utilizing food-based carbon sources such as rice, sugar cane, maize, barley, and other grains. However, increasing energy production at the expense of food crops can cause a food crisis.

Rice straw is produced as a by-product of rice production during harvest, and it is a rich source of lingo-cellulosic material because of its high lignin concentration. It has become increasingly popular over the last decade to burn rice straw in open fields as a post-harvest management tool for paddy fields; however, it contributes to air pollution as a result of this practice. According to the International Rice

**Corresponding author*: sanjay0193@gmail.com

Sanjay Kumar, Narendra Kumar and Shahid-ul-Islam (eds.) Role of Microbes in Industrial Products and Processes, (195–222) © 2023 Scrivener Publishing LLC

Research Institute, the annual production of rice straw (Parali) ranges between 800–1000 million tons, and it has the capacity to produce approximately 205 billion liters of bioethanol per year. After being treated for the physical, chemical, and biological components, lingo-cellulosic waste created from paddy fields can be used as a sustainable source of bio-ethanol production as an alternative. The following article includes thorough information on recent breakthroughs in the pre-treatment of lingo-cellulosic materials, sustainable bio-ethanol production from parali, bioethanol yields, properties, and possibilities of bioethanol as an alternative to fossil fuels.

Keywords: Bioethanol, lignocellulose biomass (LCB), pre-treatment, fermentation, recovery

8.1 Introduction

With each passing day, the world's consumption of energy (crude oil) related to transportation grows more and more, contributing to the world's problem of fossil fuel depletion. As a result of these kinds of issues, we have found allies in the search for new sources of artificial energy that can supply the world's energy needs in the long term. This is because biomass is the most frequent source of energy and has the potential to be used in the creation of alternative fuel energy in the not-too-distant future. Many professionals in the bio-industry feel that biofuel has the potential to be the fuel of the future [1].

Biofuels are usually made from waste from the food industry, like corn, wheat, rice, and sugarcane residue [2]. Bioethanol is the most eco-friendly, economical, and renewable fuel because it has the same properties as gasoline [1]. In Bioethanol production, the main ingredient is lignocellulose and hemicellulose, which can be found in wheat and rice residue [3]. Because India grows wheat, rice, and sugarcane second only to the United States, it is one of the world's top three producers. However, straw is taken away from the grains during the harvest season. This causes the straw to pile up and spread out. Feedstock that contains sugar can be used to make Bioethanol, like sugar beets or sugarcane. Starchy material like wheat and barley can also be used and lignocellulose biomass like rice husk/straw and wood.

Rice is a substantial source of food for 40% of the world's people. In India alone, straw is made up of 130 million tons of straw every year. Almost half of the straw is utilized for animal fodder. At the same time, the remaining is primarily burned in the fields, which results in the loss of the majority of the energy in the straw and the release of carbonaceous particles into the air, which can cause a variety of respiratory illnesses. Even though a lot of rice straw is made, farmers in this country need to learn about the value of agro-waste and how to use rice straw for bioethanol so that it can be utilized as an alternative to the source of biofuel production [4].

Rice straw is mainly made up of cellulose, hemicellulose, lignin, silica, and some ash [3]. The lignocellulosic material utilized in the synthesis of bioethanol contains hexose and pentose sugar. However, cellulose in lignocellulosic substrates is contained in the cell wall, which is surrounded by lignin and hemicellulose matrix, making cellulose accessibility problematic. As a result, pretreatment of lignocellulosic material is required for cellulose accessibility so that it can be converted into glucose [5].

8.2 Production Process of Bioethanol from Lignocellulosic Waste

The different steps are followed for ethanol production using lignocellulosic biomass as a carbon source. These steps are described below (Figure 8.1) for a better understanding of the process.

8.2.1 Lignocellulosic Waste as a Raw Material

Rice straw is a primary lignocellulosic waste that is a preferred raw material for bioethanol production because it is available round the year. Rice straw disposal is a limited and even costly process. A small percentage of paddy straw is used as fodder for livestock, while the rest of it is cleared from the agricultural land by burning, which releases specific amounts of dangerous chemicals, causing air pollution and even affecting the health of human beings [6]. The Table 8.1 displays the structural and chemical makeup of the raw materials.

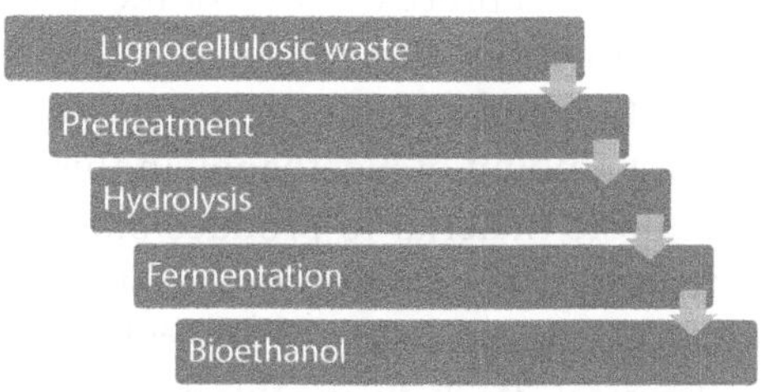

Figure 8.1 Process overview of bioethanol production from lignocellulosic waste.

Table 8.1 Structure and chemical composition of raw material i.e. rice straw [7].

	Cellulose	Hemicellulose	Lignin
Structural unit	D-Glucopyranose	Mannose, D-Xylose, Galactose, L-Arabinose, Glucuronic acid	Syringil, Guaiacyl, Para-Hydroxy-Phenyl
Bonding	Beta-1-4 Glycosidic linkage	Beta-1-4 Glycosidic linkage, Beta-1-2(3-6) Glycosydic linkage	C-C, R-O-R'
Polymeric level	1000-10,000	<=200	4000
Polymer	Beta-1-4-Glucan	Galacto-gluco-mannan, Glucomannan, Xylan	G-, GS-, GSH- type
Structure	Crystalline and amorphous area	Few portions of its structure is a crystalline area, while the majority of its structure is amorphous	Amorphous, Nonlinear, Non-Uniform, 3D polymer
Binding forces	Hydrogen bond	Chemical bond	Chemical bond

8.2.2 Pre-Treatment Methods

Pre-treatment procedures are used to dissolve and separate the components of lignocellulosic biomass. This step alone consumes around 33% of the overall value of lignocellulosic bioethanol production, therefore selecting the suitable technique for this process is one of the most crucial decisions to be made. The pre-treatment method of choice improves the efficiency of the enzyme saccharification and enzyme digestibility of the straw, allowing it to produce the highest amount of bioethanol. In this case, paddy straw is utilized as a lignocellulosic material to produce bioethanol. The lignocellulosic complex suffers from a reduction in crystallinity when the matrix is destroyed, and the increased fraction of amorphous cellulose renders it more vulnerable to chemical attack during this process. The simplified effect that pretreatment has on lignocellulosic material, depicted in Figure 8.2.

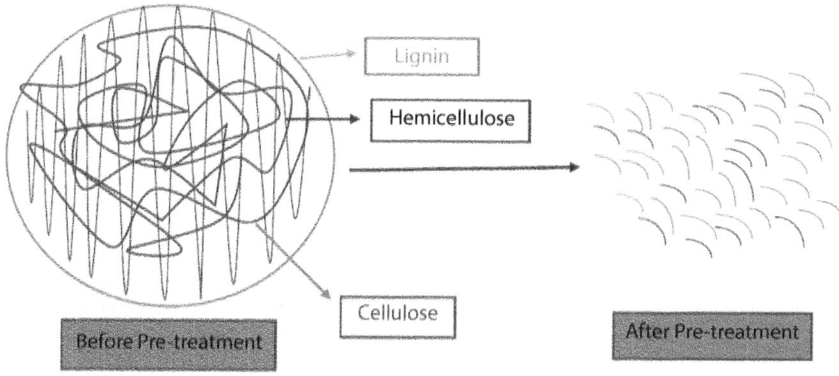

Figure 8.2 Simplified impact of pre-treatment on lignocellulosic material [9].

The lignocellulosic biomass is processed for hydrolysis in order to maximize monomeric sugar yields. Pre-treatment aims to [8]:

- Sugars are formed either directly or by hydrolysis.
- Avoid the breakdown of sugars that are made.
- Inhibitory products are formed in a limited amount
- Reduce energy requirements
- Minimization of costs

8.2.2.1 *Physical Treatment*

Grinding and milling, steaming, irradiation, pressure, and heating are examples of physical pre-treatment procedures that increase the available surface area and size of holes while decreasing the crystallinity of cellulose [11]. Steam pre-treatment, size reduction, and microwave or ultrasonic pre-treatment are used in the physical treatment. The rice straw steam pre-treatment procedure takes around 60 minutes at a temperature of 121°C and pressure of 15 psi [8].

Advantages of using physical pre-treatment are [10, 12]:

- Beneficial to get desired particle size by increasing the surface area.
- Effective in reducing cellulose crystallinity.
- Help enzymatic hydrolysis.

The following Figure 8.3 illustrates the pretreatment of lignocellulosic biomass.

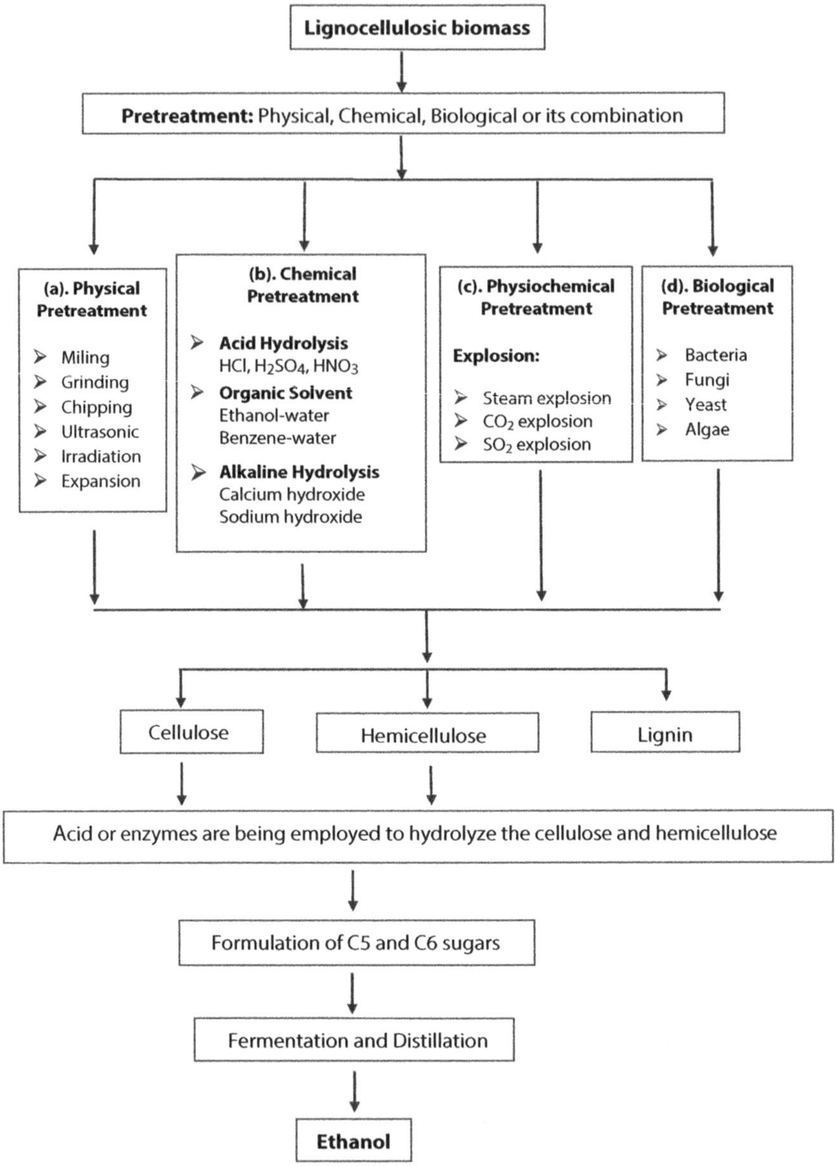

Figure 8.3 Pre-treatment of lignocellulosic biomass [10].

8.2.2.2 Chemical Treatment

Chemical pretreatment is crucial since other treatment processes cannot properly convert lignocelluloses to soluble sugars in the absence of effective chemical pretreatment. There are certain potential chemicals, such as acidic and alkaline chemicals [13]. Rice straw is alkaline pre-treated for 60 minutes with any of the following chemicals: $NaHCO_3$, $CaCO_3$, and NH_3 at 121°C and 15 psi [8]. Acidic pre-treatment of rice straw can be done with varying concentrations of various acids, such as H_2SO_4, lactic acid, and citric acid, at 121 °C, 15 psi for 60 minutes, as well as at room temperature. The acid treatment is capable of breaking the most significant number of polysaccharide-lignin connections, which aids in the recovery of monomeric sugars [14].

The advantage of using chemical pre-treatment is [10, 12]:

- ➢ Reducing biomass size by disrupting lignocellulosic components, primarily hemicellulose.
- ➢ Acid concentration has a critical role in pretreatment. Pretreatment with dilute acids removes the most hemicellulose, increasing sugar yield and even modifying the lignin structure, whereas strong acids can only hydrolyze cellulose. Certain acids, such as H_2SO_4 and HCl, are pretty inexpensive.
- ➢ Alkali solutions can remove the lignin from biomass, uncovering the polysaccharides. They can also break down the crystalline cellulose, increasing surface area and speeding up the hydrolysis process.

8.2.2.3 Physiochemical Treatment

The lignocellulose structure can be decomposed by physicochemical treatment. CO_2 explosion, steam explosion, SO_2 explosion, liquid hot water, and wet oxidation are some of the methods used [15].

Advantages of physiochemical pre-treatment are [10, 12]:

- ➢ Low chemical and energy expenditure.
- ➢ Hemicellulose and lignin decomposition.
- ➢ Acids assist in hydrolysis.
- ➢ Increased surface area exposure.

8.2.2.4 *Biological Treatment*

Biological pre-treatments are generally non-toxic and eco-friendly, use little energy, and do not produce inhibitors for downstream processes. To hydrolyze cellulose and remove lignin, bacterial or fungal pretreatments can be utilized. Biological approaches involve using fungus (brown fungi, white fungi, and soft rot fungi), bacteria, and enzymes to break down hemicellulose and lignin. Microbes perform best in optimal conditions; hence, the conditions for biological pre-treatments must be favorable. They are influenced by physical (such as temperature and particle size), chemical (such as pH), and biological (strain of bacteria or fungi) factors [15].

Advantage of biological pre-treatment are [10, 12]:

- Eco-friendly
- Lower requirement of energy.
- Very low cost
- Zero chemical required.
- Helpful in hydrolysis process of hemicellulose, cellulose, and lignin.

The various types of pretreatment are summarised in Table 8.2.

Table 8.2 Methods of lignocellulosic pre-treatment [16].

Type of method	Type of pre-treatment	Mechanism of action
Physical	Fragmentation (Milling, Grinding, Rolling)	Fragmentation of the lignocellulosic chain into smaller segments, exposing the lignocellulosic structure to the reagents used in subsequent treatment processes.
	Microwave Radiation	Reduction of cellulose crystal structure.
	Sonication	Rapture of hydrogen bond in the lignocellulosic structure.
	Gamma radiation	Rapture of beta-1-4-Glycosidic linkage.

(*Continued*)

Table 8.2 Methods of lignocellulosic pre-treatment [16]. (*Continued*)

Type of method	Type of pre-treatment	Mechanism of action
Chemical	Acid hydrolysis	Hemicellulose decomposition and lignin dissolution
	Alkali pretreatment	Saponification of lignocellulose and changes to the structure of lignin.
	Oxidation	Lignin and hemicellulose dissolution and cellulose crystal separation
Physiochemical	Steam explosion	Hemicellulose dissolution at 150 degrees C, whereas lignin dissolution at 180 degrees or above.
	CO_2/SO_2 explosion	Decomposition of hemicellulose and lignin.
Biological	Fungi	Decomposition of lignin and hemicellulose.
	Bacteria	Decomposition of lignin and hemicellulose.
	Enzymatic treatment	Decomposition of hemicellulose and cellulose.
Combined	Combination of two or more treatments such as **1.** Hydrothermal and wet disk milling **2.** supercritical CO_2 and ultrasonic pre-treatment **3.** Steam-pre-treatment combined with hydrogen peroxide etc.	Easy digestion of lignocellulosic materials High yield of lignin and hemicelluloses from lignocellulosic materials

8.2.2.5 Combined Pre-Treatment

Alkali, an oxidative de-lignification, a biological pre-treatment, and steam explosion can all easily remove lignin from lignocellulosic materials. Acid, alkali, and steam explosion pre-treatments have some effect on the removal of hemicelluloses. It can be done with physical methods like grinding or milling or radiation or chemical methods like a steam explosion or supercritical CO2 pre-treatment to make more of the surface area available to be cleaned or changed. They make it easier for lignocellulosic materials to be digested and get more lignin and hemicelluloses out of them, which leads to more ethanol being made. Pre-treatments that combine with each other usually have a good effect on the biomass, and they can even help with enzymatic hydrolysis. This means that the combined pre-treatments do more than just make lignocellulosic materials break down faster; they also get more of the valuable parts of them, which could be used to make more bioethanol and its by-products, which could be used on a large scale [17]. The combined pretreatment of lignocellulosic biomass is depicted in the Figure 8.4.

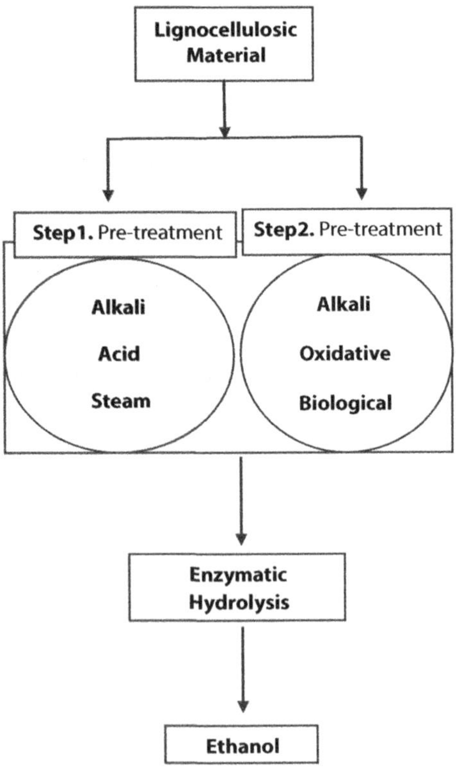

Figure 8.4 Combined pre-treatment [17].

8.2.3 Hydrolysis

Pretreated biomass still contains some cellulose and even moisture, which hydrolysis can remove [18]. The hydrolysis process can be carried out using either acid or enzymes. Begin with acid hydrolysis, which can be done with either a dilute or strong acid form. Dilute acid hydrolysis requires a low acid concentration and is performed at a higher temperature. In contrast, concentrated acid hydrolysis requires a high acid concentration but is performed at a lower temperature [19]. As a result, dilute acid hydrolysis is more efficient and breaks more bonds than concentrated acid hydrolysis [20]. The primary goal of performing this treatment with weak acid concentration is to remove as much hemicellulose as possible from the lignocellulosic material. For 90 minutes, a temperature range of 120°C–220°C and a concentration of roughly 1% sulphuric acid are used.

The mechanism of action of acid hydrolysis includes [21]:

a. **1st step:** Disruption of the crystalline structure of cellulose and fiber dissolution with concentrated acids.
b. **2nd step:** Diluting acid causes glycosidic bond hydrolysis, resulting in mono-saccharides.

Hydrolysis of pretreated biomass can also be accomplished using specific enzymes; the biomass is added with 0.1 M Na-acetate buffer at a pH of 5.0 and inoculated with mold-like bacteria or fungi 124 at 60°C at 150 rpm for 6 hours. After centrifuging the samples at 10,000 rpm for 10 minutes, the supernatant was collected and used to estimate reducing sugars with the dinitro salicylic (DNS) agent [22].

8.2.4 Fermentation

Using microorganisms (Table 8.3) such as yeast, fungus, and bacteria, fermentation can be used to transform pre-treated biomass into bioethanol. These microbes consume the fermentable sugars in the biomass and produce ethyl alcohol and other byproducts [19]. After the biomass has been pre-treated, it must be hydrolyzed and fermented, which can be done separately or concurrently. When it comes to the synthesis of bioethanol, three processes are commonly used: separate hydrolysis followed by fermentation (SHF), simultaneous saccharification followed by fermentation (SSF), and simultaneous saccharification followed by co-fermentation (SSCF) [20]. Separate hydrolysis of pretreated biomass is performed in SHF, followed by fermentation. Enzymatic hydrolysis and fermentation occur concurrently

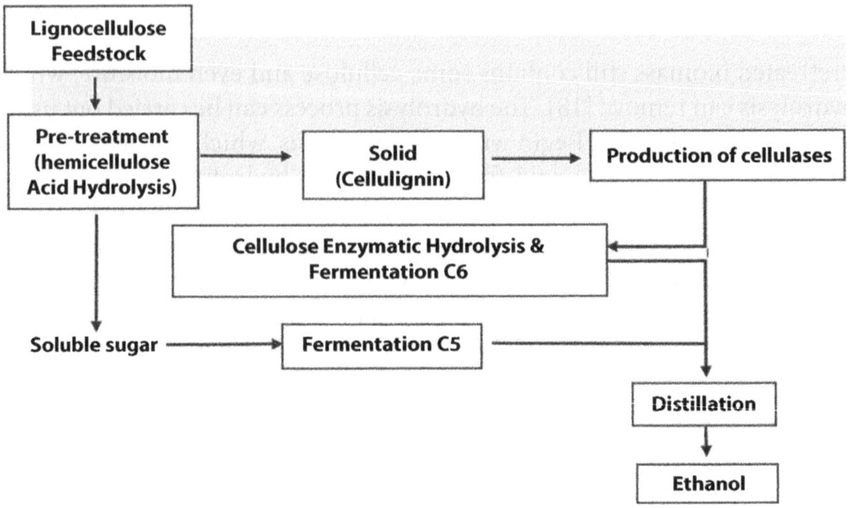

Figure 8.5 The simultaneous saccharification and fermentation (SSF) schematic process [24].

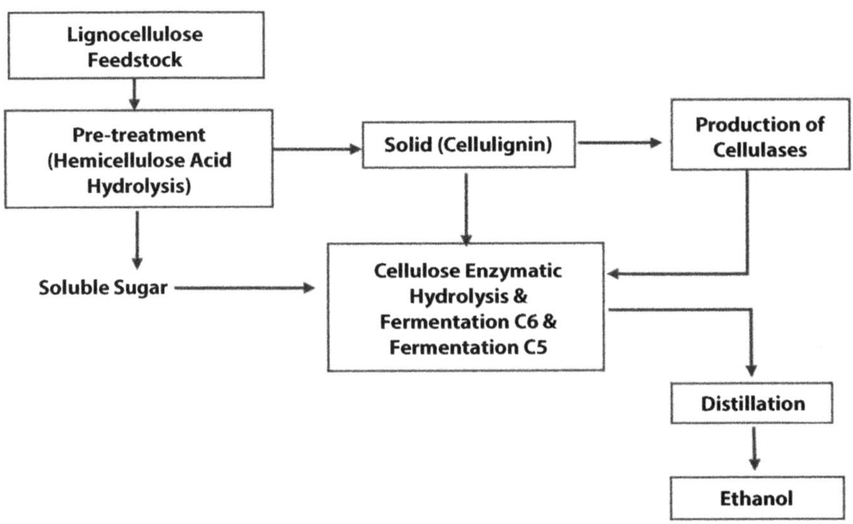

Figure 8.6 The schematic of the process called "Simultaneous Saccharification and Co-fermentation (SSCF)" [24].

in the SSF and SSCF processes to keep the concentration of glucose low. In SSF, the fermentation of glucose and pentose is separated, but in SSCF, the fermentation of glucose and pentose is combined. Both the SSF and SSCF processes are preferred over the SHF method due to advantages such

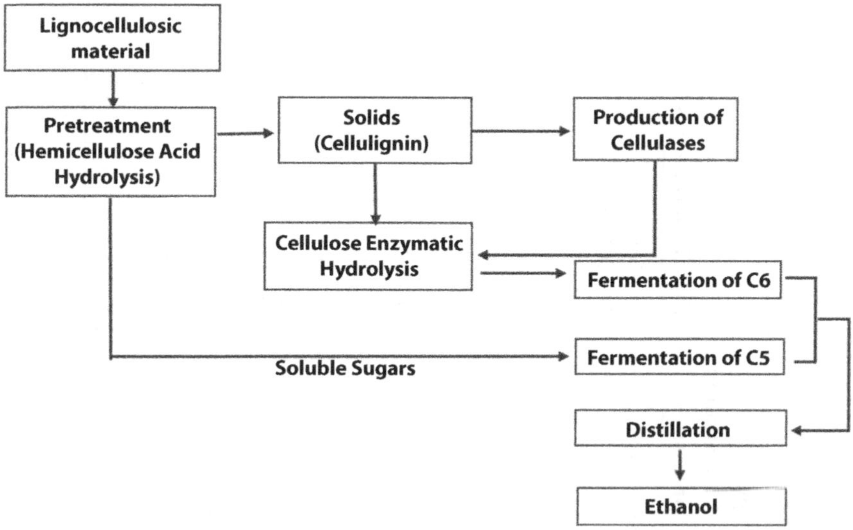

Figure 8.7 Separate hydrolysis and fermentation process (SHF) schematic process [24].

as reduced costs, higher ethanol yield, operation in the same tank, and shorter processing time [23]. Figures 8.5, 8.6, and 8.7, respectively, present an overview of the processes referred to as SSF, SSCF, and SHF.

8.3 Environmental Issues which can be Mitigated by the Production of Bioethanol Using Rice Straw

Rice straw incineration benefits farmers but disadvantages from an environmental perspective [59]. Because of paddy field burning, India has the most polluted cities globally. Open-field burning has several advantages, including being effective, dependable, affordable, requiring little labor, and requiring no expensive equipment from the farmer's perspective [60]. Rice straw wastes that are burned openly pollute the air and contribute to greenhouse gas emissions such as methane, carbon dioxide, ozone, nitrous oxide, and water vapors [59], as well as affecting soil fertility, biodiversity losses, and human health hazards [61]. These gases and aerosols contain carbonaceous matter, which affects global climate change and may result in regional increases in aerosols, acid deposition, tropospheric ozone rise, and stratospheric ozone layer depletion. Black carbon emissions also play a significant impact in influencing global warming [62]. Hazardous emissions from field burning are coupled with gases emitted by coal-fired power plants and cement manufacturing plants, resulting in severe respiratory

Table 8.3 Micro-organism's for the production of bioethanol using LCB as a feedstock.

S.N.	Microorganism	Feedstock	Pre-treatment method	Type of fermentation process	Kinetic parameters	Ref.
1.	*Pichia stipitis* BCC151	Bagasse (5%, w/v) hydrolysate	Mechanical pretreatment by ball milling,	SHF(24h)	P = 8.4 g/l Y = 0.29 g/g Q = 0.35 g/l/h	[25]
2.	*Pichia stipitis* BCC151	Bagasse (5%, w/v) hydrolysate	Mechanical pretreatment by ball milling,	SSF(72h)	P = 8 g/l Y = 0.27 g/g Q = 0.11 g/l/h	[25]
3.	*Saccharomyces cerevisiae*	Corn stover	Co-solvent	SSF	Y = 89.2%	[26]
4.	*Zymomonasmobilis* (ZMA7-2)	Food waste	Biological pretreatment	SHF	P = 99.78g/l Y = 0.50g/g Q = 2.35g/l/h	[27]
5.	*S. cerevisiae*	Ethanol-extracted (EECB) cane bagasse	Steam explosion	SHF	P = 27.1g/l Y = 46.7% Q = 0.31g/l/h	[28]

(*Continued*)

Table 8.3 Micro-organism's for the production of bioethanol using LCB as a feedstock. (*Continued*)

S.N.	Microorganism	Feedstock	Pre-treatment method	Type of fermentation process	Kinetic parameters	Ref.
6.	*Saccharomyces cerevisiae* (thermotolerant strain PYCC 2613)	Alkali extracted *Cistus ladanifer* (rockrose)	Steam explosion	SSF	P = 16.1g/l Y = yield of 22.1 g ethanol/100 g R-SE-OH. Q = 0.2g/l/h	[29]
7.	*Saccharomyces cerevisiae*	Sugarcane bagasse	Alkaline type	SSF	P = 7.27 g/l Y = 93.8%	[30]
8.	*S. cerevisiae* Thermossac	Sugarcane bagasse	Steam explosion	NA	P = 25.6 g/l Y = 44.4% Q = 2.1g/l/h	[31]
9.	*S. cerevisiae* Thermossac	Sugarcane bagasse	Sono-assisted alkali-washed	NA	P = 23.4g/l Y = 46.6% Q = 2g/l/h	[31]
10.	*Saccharomyces cerevisiae* LBM-1	Sugarcane bagasse	Alkaline pretreatments	SSSF	P = 8.8g/l Q = 0.88g/l/h	[32]
11.	*Saccharomyces cerevisiae* LBM-1	Eucalyptus	Alkaline pretreatment	SSSF	P = 3.4g/l Q = 0.34g/l/h	[32]

(*Continued*)

Table 8.3 Micro-organism's for the production of bioethanol using LCB as a feedstock. (*Continued*)

S.N.	Microorganism	Feedstock	Pre-treatment method	Type of fermentation process	Kinetic parameters	Ref.
12.	*Saccharomyces cerevisiae* LBM-1	Sugarcane Straw	Alkaline pretreatments	SSSF	P = 6.4g/l Q = 0.64g/l/h	[32]
13.	*Saccharomyces cerevisiae*	Rice straw	Twin gear reactor (TGR) pretreatment	SSF	Y = 0.032 g/g of dry mass	[33]
14.	*Saccharomyces cerevisiae* CCUG 53310	Rice straw	Alkaline pretreatments	DSSF	P = 83.1 g/l	[34]
15.	*Saccharomyces cerevisiae* SR8u	Wheat straw (WS) (*Triticum aestivum* L.)	Alkaline pretreatments	SSCF	P = 31.1g/l	[35]
16.	*Saccharomyces cerevisiae* SR8 u	Wheat straw	Alkaline pretreatment	NA	P = 65g/l	[36]
17.	*Escherichia coli* WDHL	Wheat straw	Dilute acid pretreatment	NA	P = 8.9g/l Y = 0.38g/g Q = 0.051g/l/h	[37]

(*Continued*)

Table 8.3 Micro-organism's for the production of bioethanol using LCB as a feedstock. (*Continued*)

S.N.	Microorganism	Feedstock	Pre-treatment method	Type of fermentation process	Kinetic parameters	Ref.
18.	*S. cerevisiae*	Rapeseed straw	Dilute acid pretreatment	NA	Y = 88–90%	[38]
19.	*S. cerevisiae*	Rye stillages	Dilute acid pretreatment	NA	P = 20.63g/l	[39]
20.	*S. cerevisiae*	Wheat stillages	Dilute acid pretreatment	NA	P = 17.86g/l	[39]
21.	*S. cerevisiae*	Maize stillages	Dilute acid pretreatment	NA	P = 35.44g/l	[39]
22.	*Saccharomyces cerevisiae* PE-2	Corn residues	Microwave-assisted pretreatment	SSF	P = 33.88 g/L	[40]
23.	Dry yeast	Cotton stalks	Sequential combination of organosolvent and hydrothermal pretreatment	PSSF	P = 32.3 g/L Q = 0.49 g/l/h	[41]
24.	Dry yeast	Cotton stalks	Microwave-assisted acid pretreatment	PSSF	P = 15.9g/l Q = 0.33 g/l/h	[41]

(*Continued*)

Table 8.3 Micro-organism's for the production of bioethanol using LCB as a feedstock. (*Continued*)

S.N.	Microorganism	Feedstock	Pre-treatment method	Type of fermentation process	Kinetic parameters	Ref.
25.	Dry yeast	Cotton stalks	Alkali treatment	PSSF	P = 15.9g/l Q = 0.33g/l/h	[41]
26.	Dry yeast	Cotton stalks	Organosolv	PSSF	P = 22.1g/l Q = 0.52g/l/h	[41]
27.	*Pichia kudriavzevii* HOP-1	Cotton stalks	Alkali pretreatment	PSSF (12 h) [1]	P = 19.48g/l Y = 0.21 g/g Q = 0.41g/l/h	[42]
28.	*P. kudriavzevii* HOP-1	Cotton stalks	Ozone pretreatment	SHF	P = 10.96g/l Q = 0.46g/l/h	[42]
29.	*Saccharomyces cerevisiae* NRRL Y-132	Cotton stalks	Ionic liquid, 1-ethyl-3-methylimidazolium acetate (EMIMAc),	SHF	P = 22.9g/l Q = 1.91g/l/h	[43]
30.	*S. cerevisiae* NRRL Y-132	Cotton stalks	Alkaline pretreatment	SHF	P = 19.8g/l Q = 1.65g/l/h	[43]

(*Continued*)

Table 8.3 Micro-organism's for the production of bioethanol using LCB as a feedstock. (*Continued*)

S.N.	Microorganism	Feedstock	Pre-treatment method	Type of fermentation process	Kinetic parameters	Ref.
31.	*Clavispora* NRRL Y-50464	Rice straw	Choline chloride-based solvent(Green solvents)	SSF	P = 36.7g/l	[44]
32.	(a). *S. cerevisiae* MTCC 174 (b). *Zymomonasmobilis* subsp. *mobilis* MTCC 91	Wheat straw	Microwave-assisted NaOH pretreatments	NA	Y = 68.2%	[45]
33.	*B. subtilis* NA15	Rice straw	Microwave-alkali-acid pretreatment	SSF	P = 25.2 g/L (titer value) Y = 0.38 g/g Q = 1.05 g/L/h	[46]
34.	*S. cerevisiae* YC-097	Wheat straw	Microwave-assisted alkali pretreatment	SSF	P = 34.3 g/l Y = 69.3%	[47]
35.	*Saccharomyces Cerevisiae*	Sugarcane bagasse	Ultrasound with acid and Ultrasound with alkaline	NA	P = 0.0.911 g/L	[48]

(*Continued*)

Table 8.3 Micro-organism's for the production of bioethanol using LCB as a feedstock. (*Continued*)

S.N.	Microorganism	Feedstock	Pre-treatment method	Type of fermentation process	Kinetic parameters	Ref.
36.	*Saccharomyces cerevisiae* strain Lg8-1	Wheat (*Triticum aestivum* L.) straw	Alkaline pre-extraction followed by steam pretreatment	NA	P = 54.5 g/L	[49]
37.	*Saccharomyces cerevisiae* Ethanol Red*	Triticale straw	Steam-explosion	SSF	P = 29.31 g/l Y = 84.7%	[50]
38.	*S. cerevisiae*	Barley straw	Sequential Steam Explosion and Reactive Extrusion Pretreatment	SSF	P = 31.7 g/l Y = 0.40 g/g Q = 0.96 g/l/h	[51]
39.	*Saccharomyces cerevisiae*	Sweet sorghum	Steam treatment	NA	Y = above 90%	[52]
40.	*Scheffersomycesstipitis* NRRL Y-7124	Rice straw	Biological treatment by baker›s yeast (*Saccharomyces cerevisiae*)	NA	P = 23 g/l Y = 0.27 g/g Q = 0.52 g/l/h	[53]

(*Continued*)

Table 8.3 Micro-organism's for the production of bioethanol using LCB as a feedstock. (*Continued*)

S.N.	Microorganism	Feedstock	Pre-treatment method	Type of fermentation process	Kinetic parameters	Ref.
41.	*Saccharomyces cerevisiae* strain LN	Paddy straw	*Trametes hirsuta* MTCC 136 used for biological pretreatment	NA	P = 0.86 g/l	[54]
42.	*S. cerevisiae* and *C. tropicalis* NSW-NW	Rice straw	*T. reesei* Aq-5b and *T. viride* NSW-XM for biological treatment	NA	P = 4.97–5.05 g/L Q = 2.17 g/l/h	[55]
43.	Dry yeast	Wheat straw	Phosphoric acid and hydrogen peroxide	SSF	P = 69.9 g/L Y = 0.155 g/g	[56]
44.	*Saccharomyces cerevisiae*	Dried sugarcane bagasse	Liquid hot water (LHW) pretreatment and wet disk milling	SScF	Y = 93.8%	[57]
45.	*Saccharomyces cerevisiae* type II distiller's yeast	Triticale straw	1-ethyl-3-methylimidazolium acetate	NA	P = 10.64 g/l	[58]

Table 8.4 Rank of polluted cities in the world [60].

Rank	Countries	Cities	Score
1.	Egypt	Cairo	95.8361
2.	India	Delhi	86.7024
3.	China	Beijing	76.4648
4.	Russia	Moscow	75.5634
5.	Turkey	Istanbul	72.9714

ailments and various catastrophic health conditions [63]. Another problem is dealing with rice straw's high ash content as well as alkali elements like sodium and potassium [64]. This is a major cause of air pollution and global warming, affecting many countries. Table 8.4 shows such countries.

8.4 Modern Usage: An Hour's Requirement

In India, fossil fuel consumption should be reduced and a shift to renewable fuels is required to meet expanding energy demands in the transportation sector and growing environmental concerns. The country's national biofuel rules made it lawful to incorporate bioethanol into gasoline [65]. Bioethanol has the potential to be an alternative source of energy due to its outstanding properties, including cleaner combustion due to the presence of oxygen in its composition. Now that bioethanol is recognized as a fuel for direct ethanol fuel cells and biofuel cells, it is available for purchase. Bioethanol is a transparent liquid that can be used in place of gasoline in cars [66]. As the transition from fossil fuels to biofuels is a significant and crucial phase. Increased consumption of biofuels is a significant step toward environmental conservation [67]. Due to its high-octane number, bioethanol fits all of the standards for gasoline mixing, but due to its low specific gravity and high heat of vaporization, it is incompatible with self-ignition in diesel engines. Thus, when diesel-bioethanol blended fuel is utilized, self-ignition is enhanced using ignition improvers, glow plugs, surface ignition, and pilot injection [68]. A comprehensive policy framework is a necessary condition for improving the country's ethanol blending situation, as it encompasses several critical elements, such as [69, 70]:

- EBPP is inextricably related to agricultural policy and markets.
- Other sectors' need for ethanol (nonblending) will continue to rise.
- The nature of the issues that the sector is confronting is so intertwined that they serve to strengthen one another.
- Only a limited amount of flexibility can be accommodated at the state level under the centrally planned strategy.

8.5 Conclusion

Rice straw, or parali, is widely available in bulk worldwide and is a significant source of environmental and human health concerns (air pollution caused by openly burning rice leftovers) (respiratory illness because of increased levels of air pollution). Agricultural wastes such as rice straw and wheat straw are high in lignocellulose material and may be used as raw materials in bioethanol manufacturing. They thrive worldwide and require no more land, water, or energy. As of now, numerous researchers have made significant efforts to develop technologies for the production of biofuels and bioenergy, taking into account their use and potential negative environmental impacts. Bioethanol is a renewable biofuel produced using a biotechnological method that is environmentally friendly. Bioethanol is a renewable energy source that can be utilized in place of conventional fossil fuels. Bioethanol is a type of alcohol that is produced by fermenting carbohydrates. Bioethanol can be combined with gasoline and utilized as a transportation fuel with no issue.

References

1. Singh, A., Prajapati, P., Vyas, S., Gaur, V. K., Sindhu, R., Binod, P., ... & Varjani, S., A comprehensive review of feedstocks as sustainable substrates for next-generation biofuels. *Bioenergy Res.*, 1–18, 2022.
2. María, A., Pabón, A., Felissia, F.E., Mendieta, C.M., Chamorro, E., Area, M.C., Improvement of bioethanol production from rice husks. *Cell. Chem. Technol.*, 54, 8, 689–698, 2020.
3. Swain, M.R., Singh, A., Sharma, A.K., Tuli, D.K., Bioethanol production from rice-and wheat straw: An overview, in: *Bioethanol Production From Food Crops*, pp. 213–231, 2019.

4. Jin, X., Song, J., Liu, G.Q., Bioethanol production from rice straw through an enzymatic route mediated by enzymes developed in-house from *Aspergillus fumigatus*. *Energy*, 190, 116395, 2020.
5. Wi, S.G., Choi, I.S., Kim, K.H., Kim, H.M., Bae, H.J., Bioethanol production from rice straw by popping pretreatment. *Biotechnol. Biofuels*, 6, 1, 1–7, 2013.
6. Wi, S.G., Choi, I.S., Kim, K.H., Kim, H.M., Bae, H.J., Bioethanol production from rice straw by popping pretreatment. *Biotechnol. Biofuels*, 6, 1, 1–7, 2013.
7. Chen, H., Liu, J., Chang, X., Chen, D., Xue, Y., Liu, P., Han, S., A review on the pretreatment of lignocellulose for high-value chemicals. *Fuel Process. Technol.*, 160, 196–206, 2017. https://doi.org/10.1016/j.fuproc.2016.12.007.
8. Kumar, A., Jain, K.K., Singh, B., Process optimization for chemical pretreatment of rice straw for bioethanol production. *Renew. Energy*, 156, 1233–1243, 2020.
9. Veluchamy, C. and Kalamdhad, A.S., Enhancement of hydrolysis of lignocellulose waste pulp and paper mill sludge through different heating processes on thermal pretreatment. *J. Clean. Prod.*, 168, 219–226, 2017.
10. Sharma, H.K., Xu, C., Qin, W., Biological pretreatment of lignocellulosic biomass for biofuels and bioproducts: An overview. *Waste Biomass Valorization*, 10, 2, 235–251, 2019.
11. Mohanty, B. and Abdullahi, I., II, Bioethanol production from lignocellulosic waste-a review. *Biosci. Biotechnol. Res. Asia*, 13, 2, 1153, 2016.
12. Maurya, D.P., Singla, A., Negi, S., An overview of key pretreatment processes for biological conversion of lignocellulosic biomass to bioethanol. *3 Biotech.*, 5, 5, 597–609, 2015.
13. Khedr, F.G., Tohamy, E.Y., El-Gamal, A.D., Abouelwafa, A.M., Bioconversion of rice straw into bioethanol by enzymatic hydrolysis of bacillus subtilis. *IOSR J. Pharm. Biol. Sci.*, 14, 09–29, 2019.
14. Rezania, S., Oryani, B., Cho, J., Talaiekhozani, A., Sabbagh, F., Hashemi, B., Mohammadi, A.A., Different pretreatment technologies of lignocellulosic biomass for bioethanol production: An overview. *Energy*, 199, 117457, 2020.
15. Tu, W.C. and Hallett, J.P., Recent advances in the pretreatment of lignocellulosic biomass. *Curr. Opin. Green Sustain. Chem.*, 20, 11–17, 2019.
16. Kucharska, K., Rybarczyk, P., Hołowacz, I., Łukajtis, R., Glinka, M., Kamiński, M., Pretreatment of lignocellulosic materials as substrates for fermentation processes. *Molecules*, 23, 11, 2937, 2018.
17. Sun, S., Sun, S., Cao, X., Sun, R., The role of pretreatment in improving the enzymatic hydrolysis of lignocellulosic materials. *Bioresour. Technol.*, 199, 49–58, 2016.
18. Sharma, S., Nandal, P., Arora, A., Ethanol production from NaOH pretreated rice straw: A cost-effective option to manage rice crop residue. *Waste Biomass Valorization*, 10, 11, 3427–3434, 2019.
19. Abdu Yusuf, A. and Inambao, F.L., Bioethanol production techniques from lignocellulosic biomass as alternative fuel: A review. *Int. J. Mech. Eng. Technol.*, 10, 6, 2019.

20. Azhar, S.H.M., Abdulla, R., Jambo, S.A., Marbawi, H., Gansau, J.A., Faik, A.A.M., Rodrigues, K.F., Yeasts in sustainable bioethanol production: A review. *Biochem. Biophys. Rep.*, 10, 52–61, 2017.
21. Zhou, Z., Liu, D., Zhao, X., Conversion of lignocellulose to biofuels and chemicals via sugar platform: An updated review on chemistry and mechanisms of acid hydrolysis of lignocellulose. *Renew. Sust. Energ. Rev.*, 146, 111169, 2021.
22. Ashoor, S. and Sukumaran, R.K., Mild alkaline pretreatment can achieve high hydrolytic and fermentation efficiencies for rice straw conversion to bioethanol. *Prep. Biochem. Biotechnol.*, 50, 8, 814–819, 2020.
23. Canilha, L., Chandel, A.K., Antunes, F.A.F., Luiz da Costa Freitas, W., das Graças Almeida Felipe, M., da Silva, S.S., Bioconversion of sugarcane biomass into ethanol: An overview about composition, pretreatment methods, detoxification of hydrolysates, enzymatic saccharification, and ethanol fermentation. *J. Biomed. Biotechnol.*, 2012, 2012.
24. Sebayang, A.H., Masjuki, H.H., Ong, H.C., Dharma, S., Silitonga, A.S., Mahlia, T.M., II, Aditiya, H.B., A perspective on bioethanol production from biomass as alternative fuel for spark ignition engine. *RSC Adv.*, 6, 18, 14964–14992, 2016.
25. Buaban, B., Inoue, H., Yano, S., Tanapongpipat, S., Ruanglek, V., Champreda, V., Eurwilaichitr, L., Bioethanol production from ball milled bagasse using an on-site produced fungal enzyme cocktail and xylose-fermenting Pichia stipitis. *J. Biosci. Bioeng.*, 110, 1, 18–25, 2010.
26. Nguyen, T.Y., Cai, C.M., Osman, O., Kumar, R., Wyman, C.E., CELF pretreatment of corn stover boosts ethanol titers and yields from high solids SSF with low enzyme loadings. *Green Chem.*, 18, 6, 1581–1589, 2016.
27. Ma, K., Ruan, Z., Shui, Z., Wang, Y., Hu, G., He, M., Open fermentative production of fuel ethanol from food waste by an acid-tolerant mutant strain of Zymomonasmobilis. *Bioresour. Technol.*, 203, 295–302, 2016.
28. Neves, P.V., Pitarelo, A.P., Ramos, L.P., Production of cellulosic ethanol from sugarcane bagasse by steam explosion: Effect of extractives content, acid catalysis and different fermentation technologies. *Bioresource Technol.*, 208, 184–194, 2016.
29. Ferro, M.D., Fernandes, M.C., Paulino, A.F., Prozil, S.O., Gravitis, J., Evtuguin, D.V., Xavier, A.M., Bioethanol production from steam explosion pretreated and alkali extracted Cistus ladanifer (rockrose). *Biochem. Eng. J.*, 104, 98–105, 2015.
30. Nosratpour, M.J., Karimi, K., Sadeghi, M., Improvement of ethanol and biogas production from sugarcane bagasse using sodium alkaline pretreatments. *J. Environ. Manage.*, 226, 329–339, 2018.
31. da Silva, G.T., Chiarello, L.M., Lima, E.M., Ramos, L.P., Sono-assisted alkaline pretreatment of sugarcane bagasse for cellulosic ethanol production. *Catal. Today*, 269, 21–28, 2016.

32. de Carvalho, D.M., de Queiroz, J.H., Colodette, J.L., Assessment of alkaline pretreatment for the production of bioethanol from eucalyptus, sugarcane bagasse and sugarcane straw. *Ind. Crops Prod.*, 94, 932–941, 2016.
33. Ahmed, M.A., Rehman, M.S.U., Terán-Hilares, R., Khalid, S., Han, J., II, Optimization of twin gear-based pretreatment of rice straw for bioethanol production. *Energy Convers. Manage.*, 141, 120–125, 2017.
34. Molaverdi, M., Karimi, K., Mirmohamadsadeghi, S., Improvement of dry simultaneous saccharification and fermentation of rice straw to high concentration ethanol by sodium carbonate pretreatment. *Energy*, 167, 654–660, 2019.
35. Yuan, Z., Wen, Y., Li, G., Production of bioethanol and value added compounds from wheat straw through combined alkaline/alkaline-peroxide pretreatment. *Bioresour. Technol.*, 259, 228–236, 2018.
36. Yuan, Z., Wen, Y., Kapu, N.S., Beatson, R., Evaluation of an organosolv-based biorefinery process to fractionate wheat straw into ethanol and co-products. *Ind. Crops Prod.*, 121, 294–302, 2018.
37. Lopez-Hidalgo, A.M., Sánchez, A., De León-Rodríguez, A., Simultaneous production of bioethanol and biohydrogen by Escherichia coli WDHL using wheat straw hydrolysate as substrate. *Fuel*, 188, 19–27, 2017.
38. Kuglarz, M., Alvarado-Morales, M., Dąbkowska, K., Angelidaki, I., Integrated production of cellulosic bioethanol and succinic acid from rapeseed straw after dilute-acid pretreatment. *Bioresour. Technol.*, 265, 191–199, 2018.
39. Mikulski, D. and Kłosowski, G., Efficiency of dilute sulfuric acid pretreatment of distillery stillage in the production of cellulosic ethanol. *Bioresour. Taechnol.*, 268, 424–433, 2018.
40. Aguilar-Reynosa, A., Romaní, A., Rodríguez-Jasso, R.M., Aguilar, C.N., Garrote, G., Ruiz, H.A., Comparison of microwave and conduction-convection heating autohydrolysis pretreatment for bioethanol production. *Bioresour. Taechnol.*, 243, 273–283, 2017.
41. Dimos, K., Paschos, T., Louloudi, A., Kalogiannis, K.G., Lappas, A.A., Papayannakos, N., Mamma, D., Effect of various pretreatment methods on bioethanol production from cotton stalks. *Fermentation*, 5, 1, 5, 2019.
42. Kaur, U., Oberoi, H.S., Bhargav, V.K., Sharma-Shivappa, R., Dhaliwal, S.S., Ethanol production from alkali-and ozone-treated cotton stalks using thermotolerant Pichia kudriavzevii HOP-1. *Ind. Crops Prod.*, 37, 1, 219–226, 2012.
43. Haykir, N., II and Bakir, U., Ionic liquid pretreatment allows utilization of high substrate loadings in enzymatic hydrolysis of biomass to produce ethanol from cotton stalks. *Ind. Crops Prod.*, 51, 408–414, 2013.
44. Kumar, A.K., Parikh, B.S., Shah, E., Liu, L.Z., Cotta, M.A., Cellulosic ethanol production from green solvent-pretreated rice straw. *Biocatal. Agric. Biotechnol.*, 7, 14–23, 2016.

45. Tsegaye, B., Balomajumder, C., Roy, P., Optimization of microwave and NaOH pretreatments of wheat straw for enhancing biofuel yield. *Energy Convers. Manage.*, 186, 82–92, 2019.
46. Akhtar, N., Goyal, D., Goyal, A., Characterization of microwave-alkali-acid pre-treated rice straw for optimization of ethanol production via simultaneous saccharification and fermentation (SSF). *Energy Convers. Manage.*, 141, 133–144, 2017.
47. Zhu, S., Wu, Y., Yu, Z., Zhang, X., Wang, C., Yu, F., Jin, S., Production of ethanol from microwave-assisted alkali pretreated wheat straw. *Process Biochem.*, 41, 4, 869–873, 2006.
48. Kandasamy, M., Hamawand, I., Bowtell, L., Seneweera, S., Chakrabarty, S., Yusaf, T., Eberhard, F., Investigation of ethanol production potential from lignocellulosic material without enzymatic hydrolysis using the ultrasound technique. *Energies*, 10, 1, 62, 2017.
49. Yuan, Z., Li, G., Hegg, E.L., Enhancement of sugar recovery and ethanol production from wheat straw through alkaline pre-extraction followed by steam pretreatment. *Bioresour. Technol.*, 266, 194–202, 2018.
50. Kossatz, H.L., Rose, S.H., Viljoen-Bloom, M., van Zyl, W.H., Production of ethanol from steam exploded triticale straw in a simultaneous saccharification and fermentation process. *Process Biochem.*, 53, 10–16, 2017.
51. Oliva, J.M., Negro, M.J., Manzanares, P., Ballesteros, I., Chamorro, M. Á., Sáez, F., Moreno, A.D., A sequential steam explosion and reactive extrusion pretreatment for lignocellulosic biomass conversion within a fermen ation-based biorefinery perspective. *Fermentation*, 3, 2, 15, 2017.
52. Damay, J., Boboescu, I.Z., Duret, X., Lalonde, O., Lavoie, J.M., A novel hybrid first and second generation hemicellulosic bioethanol production process through steam treatment of dried sorghum biomass. *Bioresour. Technol.*, 263, 103–111, 2018.
53. Fonseca, B.G., Mateo, S., Moya, A.J., Roberto, I.C., Biotreatment optimization of rice straw hydrolyzates for ethanolic fermentation with Scheffersomycesstipitis. *Biomass Bioenergy*, 112, 19–28, 2018.
54. Arora, A., Priya, S., Sharma, P., Sharma, S., Nain, L., Evaluating biological pretreatment as a feasible methodology for ethanol production from paddy straw. *Biocatal. Agric. Biotechnol.*, 8, 66–72, 2016.
55. Wu, X., Zhang, J., Xu, E., Liu, Y., Cheng, Y., Addy, M., Ruan, R., Microbial hydrolysis and fermentation of rice straw for ethanol production. *Fuel*, 180, 679–686, 2016.
56. Qiu, J., Tian, D., Shen, F., Hu, J., Zeng, Y., Yang, G., Zhang, J., Bioethanol production from wheat straw by phosphoric acid plus hydrogen peroxide (PHP) pretreatment via simultaneous saccharification and fermentation (SSF) at high solid loadings. *Bioresour. Technol.*, 268, 355–362, 2018.
57. Wang, Z., Dien, B.S., Rausch, K.D., Tumbleson, M.E., Singh, V., Fermentation of undetoxified sugarcane bagasse hydrolyzates using a two stage

hydrothermal and mechanical refining pretreatment. *Bioresour. Technol.*, 261, 313–321, 2018.
58. Smuga-Kogut, M., Walendzik, B., Szymanowska-Powalowska, D., Kobus-Cisowska, J., Wojdalski, J., Wieczorek, M., Cielecka-Piontek, J., Comparison of bioethanol preparation from triticale straw using the ionic liquid and sulfate methods. *Energies*, 12, 6, 1155, 2019.
59. Romasanta, R.R., Sander, B.O., Gaihre, Y.K., Alberto, M.C., Gummert, M., Quilty, J., Wassmann, R., How does burning of rice straw affect CH4 and N2O emissions? *A Comp. Experiment Different On-Field Straw Manage. Practices. Agric. Ecosyst. Environ.*, 239, 143–153, 2017.
60. El Safty AMK Department of Occupational and Environmental Medicine. Cairo University, Egypt,. *Egypt. J. Occup. Med.*, 44, 3, 679–708, 2020.
61. Bhattacharyya, P., Bisen, J., Bhaduri, D., Priyadarsini, S., Munda, S., Chakraborti, M., Nimbrayan, P., Turn the wheel from waste to wealth: Economic and environmental gain of sustainable rice straw management practices over field burning in reference to India. *Sci. Total Environ.*, 775, 145896, 2021.
62. Lohan, S.K., Jat, H.S., Yadav, A.K., Sidhu, H.S., Jat, M.L., Choudhary, M., Sharma, P.C., Burning issues of paddy residue management in north-west states of India. *Renew. Sust. Energ. Rev.*, 81, 693–706, 2018.
63. Athira, G., Bahurudeen, A., Appari, S., Sustainable alternatives to carbon intensive paddy field burning in India: A framework for cleaner production in agriculture, energy, and construction industries. *J. Clean. Prod.*, 236, 117598, 2019.
64. Logeswaran, J., Shamsuddin, A.H., Silitonga, A.S., Mahlia, T.M., II, Prospect of using rice straw for power generation: A review. *Environ. Sci. Pollut. Res.*, 27, 21, 25956–25969, 2020.
65. Roy, M.M. and Chandra, A., Promoting biofuels: The case of ethanol blending initiative in India. *Clean Technol. Environ. Policy*, 21, 5, 953–965, 2019.
66. Best Use of Bioethanol: Fuel or Feedstock? techooid.com. https://techooid.com/best-use-of-bioethanol-fuel-or-feedstock
67. REPN, *Renewables 2015–Global Status Report*, Renewable Energy Policy Network for the 21st Century, REN21 Secretariat, Paris, France, 2016, http://www.ren21.net/wp-content/uploads/2015/07/REN12-GSR2015_Onlinebook _low1.pdf.
68. Balat, M., Balat, H., Oz, C., Progress in bioethanol processing. *Prog. Energy Combust. Sci.*, 34, 5, 551–573, 2008.
69. CSTEP, *Fuel Blending in India: Leanings and Way Forward*, Centre for Study of Science, Technology and Policy, Bangalore, 2016.
70. TERI, *Energy Security Outlook—Defining A Secure And Sustainable Energy Future For India*, The Energy Research Institute, New Delhi, 2015.

9

Modeling and Optimization of Microbial Production of Xylitol

Sharon Mano Pappu J.[1]*, Sathyanarayana N. Gummadi[2] and Tamilmani Jayabalan[1]

[1]*Department of Biotechnology, National Institute of Technology, Andhra Pradesh, India*
[2]*Applied and Industrial Microbiology Laboratory, Bhupat and Jyoti Mehta School of Biosciences, Department of Biotechnology, Indian Institute of Technology Madras, Chennai, India*

Abstract

The application of modeling approaches with computational simulation and analysis has an insightful effect on the process development of microbial products. This chapter aims to cover the trending strategies employed for modeling the biotechnological process and optimization of process conditions to achieve enhanced yield in biotechnological production. A case study is presented to understand the modeling and optimization techniques – microbial production of xylitol. The basics of modeling biotechnological processes, including the classification of mathematical models were described. Various works on xylitol production related to mathematical modeling, such as kinetic models for growth, substrate uptake, and product formation were assessed. The different types of modeling, including unstructured, structured, and cybernetic, were discussed. This chapter also discusses the optimization techniques used in the microbial production process with xylitol production as a case study. Optimization techniques including one-factor-at-a-time, design of experiments, namely full factorial design, fractional factorial design, Taguchi design, Central composite design, and Box-Behnken designs were discussed along with the analysis part in relation to process conditions such as temperature, substrate concentration, pH, agitation, and aeration. This chapter will provide an overview of various modeling approaches for the biotechnological process and will familiarize the reader with different optimization techniques.

***Keywords*:** Modeling, optimization techniques, design of experiments, Xylitol

**Corresponding author*: sharonmano@gmail.com

Sanjay Kumar, Narendra Kumar and Shahid-ul-Islam (eds.) Role of Microbes in Industrial Products and Processes, (223–254) © 2023 Scrivener Publishing LLC

9.1 Introduction

By the turn of the 21st century, renewable energy and value-added chemicals appeared as powerful competitors for feasible replacement of conventional fossil fuels and petrochemicals. Nevertheless, no outstanding work has been carried out in this area of research till the end of the 20th century, possibly due to a lack of formal guidelines and a futuristic approach for developing green and sustainable technologies. Thus, the US department of energy (DoE) considered the possibility of such technologies extensively in 2004 and proposed "top 12" building block platform chemicals primarily derived from the sugar content of the biomass, followed by another report in 2007 for the development of lignin-based biorefineries. Among which is xylitol, a sugar alcohol with the same sweetness as that of sucrose but with 40% fewer calories. The limitations in the microbial production of xylitol such as low productivity and low yields can be overcome by developing mathematical models. In addition to the invention of supercomputers and graphical user interface software/packages, the vast knowledge of biochemical pathways of microbes has changed the way of process development and system operations. It is achievable to improve xylitol production by considering the factors that affect the metabolism of the organism involved in the production process.

The proficiency in forecasting the microbial activities in the fermentation process augments the option of optimizing their performance. Mathematical modeling of bioprocess systems has become an indispensable tool in the design and optimization of the process. Improvement in technology and the recent development of several computational softwares have made the modeling approach more valuable than previous attempts. Mathematical models can be employed as a prediction tool for the development of the fermentation process and optimization of growth conditions. Optimization of bioprocess based on mathematical modeling provides a quantitative and systematic framework to increase the process profitability. The information gained from modeling will be helpful in directing the process control and modification. Simulations based on the developed model will intend to predict product concentration and control of process parameters during fermentation. The building of mathematical models of fermentation systems has the ability to (i) To minimize the experimental work; (ii) To assess the reliability of experimental results; (iii) To predict the behavior of the fermentation processes; (iv) To analyze the behavior of microbial culture and (v) To optimize the bioreactor performance.

This chapter deals with the applications of xylitol and its methods of production, emphasizing the modeling approaches and optimization techniques in the microbial production of xylitol. A section will follow on modeling of fermentation system with a focus on developing unstructured, structured, and cybernetic models with microbial xylitol production as a case study. Another purpose of the chapter is to highlight potential optimization techniques, conventional and design of experiments to maximize productivity, yield, and titer. This chapter gives an overview of modeling approaches and optimization techniques for biotechnological processes.

9.2 Xylitol

9.2.1 Applications and Demand

Xylitol ($C_5H_{12}O_5$), a five-carbon polyol (molecular weight – 152.15 g mol^{-1}), is considered as a naturally occurring polyalcohol in which a hydroxyl group is attached to each carbon atom. The pentahydroxy pentane was first discovered by German chemist Emil Fischer and his team in the year 1891, who obtained xylitol in the syrup form as a result of the reaction between xylose and sodium amalgam [1]. It has high sweetening power similar to sucrose while exhibiting one-third of calories lesser than sucrose (2.4 for xylitol vs. 4 cal g^{-1} for sucrose) [2]. The other significant characteristics include high endothermic heat (34.8 cal g^{-1}) [3]. As it does not undergo a Maillard reaction, the nutritional value of proteins remains the same [4].

Due to its sweetening property, xylitol has been extensively used in dietary food products and in the production process of caramels, chocolates, candies, jellies, marmalades, and soft drinks. Additionally, because of its high negative heat of solution, it imparts a pleasing fresh and cool sensation and is therefore predominantly used in the formulation of chewing gums, mouthwash, and toothpaste [5]. Being a low caloric sweetener and possessing insulin-independent metabolism, this has been transpired as a propitious sucrose substitute for non-insulin diabetic patients [6]. Ingestion of xylitol causes a limited increase in blood glucose and insulin levels compared to the increase in sucrose and glucose levels [7]. This property makes xylitol to be applicable in post-traumatic or post-operative states, the condition at which the secretion of excessive stress hormones results in insulin resistance [8]. Furthermore, since the metabolism of xylitol does not depend on glucose-6-phosphate dehydrogenase, it can be used as a potential sweetener for glucose-6-phosphate dehydrogenase deficient patients [8].

Several studies have also been reported the application of xylitol in the effective treatment of acute otitis media in children [9] and osteoporosis [10]. Due to its unique properties, such as high solubility [11], high negative heat of dissolution, low caloric sweetener, low glycemic rate [12], anti-cariogenicity, and cariostatic properties, xylitol has wide commercial properties applications in food, pharmaceutical, and health industries. Due to health and weight consciousness, the rise in consumers inclining towards low caloric and sugar-free food products has increased demand for xylitol and resulted in a 12% market share of the total polyols.

Half of the world's total xylitol is produced by Asia, and the remaining half of the production is by the United States, Europe, and Australia [13]. It was reported that 21% of the world's total xylitol (35,000 tons) was produced by a specific Chinese company, ensuing China's most significant market [14]. In the year 2013, xylitol had an enormous market containing an annual consumption of approximately 161.5 thousand metric tons that correspond to the market value of US$ 670 million, and the demand for xylitol is expected to reach 242 thousand metric tons. The global consumption of xylitol was approximately is expected to reach 256 thousand metric tons, 1.5 times more than its current requirement, by 2022, which corresponds to around US $16 billion. The demand for xylitol keeps increasing year by year steadily at a rate of 6% per year and to meet the world's ever-rising demand, it is imperative to produce xylitol in large quantities.

9.2.2 Methods of Xylitol Production

Xylitol, found in some fruits and vegetables, can be extracted by treating the vegetable matter with oxalic acid or can be extracted from beechwood or jute sticks [6]. Nevertheless, the extraction process becomes challenging and improvident due to very less xylitol content and high cost. Hence, the mass production of xylitol has been carried out by chemical hydrogenation of pure xylose under high pressure and temperature in the presence of catalysts on an industrial scale. The methods of xylitol production can be classified broadly into two types: (i) chemical method (ii) biotechnological method. The biotechnological method is further classified into two (i) microbial method, where microorganisms are employed for the bioconversion of xylose to xylitol, and (ii) enzymatic method, where enzymes are involved in the production process.

9.2.2.1 Chemical Method of Xylitol Production

Conventionally, xylitol is produced by hydrogenation or chemical reduction of pure xylose or xylose derived from hemicellulosic hydrolysates. The conversion process of xylan-rich hydrolysates to xylitol comprises of four steps: (i) acid hydrolysis of xylans, which results in monomeric sugar units (xylose, glucose, arabinose, mannose), (ii) Purification to recover or enrich xylose rich hydrolysate, (iii) hydrogenation of xylose to xylitol at high pressure and temperature in the presence of toxic metal catalysts (nickel, ruthenium) [15], (iv) purification and crystallization of produced xylitol [16].

The first step in the chemical method, acid hydrolysis of xylans, results in a mixture of pentoses and hexoses. The resulting hemicellulosic hydrolysate has to be purified to obtain pure xylose by removing other sugars and impurities. Currently, chromatographic techniques that are employed to remove inorganic salts and charged products [17] do not accomplish the complete separation or removal of other sugars. Consequently, xylitol yield is about 50–60% of the raw material used for the production process [6]. The significant drawbacks of the conventional chemical method of xylitol production involve extensive purification and raucous operating conditions, which are energy-intensive and waste generating. The aforementioned shortcomings of the chemical method motivated researchers to make a considerable effort in exploring alternative methods, and the microbial production of xylitol emerged to be an intriguing method.

9.2.2.2 Microbial Method of Xylitol Production

Being less expensive, microbial production of xylitol is considered as a potential alternative to the chemical production method. Unlike the chemical method, microbial production occurs under mild operating conditions and can work on heterogeneous hemicellulose hydrolysates without purification. The high-quality product and cost-effective steps project microbial processes to be a promising method for xylitol production. The comparison of factors involved in chemical and micro methods of xylitol production has been tabulated in Table 9.1.

Microbial conversion of xylose to xylitol can be accomplished by employing bacteria, fungi, and yeast. In addition to several studies focusing on the search of alternative raw materials as carbon sources, numerous species of microorganisms have also been evaluated for xylitol production by the microbial method. The fermentation conditions and production of xylitol by microorganisms such as bacteria, fungi, and yeast are tabulated in Table 9.2.

Table 9.1 Comparison of factors involved in chemical and microbial methods of xylitol production.

Parameters	Chemical method	Microbial method
Source	Pure xylose or xylose from lignocellulosic hydrolysates	Xylose from lignocellulosic hydrolysates does not require pure xylose
Process	Hydrogenation	Bioconversion by microorganisms (bacteria, yeast, and fungi)
Operating conditions	High temperature and pressure	Mild operating conditions
Catalyst	Nickel or ruthenium	No catalyst
Steps involved	Hydrolysis by acid treatment Purification of hydrolysate to remove impurities Hydrogenation of xylose Crystallization	Hydrolysis by acid or enzymatic treatment Detoxification Fermentation under optimal conditions Purification and crystallization
Energy required	High energy and laborious	Low energy

9.3 Modeling of Biotechnological Processes

Information required to analyze any biotechnological processes and predict their performance can be achieved by developing mathematical models. Many phenomena in biotechnological processes will be understood by writing kinetic equations that describe the growth of the microorganism, consumption of substrate, and enzyme responsible for the metabolism of substrate and formation of product. It is significant to obtain expertise in the area of process operation and the parameters involved in detecting the organism's behavior [31]. A suitable kinetic model of the process is essential in effective designing and scaling up the bioreactor. The mathematical description of various processes such as kinetics of biomass formation and product formation in the bioreactor and the steady-state behavior of

Table 9.2 Production of xylitol by microorganisms (bacteria, fungi, and yeast) and fermentation conditions.

Microorganism	Fermentation conditions	Initial xylose ($g\ L^{-1}$)	Xylitol (g L-1)	Yield ($g\ g^{-1}$)	Xylitol productivity ($g\ L^{-1}\ h^{-1}$)	References
Bacteria						
Enterobactorliquefaciens	30°C, pH – 7.0, 200 rpm, 96 h, shake flask	100	33.3	0.33	0.35	[18]
Corynebacterium sp. B-4247	30°C, pH – 7.2, 130 rpm, 50 h, shake flask	100	40	0.38	0.8	[19]
Gluconobacteroxydans	Enzymatic method	52.4 (Arabitol)	51.4	Molar yield (98%)	1.9	[20]
Mycobacterium smegmetis	Cell immobilization method	20 (Xylulose)	14.81	Conversion ratio 74.2%	0.61	[21]
Fungus						
Petromycesalbertensis	28°C, pH – 6.0 and 7.0, 200 rpm, 240 h, shake flask	100	39.8	0.4	0.166	[22]

(Continued)

Table 9.2 Production of xylitol by microorganisms (bacteria, fungi, and yeast) and fermentation conditions. (*Continued*)

Microorganism	Fermentation conditions	Initial xylose (g L^{-1})	Xylitol (g L-1)	Yield (g g^{-1})	Xylitol productivity (g L^{-1} h^{-1})	References
Penicillium crustosom	30°C, pH – 6.0, 200 rpm, 96 h, shake flask	11.5	0.52		0.005	[23]
Aspergillus niger	30°C, pH – 6.0, 200 rpm, 96 h, shake flask	11.5	0.36		0.004	[24]
Yeast						
Candida guilliermondi FTI-20037	30°C, pH – 7.0, 200 rpm, 78 h, shake flask	104	77.2	0.742	0.99	[24]
Candida tropicalis	30°C, pH – 5.0, 220 rpm, 78 h, shake flask	150	131	0.873	2.91	[25]
Candida boidinii NRRL Y-17213	30°C, pH – 5.5, 150 rpm, 336 h, reactor	150	53.1	0.354	0.16	[26]

(*Continued*)

Table 9.2 Production of xylitol by microorganisms (bacteria, fungi, and yeast) and fermentation conditions. (*Continued*)

Microorganism	**Fermentation conditions**	**Initial xylose ($g\ L^{-1}$)**	**Xylitol (g L-1)**	**Yield ($g\ g^{-1}$)**	**Xylitol productivity ($g\ L^{-1}\ h^{-1}$)**	**References**
Candida sp.559-9	30°C, pH – 5.0, 240 rpm, 120 h, shake flask	200	173	0.865	1.44	[27]
Pichia sp.	28°C, pH – 6.0, 250 rpm, 72 h, shake flask	40	25	0.625	0.5	[28]
Debaryomyceshansenii UFV-170	30°C, pH – 6.0, 300 rpm, 112 h, reactor (semi-aerobic)	100	76.6	0.73	0.68	[29]
Hansenula polymorpha	30°C, pH – 6.0, 150 rpm, 120 h, shake flask	125	58	0.62	0.6	[30]

substrate utilization and inhibition provides quantitative information in designing the process. Several studies have been reported on the development of models that describe the effects of various physical and operational parameters and media composition on the biotechnological production process.

9.3.1 Classification of Mathematical Models

Different bioprocess models are classified according to their probable biological explanation or their intricacy level. To make it simple, models of the fermentation process are classified based on the level of understanding and development method into two types: mechanistic and empirical models (Figure 9.1). In mechanistic models, the constituent part and working mechanism of the internal operation of a fermentation system are formulated by mathematical equations. The future system behavior can be predicted by developing mechanistic models which are developed based on deterministic principles [32]. Models that represent the relationship between input and output parameters in a data set without accounting for the underlying mechanism or basic principles are termed an empirical model [33]. But the prediction capability of the empirical model is confined to the range of data that are considered for developing the model [34]. Mechanistic models are preferable in the case when more knowledge about a process is available. Mechanistic models are developed for fermentation by writing equations for mass, heat, and momentum balances. The complex mechanism of a biological system is often modeled using the empirical equation, for example, the microbial growth kinetics is

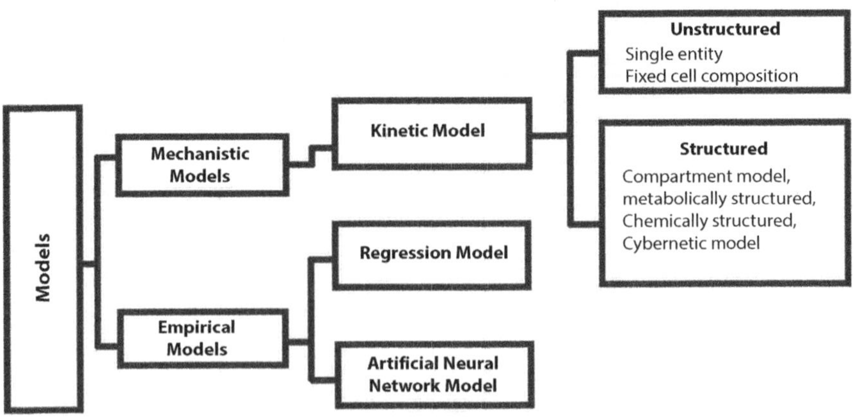

Figure 9.1 Classification of mathematical models of fermentation processes.

modeled by Monod expression. Mechanistic models are further classified into unstructured and structured kinetic models and the empirical models or data-driven models are classified into regression and neural network models as shown in Figure 9.1.

9.3.2 Modeling of Microbial Production of Xylitol

The effect of physical and operational parameters on the production of xylitol has been carried out by many research groups. The optimization of operational parameters such as pH, temperature, inoculum age and concentration, components of the media, and its effect on kinetic parameters were conducted in conical flasks [35–37]. Very few studies were conducted in a bioreactor to evaluate the effect of process parameters on reaction kinetics [38, 39]. However, to understand the biochemical reaction mechanism to enhance the xylitol production and the process design, the development of a kinetic model becomes an essential part.

9.3.2.1 Unstructured Modeling

The microbial growth, substrate consumption, and product formation represented by simple kinetic equations are termed as an unstructured model. Unstructured models are developed based on the postulation that the cell composition is constant, which is parallel to assume the existence of balanced growth [40, 41]. This assumption makes the model to be valid under steady-state and exponential phase culture and fails to describe the transient culture. Unstructured models are applicable only when the cell responds fast with change in process conditions and for small changes, which do not deviate much from the steady-state conditions [42]. Aranda *et al.* [43] have developed a stoichiometric model by considering the reaction rate and analyzing the metabolic fluxes of *Candida parapsilosis* under aerobic. Kinetic behavior of growth of the yeast *Candida parapsilosis ATCC 28474* and xylitol production in oxygen-limited conditions [44] were investigated and an unstructured model was developed to explain the kinetic behavior of the yeast. A model was developed to assess the growth in the presence of a co-substrate (glucose) to expedite the xylitol production [45] and a growth model was proposed to predict the oxygen transfer rate in the production process of xylitol by *Candida guilliermondii* [46].

The change in cell composition and the cell morphology with reference to environmental perturbations cannot be explained by unstructured models since these models consider the cell as a single entity and do not recognize the internal structure of the cell [47]. The ease in building up the unstructured models makes them more versatile in steady-state simulation studies.

A. Kinetic Models for Microbial Growth

Various models have been used for studying growth kinetics. The growth of cells is modeled by the simple Malthusian model, which is also known as exponential law [48]. The biomass concentration increases proportionally with the increase in time in a batch fermentation operated under ideal process conditions. Thereafter, all the cells in the reactor multiply with the same probability. The overall rate of biomass formation is directly proportional to the biomass and can be given as:

$$\frac{dX}{dt} = \mu X \tag{9.1}$$

where μ is the specific growth rate (h^{-1}), X is the cell dry weight concentration ($g\ L^{-1}$) and t is the fermentation time (h). The increase in biomass concentration is given by:

$$X = X_0 e^{\mu t} \tag{9.2}$$

Table 9.3 Different kinetic model equations for microbial growth.

Model	**Mathematical representation**
Monod model	$\mu = \frac{\mu_{max} S}{K_s + S}$
Logistic model	$\frac{dX}{dt} = \mu_{max} X \left(1 - \frac{X}{X_{max}}\right)$
Gompertz model	$\frac{dX}{dt} = \mu_{max} X \ \ln\left(\frac{X_{max}}{X}\right)$

S – substrate concentration ($g\ L^{-1}$), saturation constant ($g\ L^{-1}$), X – Cell dry weight concentration ($g\ L^{-1}$), μ_{max} – Maximum specific growth rate (h^{-1}) and X_{max} – Maximum cell dry weight concentration ($g\ L^{-1}$).

The drawback of Malthius model is that it relates to biomass concentration and ignores the substrate required for the growth. Other unstructured models that relate substrate and cell concentration are Contois, Moser, Tessier, and Verlhurst models [49]. Another model which incorporates substrate inhibition was proposed by Andrew [50].

The growth of microorganisms can be evaluated by the most commonly used unstructured models - Monod, Logistic and Gompertz models and the equations for these models are tabulated in Table 9.3 [50].

B. Kinetic Models for Substrate Utilization

The kinetics of substrate utilization can be given as follows:

$$\frac{ds}{dt} = -\frac{1}{Y_{X/S}}\frac{dX}{dt} - \frac{1}{Y_{P/S}}\frac{dP}{dt} + m_s\frac{dX}{dt} \tag{9.3}$$

where $Y_{X/S}$ and $Y_{P/S}$ represent the yield coefficient of biomass and product formed with respect to substrate consumption and m_s represents the maintenance coefficient accredited to the maintenance functions of the cell such as motility and maintenance of concentration gradient.

C. Kinetic Models of Product Formation

Product formation kinetics can be written based on the stoichiometric connection between substrate utilization and product formation or biomass formation. The substrate concentration decreases as the fermentation progresses and the rate of product formation parallels the rate of cell growth. Luedeking-Piret model (Equation 9.4) is the commonly used model to describe product formation and can explain both the growth and non-growth associated product formation.

$$\frac{dP}{dt} = \alpha\frac{dX}{dt} + \beta \tag{9.4}$$

9.3.2.2 *Structured Modeling*

Unlike unstructured models, the structured model divides the cell into different compartments [51]. The model could predict cell behavior and the change in cell composition with respect to the change in the external environment. The structured model would be able to describe the complete growth kinetics of the cell as this model separates the cell into different compartments that are distinct in their physical or chemical nature [52]. In addition to mass and energy balance equations, the structured model constitutes of stoichiometric equations and is able to explain the change in the composition of one compartment with reference to another compartment and change in compartment composition with respect to the change in process conditions [53]. A stoichiometric model deliberating the metabolic fluxes involved in xylitol production under microaerobic conditions was developed. In order to ascertain a stoichiometric reaction rate analysis of the network, the metabolic reactions in xylitol production by *Candida parapsilosis* were incorporated. It was shown that the developed stoichiometric model predicts the xylitol yield in several xylose-assimilating yeasts with a mean error of 12.8%.

Compared with the unstructured model, the predicting capability and extrapolation capacity of structured models are superior and can be applied in broader domains. The information about the physiological state of the cell, its composition, and regulatory adaptation to the environment can be provided by developing structured models [54]. The structured model may be a multi-compartment or chemically structured or genetically structured model [55, 56].

9.3.2.3 Cybernetic Modeling

Microorganisms require several nutrients for growth and maintenance and the media components supply this nutritional requirement of the cell. The behavior of microorganisms in the presence of multi substrates can be addressed by accounting for the regulatory mechanism of the cells. A cybernetic modeling approach can be employed by considering the optimal nature of the regulatory process of microorganisms. The cybernetic or goal-seeking approach was used to describe the metabolic regulation of the cell by applying the optimal behavior of microorganisms when cultured in a substitutable carbon source [57]. This dynamic model comprises of a set of ordinary differential equations that could describe the enzyme levels and concentration of metabolites within the cell and its external environment [58]. This model aids a simple way to identify the control structure of complex biochemical pathways or reactions involved in the cell. Cybernetic models assist in the prediction of microbial response when the growth rate is restricted by one or more carbon substrates [59] and are also focused on metabolite production [60]. The cellular regulatory processes of induction or repression and inhibition or activation are represented by cybernetic variables u_i and v_i. The inclusion of cybernetic variables allows the model to predict the profile of biomass, metabolite concentration, and enzyme levels during fermentation. Cybernetic models are initially applied to predict the diauxic growth patterns in the presence of multiple substrates. Cybernetic models are also helpful in predicting the dynamic cell response in reference to environmental changes [61]. A cybernetic mathematical model examined the diauxic growth pattern of *Debaryomycesnepalensis* on mixed substrates in batch culture [62], which could effectively portray the diauxic growth, sequential utilization of substrates, and production of xylitol under the different ratio of co-substrate.

Due to many assumptions made during the development of mechanistic models, some constraints and shortcomings exist that limit the use of mechanistic models. Empirical models such as statistical regression techniques and artificial neural networks aid in establishing input-output parameter relationships and also in process optimization.

9.3.3 Optimization Techniques in Biotechnological Processes – Xylitol Production as a Case Study

The efficiency of the fermentation process can be improved significantly by optimizing the media components and process conditions using techniques that help identify significant factors and determine the optimal value of critical process controlling factors. The optimization process obtains the desired response or output with acceptable variations [83]. Conventionally, optimization of fermentation parameters was done by traditional approaches that were time-consuming, expensive, and involved many experiments with limited precision. Nonetheless, with the introduction of statistical and contemporary mathematical methods, optimization of the fermentation process has become more effective and economical, as statistical methods involve relatively fewer experiments. In order to obtain the effect of factors on output response, the significant factors have to be selected and varied in a determined pattern followed by an investigation of experimental results.

Experimental design can be categorized into two types based on the number of factors considered for investigations: (i) one-factor-at-a-time and (ii) Multifactor design (factorial design). Designing the experiments is of utmost importance in optimizing the productivity and yield of a fermentation process, as any microbial fermentation process is influenced by several factors such as organism of choice, carbon and nitrogen substrates, inorganic nutrients, and process operational conditions. The selection of an appropriate experimental design helps identify the significant factors,

Table 9.4 One-factor-at a time design strategy for improving the productivity and xylitol yield using pure substrates and hemicellulosic hydrolysates.

S. no.	Organism used	Media	Fermentation conditions	Mode of operation	Design	Analysis	Software used	Independent variables (parameter optimized)	Dependent variable (optimum values)	References
1	*Candida tropicalis* ATCC 13803	Semi-synthetic	30°C, 500 rpm, 1.0 vvm	Batch reactor, Total volume -3.5 L, Working volume – 1.0 L	One factor at a time	-	-	Batch: Xylose – 100 g L^{-1}; Fed-batch: Fermentation time – 14 h, Dilution rate – 0.01 h^{-1}; Feed -119 g L^{-1}.	Xylitol – 187 gL^{-1}; $Y_{P/S}$ – 0.75, Q_p- 3.9	[63]
2	*Candida tropicalis* CICC 1779	Cotton stalk HCH (Detoxified)	30°C, 200 rpm,	Batch, Shake flask (100 mL)	One factor at a time, RSM	RA	Design Expert (8.0), Origin 7.0, DPS – 9.0	Reducing sugar concentration – 94 g L^{-1}, Inoculum ratio – 5.7% (w/v), initial pH – 5.1, Fermentation time, Agitation rate	Xylitol – 12.73 g L^{-1}	[64]
3	*Candida tropicalis*	Synthetic xylose & corncob HCH	30°C, pH – 4.5, 400 rpm, 0.7 vvm	Total volume – 10 L, Working volume – 5 L	One factor at a time	-	-	Treatment temperature – 30°C, activated charcoal – 15 g L^{-1}, xylitol saturation concentration – 10 times the original concentration (637 g L^{-1}), crystallization temperature – 20°C followed by 8°C	Xylitol crystallization yield – 76.20%	[65]

study the effect of various factors on the process response, and aids in yield enhancement.

9.3.3.1 One-Factor-at-a-Time Design

One-factor-at-a-time (OFAT) design is a classical optimization technique that examines only one factor or variable at a time, whereas the other factors are maintained constant level. In order to study the effect of a factor on the output response, the factor to be inspected is altered over the desired range. After conducting the experiments, the responses are analyzed graphically or statistically to elaborate how the factor in concern affects the output of the process. OFAT has been applied widely in the optimization of fermentation parameters since it is easy to design and convenient to conduct experiments. It is used in the initial stages of medium formulation and process optimization in developing the fermentation process and also in diverse fields of science and engineering. Nevertheless, this design method has some drawbacks: (i) the interaction effect of factors is not considered, thus the optimal values determined by this method may not be the original optimal value; (ii) a large number of experiments are involved in this method, which makes it time-consuming and laborious. Several studies have employed one-factor at a time design for improvement of productivity and xylitol yield using pure substrate and hemicellulosic hydrolysate (Table 9.4).

9.3.3.2 Design of Experiments

Design of experiments (DOE) or statistical methods helps overcome the shortcomings of OFAT design. DOE can be used to study the interaction effect of factors on the process response. A basic theory of experimental design by Fisher exhibited that varying the level of many components will be more effective than the classical optimization method. The experimental design consists of a set of strategically planned and conducted experiments to obtain information about the interaction effect of more than one factor. The optimization procedure of the experimental design begins with a screening of the most significant factors and subsequent analysis. The statistical technique improves optimization efficiency and reduces experimental time and labor cost. Factorial designs assist in evaluating the effects of many factors at more than one level. A uniform framework is presented by the coded levels of the factor in a way to investigate the interaction effect in any experimental state, whereas the actual level of the factor depends on the factor in the choice to be examined. Model fitting and analysis for a factorial design can be implemented based on a coded level, making comparison easier as the resultant model coefficients are dimensionless. Factorial designs can be classified into two types: (i) Full factorial design (ii) Fractional factorial design.

A. Full Factorial Design

A full factorial design accounts for testing all the combinations of each factor level. In this design, the number of experiments for 'n' factors at 'm' levels is given by m^n. The full factorial design is mostly applied at two levels (+1(high level) and -1(low level)) for 'n' factors and the design is denoted by 2^n. An appropriate polynomial model can be used to study the effect of the factors on the output response. As the number of factors increases, the number of runs increases geometrically. For instance, when many number of factors are to be investigated simultaneously, this design results in a large number of experiment runs which makes full factorial design practically and economically not feasible.

B. Fractional Factorial Design

As the number of experimental runs is relatively large for a full factorial design, detailed information about the effect of factors can be attained by conducting only a fraction of experiments in the full factorial design. The fractional factorial design becomes an alternative choice of design when the number of runs is too high to be conducted practically. The following fractional factorial designs discussed in detail are widely applied to optimize the fermentation process.

C. Taguchi Design

Taguchi design, a partial/fractional design employs an orthogonal array which aids in evaluating the effect of many factors at two or more levels on the output response with a relatively lesser number of runs. The Taguchi's method was designed so that the performance of the design is less sensitive to change in the uncontrolled variable. It is applicable in many optimization problems due to its distribution-free and orthogonal array design. In addition to this, it helps determine the quality of a factor when it deviates from its target value. It proposes offline control features at three stages: system strategy, parameter design, and tolerance design phase. The Taguchi's design reduces experimental time and cost, thus resulting in improvement in the quality. The shortcomings associated with Taguchi designs are, (i) Orthogonal array designs are few in number and fail to deal with all the interaction effects within the proposed designs, (ii) It can provide solutions for multi-response optimization problems only based on judgmental knowledge.

Table 9.5 Statistical optimization strategies for improving the productivity and xylitol yield using pure substrates and hemicellulosic hydrolysates.

S. no.	Organism used	Media	Fermentation conditions	Mode of operation	Design	Analysis	Software used	Independent variables (parameter optimized)	Dependent variable (optimum values)	References
1	*Candida guilliermondii* FTI 20037	Rice straw HCH (Non-detoxified)	30°C, 200 rpm, 72 h, semi aerobic	Batch, Shake flask (50 mL)	2^3FD	RA, RSM	Statistica	Rice bran, Ammonium sulfate, $CaCl_2$	$Y_{P/S}$ – 0.68, Q_P- 0.54	[67]
2	*Candida guilliermondii*	Rice straw HCH (Non-detoxified)	30°C, 200 rpm, 94 h	Batch, Shake flask (50 mL)	2^{4-1}FD	RSM	Stat Graphics Program (v 6.0)	Xylose concentration – 60 g L^{-1}, Agitation speed – 200 rpm, Inoculum level – 5 g L^{-1}, in absence of nutrients	$Y_{P/S}$ – 0.65, Q_P- 0.66	[68]
3	*Candida guilliermondii* FTI 20037	Rice straw HCH (Non-detoxified)	30°C, 250 rpm, 94 h	Batch, Shake flask (50 mL)	2^2FD (CCD – 5 coded levels)	RSM	Design Expert (5.0), Statistica (5.0)	Xylose concentration – 82 g L-1, Inoculum level – 3 g L-1	Xylitol – 52 g L^{-1}; $Y_{P/S}$ – 0.85	[67]

(*Continued*)

Table 9.5 Statistical optimization strategies for improving the productivity and xylitol yield using pure substrates and hemicellulosic hydrolysates. (*Continued*)

S. no	Organism used	Media	Fermentation conditions	Mode of operation	Design	Analysis	Software used	Independent variables (parameter optimized)	Dependent variable (optimum values)	References
4	*Debaryomyceshansenii* NRRL Y-7426	Rice straw HCH(Detoxified)	30°C, 250 rpm, 72 h, pH – 5.5, semi – aerobic	Batch reactor, Total volume – 3.0 L, Working volume – 1.0 L	3^4 full FD	RA	Design Expert (6.0), Statistica (6.0)	Initial xylose concentration – 71 g L^{-1}, Oxygen mass flow rate – 4.1 mg O_2/s	Simultaneous maxima: Xylitol – 42.2 g L^{-1}; $Y_{P/S}$ – 0.71; Q_P- 0.53	[69]
5	*Candida guilliermondii* FTI 20037	HCH	30°C, 200 rpm, 116 h	Batch, Shake flask (50 mL)	2^4FD	RA	Statistica (5.0)	pH – 2.0, Agitation – 150 rpm, Contact time – 60 min, Temperature – 45°C	Lignin degradation rate – 25.8 %, Color removal – 48.9%, Xylitol – 66 gL^{-1}; $Y_{P/S}$ – 0.71; Q_P- 0.53	[70]
6	*Candida guilliermondii* FTI 20037	Rice straw HCH (Detoxified)	30°C, 200 rpm	Batch, Shake flask (50 mL)	2^3FD	RA	Design Expert (5.0), Statistica (5.0)	pH – 7.5, Initial xylose concentration – 70 g L^{-1}, Inoculum level – 3 g L^{-1}	Xylitol – 43.5 g L^{-1}; $Y_{P/S}$ – 0.74	[71]

(*Continued*)

Table 9.5 Statistical optimization strategies for improving the productivity and xylitol yield using pure substrates and hemicellulosic hydrolysates. (*Continued*)

S. no	Organism used	Media	Fermentation conditions	Mode of operation	Design	Analysis	Software used	Independent variables (parameter optimized)	Dependent variable (optimum values)	References
7	*Candida guilliermondii* FTI 20037	Brewer's spent grain hydrolysate	30°C, 200 rpm, 96 h	Batch, Shake flask (100 mL)	2^2FD (CCD)	RA	Design Expert (5.0), Statistica (5.0)	Initial xylose concentration - 70 g L^{-1}, Non-supplemented with nutrient	$Y_{P/S}$ - 0.78, Q_p- 0.58	[72]
8	*Candida tropicalis* W103	Corncob HCH	35°C, 500 rpm, 0.6 vvm, 96 h	Batch reactor, Total volume - 1.0 L, Working volume - 0.75 L	-	-	-	pH - 6.0, Acetic acid low	Xylitol - 68.4 g L^{-1}; $Y_{P/S}$ - 0.70, Q_p- 0.95	[73]
9	*Candida guilliermondii* FTI 20037	Sugarcane HCH	30°C	Fluidized bed reactor (FBR), Total volume -2.0L	2^3FD	Multi response analysis		Air flow - 200 ml min^{-1}, Fluidization flux - 57 L h^{-1}, Concentration factor - 3.7	$Y_{P/S}$ - 0.70, Q_p- 0.44	[74]

(*Continued*)

Table 9.5 Statistical optimization strategies for improving the productivity and xylitol yield using pure substrates and hemicellulosic hydrolysates. (*Continued*)

S. no	Organism used	Media	Fermentation conditions	Mode of operation	Design	Analysis	Software used	Independent variables (parameter optimized)	Dependent variable (optimum values)	References
10	*Candida tropicalis* HDY - 02	Corncob HCH (Detoxified)	35°C, 200 rpm	Batch, Shake flask (100 mL)	PB design followed by CCD	RA	Design Expert (7.02)	$(NH_4)_2SO_4$ - 5 g L^{-1}, KH_2PO_4 - 1.3 g L^{-1}, Yeast extract - 4.6 g L^{-1}, $MgSO_4 \cdot 7H_2O$ - 0.6 g L^{-1}	Xylitol - 58 g L^{-1}; $Y_{P/S}$ - 0.73	[75]
11	*Pachysolintannophilus* MTCC 1077	Corncob HCH (Detoxified)	pH - 7.0, 30°C, 200 rpm, 48 h	Batch, Shake flask (100 mL)	PB design followed by (i) Box - Behnken - Full FD (ii) CCD - full FD	RA	Design Expert, Minitab	(i) Box- Behnken: Peptone - 6.03 g L^{-1}, Xylose - 10.62 g L^{-1}, Yeast extract - 4.66 g L^{-1}, $MgSO_4.7H_2O$ - 1.39 g L^{-1}; (ii) CCD: Temperature - 36.56°C, pH - 7.27, Agitation - 194.44 rpm, inoculum size - 3.69 ml, Substrate concentration - 3.55 g L^{-1}.	$Y_{P/S}$ - 0.80	[76]

(*Continued*)

Table 9.5 Statistical optimization strategies for improving the productivity and xylitol yield using pure substrates and hemicellulosic hydrolysates. (*Continued*)

S. no	Organism used	Media	Fermentation conditions	Mode of operation	Design	Analysis	Software used	Independent variables (parameter optimized)	Dependent variable (optimum values)	References
12	*Candida mogii* ATCC 18364	Semi-synthetic	30°C, 200 rpm, 96 h	Batch reactor, Total volume -5.0 L, Working volume – 3.0 L	Genetic algorithm coupled neural network	-	-	$(NH_4)_2SO_4$ – 3.92 g L^{-1}, $(NH_4)_2HPO_4$ – 0.36 g L^{-1}, KH_2PO_4 – 6.59 g L^{-1}, Yeast extract – 4.6 g L^{-1}, Peptone – 1.46 g L^{-1}, $MgSO_4.7H_2O$ – 0.49 g L^{-1}	$Y_{P/S}$ – 0.646	[77]
13	*Debaryomyceshansenii* UFV – 170	Semi synthetic	30°C, 200 rpm,	Batch, Shake flask (100 mL)	3^3 full FD	RA	Design Expert (6.0.10) – trial version, Statistica (5.0)	Initial substrate concentration – 156 g L-1, Initial biomass concentration – 6.4 g L-1, Agitation – 280 rpm	Xylitol – 116.25 gL^{-1}; $Y_{P/S}$ – 0.77, Q_p- 1.49	[78]

(*Continued*)

Table 9.5 Statistical optimization strategies for improving the productivity and xylitol yield using pure substrates and hemicellulosic hydrolysates. (*Continued*)

S. no	Organism used	Media	Fermentation conditions	Mode of operation	Design	Analysis	Software used	Independent variables (parameter optimized)	Dependent variable (optimum values)	References
14	*Candida sp.*	Corn steep liquor	30°C, 200 rpm	Batch, Shake flask	Taguchi design (L18) (2^1 X 3^7) Orthogonal array	-	-	Temperature - 33°C, Agitation - 250 rpm, Corn steep liquor (% w/v) - 2.0, Xylose (% w/v) - 2.0, KH_2PO_4 (% w/v) - 0.5	Yield - 78.9%	[79]
15	*Debaryomycesnepalensis* NCYC 3413	Semi synthetic	30°C, 180 rpm	Shake flask	PB followed by CCD 2^3 full FD		Design Expert (8.0)	PB: Xylose - 100 g L^{-1}, K_2HPO_4 - 10.6 g L^{-1}, $ZnSO_4$ - 8.9 mg L^{-1}	Xylitol - 36 gL^{-1};	[80]
16	*Candida guilliermondii*	Semi synthetic	30°C, 200 rpm, 96 h	Batch, Shake flask (100 mL)	2^2 factorial design (CCD)	RA	Design Expert	Vanillin - 2.0 g L^{-1}	$Y_{P/S}$ - 0.84, Q_p- 0.74	[81]
17	*Debaryomycesnepalensis*	Synthetic xylose	30°C	Batch, Shake flask (100 mL)	Uniform Design,	Multi response analysis	-	pH, agitation, aeration	Xylitol - 116.25 gL^{-1}; $Y_{P/S}$ - 0.77, Q_p- 1.49	[82]

D. Central Composite and Box-Behnken Design

Instead of using a three-level full factorial design, CCD and Box-Behnken are widely applied DOE for developing a quadratic (second-order) model equation. Box and Wilson were the first to describe the central composite design (CCD) [66]. It was initially developed as a five-level fractional factorial design. It comprises of 2^n factorial design representing two levels of the factors represented as +1 and -1, center points as zero, which is the median values of factors in consideration, and star points represented as $+\alpha$ and $-\alpha$, which takes the value above and below the median value of the factorial levels. Center points are often replicated to improve the measurement precision, and star points in the design are double the number of factors considered in the DOE. Based on the factor levels and set of experiments, CCD is divided into three types: Face centered CCD (CCF), (ii) Inscribed CCD (CCI), and (iii) Circumcentered CCD (CCC). In the case of Box-Behnken design, three levels of factors are considered for the study. It can be counted as an incomplete block design except for specific factors randomly added around all combinations, while the other factors remain at their center levels.

A second-order polynomial model is typically projected to describe the effects of factors on the output response with respect to the experimental results from CCD or Box-Behnken design. A 3D surface plot or contour plot based on the estimated second-order polynomial model can be plotted by varying the levels of two factors while keeping the level of the other factor constant. It visually represents the response over the domain range, and the model equation aids in estimating optimal response. The factors that significantly affect the output response can be determined by performing an analysis of variance (ANOVA) for the model equation.

Several statistical approaches have optimized medium components and process variables affecting xylitol production. Various optimization procedures discussed in the literature are summarized in Table 9.5. One-factor-at-a-time design was used to optimize the production of xylitol by the fed-batch method and enzymatic method. The fractional factorial design was widely employed in optimizing the medium components, operational parameters in the production of xylitol, and also in the crystallization of xylitol.

9.4 Conclusions and Perspectives

Though microorganisms' biochemistry of xylitol production is well established, the enhancement of large-scale productivity requires skillful fermentation modeling and optimization. Modeling the kinetic behavior of the microorganisms such as their growth, metabolic activities, response to environmental stress, and product formation aids in understanding the underlying mechanism of product formation. On the other hand, the optimization techniques maximize the titer values, and the modern strategies namely artificial neural network approach and hybrid approach would provide a better prediction of yield and productivity at different operating conditions. The optimization of various process parameters affecting xylitol production and co-production of other high-value chemicals might reduce environmental impact and enhance overall process economics. The aforementioned optimization techniques and modeling strategies would give an insight on which approach would be well-suited for a particular process of interest.

References

1. Fan, A.M., L. O. Nabors, R. C. Gelardi, (eds), Alternative sweeteners. Marcel Dekker, New York. *J. Appl. Toxicol.*, 7, 150–150, 1987. https://doi.org/10.1002/jat.2550070216.
2. Nayak, P.A., Nayak, U.A., Khandelwal, V., The effect of xylitol on dental caries and oral flora. *Clin. Cosmet. Investig. Dent.*, 6, 89–94, 2014. https://doi.org/10.2147/CCIDE.S55761.
3. Mussatto, S.I. and Roberto, I.C., Xylitol production from high xylose concentration: Evaluation of the fermentation in bioreactor under different stirring rates. *J. Appl. Microbiol.*, 95, 331–337, 2003. https://doi.org/10.1046/j.1365-2672.2003.01990.x.
4. Parajó, J.C., Domínguez, H., Domínguez, J., Biotechnological production of xylitol. Part 2: Operation in culture media made with commercial sugars. *Bioresour. Technol.*, 65, 203–212, 1998. https://doi.org/10.1016/S0960-8524(98)00036-4.
5. Rafiqul, I.S.M. and Sakinah, A.M.M., Processes for the production of Xylitol — A review. *Food Rev. Int.*, 29, 127–156, 2013. https://doi.org/10.1080/87559129.2012.714434.
6. Nigam, P. and Singh, D., Processes of fermentative production of xylitol — A sugar substitute. *Process Biochem.*, 30, 117–124, 1995. https://doi.org/10.1016/0032-9592(95)80001-8.
7. Hassinger, W., Sauer, G., Cordes, U., Krause, U., Beyer, J., Baessler, K.H., The effects of equal caloric amounts of xylitol, sucrose and starch on insulin requirements and blood glucose levels in insulin-dependent diabetics. *Diabetologia*, 21, 37–40, 1981. https://doi.org/10.1007/BF03216221.
8. Ylikahri, R., Metabolic and nutritional aspects of xylitol. *Adv. Food Res.*, 25, 159–180, 1979.

9. Danhauer, J.L., Kelly, A., Johnson, C.E., Is mother-child transmission a possible vehicle for xylitol prophylaxis in acute otitis media? *Int. J. Audiol.*, 50, 661–672, 2011. https://doi.org/10.3109/14992027.2011.590824.
10. Mattila, P.T., Kangasmaa, H., Knuuttila, M.L.E., The effect of a simultaneous dietary administration of xylitol and ethanol on bone resorption. *Metabolism*, 54, 548–551, 2005. https://doi.org/10.1016/j.metabol.2004.11.011.
11. Elamin, K., Sjöström, J., Jansson, H., Swenson, J., Calorimetric and relaxation properties of xylitol-water mixtures. *J. Chem. Phys.*, 136, 104508, 2012. https://doi.org/10.1063/1.3692609.
12. Islam, M.S., Effects of xylitol as a sugar substitute on diabetes-related parameters in nondiabetic rats. *J. Med. Food*, 14, 505–511, 2011. https://doi.org/10.1089/jmf.2010.0015.
13. Ravella, S.R., Gallagher, J., Fish, S., Prakasham, R.S., Overview on commercial production of xylitol, economic analysis and market trends, in: *D-Xylitol*, S.S. da Silva, and A.K. Chandel (Eds.), pp. 291–306, Springer Berlin Heidelberg, Berlin, Heidelberg, 2012.
14. da Silva, S.S. and Chandel, A.K., *D-Xylitol: Fermentative Production, Application and Commercialization*, Springer, Berlin, New York, 2012.
15. Chen, X., Jiang, Z.-H., Chen, S., Qin, W., Microbial and bioconversion production of D-xylitol and its detection and application. *Int. J. Biol. Sci.*, 6, 834–844, 2010.
16. Winkelhausen, E. and Kuzmanova, S., Microbial conversion of D-xylose to xylitol. *J. Ferment. Bioeng.*, 86, 1–14, 1998. https://doi.org/10.1016/S0922-338X(98)80026-3.
17. de Mancilha, I.M. and Karim, M.N., Evaluation of ion exchange resins for removal of inhibitory compounds from corn stover hydrolyzate for xylitol fermentation. *Biotechnol. Prog.*, 19, 1837–1841, 2003. https://doi.org/10.1021/bp034069x.
18. Yoshitake, J., Ohiwa, H., Shimamura, M., Imai, T., Production of polyalcohol by a *Corynebacterium* sp.: Part I. Production of pentitol from aldopentose. *Agric. Biol. Chem.*, 35, 905–911, 1971. https://doi.org/10.1080/00021369.1971.10860014.
19. Rangaswamy, S. and Agblevor, F., Screening of facultative anaerobic bacteria utilizing D-xylose for xylitol production. *Appl. Microbiol. Biotechnol.*, 60, 88–93, 2002. https://doi.org/10.1007/s00253-002-1067-8.
20. Suzuki, S., Sugiyama, M., Mihara, Y., Hashiguchi, K., Yokozeki, K., Novel enzymatic method for the production of xylitol from D-arabitol by *gluconobacter oxydans*. *Biosci. Biotechnol. Biochem.*, 66, 2614–2620, 2002. https://doi.org/10.1271/bbb.66.2614.
21. Izumori, K. and Tuzaki, K., Production of xylitol from D-xylulose by mycobacterium smegmatis. *J. Ferment. Technol.*, 66, 33–36, 1988. https://doi.org/10.1016/0385-6380(88)90126-4.

22. Dahiya, J.S., Xylitol production by *Petromyces albertensis* grown on medium containing D-xylose. *Can. J. Microbiol.*, 37, 14–18, 1991. https://doi.org/10.1139/m91-003.
23. Sampaio, F.C., da Silveira, W.B., Chaves-Alves, V.M., Passos, F.M.L., Coelho, J.L.C., Screening of filamentous fungi for production of xylitol from D-xylose. *Braz. J. Microbiol.*, 34, 321–324, 2003. https://doi.org/10.1590/S1517-83822003000400007.
24. Barbosa, M.F.S., Medeiros, M.B., Mancilha, I.M., Schneider, H., Lee, H., Screening of yeasts for production of xylitol from D-xylose and some factors which affect xylitol yield in *Candida guilliermondii*. *J. Ind. Microbiol.*, 3, 241–251, 1988. https://doi.org/10.1007/BF01569582.
25. Oh, D.K. and Kim, S.Y., Increase of xylitol yield by feeding xylose and glucose in Candida tropicalis. *Appl. Microbiol. Biotechnol.*, 50, 419–425, 1998.
26. Vandeska, E., Amartey, S., Kuzmanova, S., Jeffries, T., Effects of environmental conditions on production of xylitol by *Candida boidinii*. *World J. Microbiol. Biotechnol.*, 11, 213–218, 1995. https://doi.org/10.1007/BF00704652.
27. Ikeuchi, T., Kiritani, R., Azuma, M., Ooshima, H., Effect of D-glucose on induction of xylose reductase and xylitol dehydrogenase in *Candida tropicalis* in the presence of NaCl. *J. Basic Microbiol.*, 40, 167–175, 2000. https://doi.org/10.1002/1521-4028(200007)40:3<167::AID-JOBM167>3.0. CO;2-Y.
28. Rao, R.S., Bhadra, B., Shivaji, S., Isolation and characterization of xylitol-producing yeasts from the gut of colleopteran insects. *Curr. Microbiol.*, 55, 441–446, 2007. https://doi.org/10.1007/s00284-007-9005-8.
29. Sampaio, F.C., Torre, P., Lopes-Passos, F.M., Perego, P., Vieira-Passos, F.J., Converti, A., Xylose metabolism in *debaryomyces hansenii* UFV-170. Effect of the specific oxygen uptake rate. *Biotechnol. Prog.*, 20, 1641–1650, 2004. https://doi.org/10.1021/bp049691j.
30. Suryadi, H., Katsuragi, T., Yoshida, N., Suzuki, S., Tani, Y., Polyol production by culture of methanol-utilizing yeast. *J. Biosci. Bioeng.*, 89, 236–240, 2000. https://doi.org/10.1016/S1389-1723(00)88825-8.
31. Okpokwasili, G.C. and Nweke, C.O., Microbial growth and substrate utilization kinetics. *Afr. J. Biotechnol.*, 5, 305–317, 2005.
32. Gernaey, K.V., Lantz, A.E., Tufvesson, P., Woodley, J.M., Sin, G., Application of mechanistic models to fermentation and biocatalysis for next-generation processes. *Trends Biotechnol.*, 28, 346–354, 2010. https://doi.org/10.1016/j.tibtech.2010.03.006.
33. Ödman, P., Johansen, C.L., Olsson, L., Gernaey, K.V., Lantz, A.E., On-line estimation of biomass, glucose and ethanol in Saccharomyces cerevisiae cultivations using *in-situ* multi-wavelength fluorescence and software sensors. *J. Biotechnol.*, 144, 102–112, 2009. https://doi.org/10.1016/j.jbiotec.2009.08.018.
34. Petrides, D., Cooney, C.L., Evans, L.B., Field, R.P., Snoswell, M., Bioprocess simulation: An integrated approach to process development. *Comput. Chem. Eng.*, 13, 553–561, 1989. https://doi.org/10.1016/0098-1354(89)85038-0.

35. Converti, A. and Domínguez, J.M., Influence of temperature and pH on xylitol production from xylose by *Debaryomyces hansenii*. *Biotechnol. Bioeng.*, 75, 39–45, 2001. https://doi.org/10.1002/bit.1162.
36. Sampaio, F.C., Chaves-Alves, V.M., Converti, A., Lopes Passos, F.M., Cavalcante Coelho, J.L., Influence of cultivation conditions on xylose-to-xylitol bioconversion by a new isolate of *Debaryomyces hansenii*. *Bioresour. Technol.*, 99, 502–508, 2008. https://doi.org/10.1016/j.biortech.2007.01.017.
37. Walther, T., Hensirisak, P., Agblevor, F.A., The influence of aeration and hemicellulosic sugars on xylitol production by *Candida tropicalis*. *Bioresour. Technol.*, 76, 213–220, 2001. https://doi.org/10.1016/S0960-8524(00)00113-9.
38. Silva, C.J.S.M., Mussatto, S.I., Roberto, I.C., Study of xylitol production by *Candida guilliermondii* on a bench bioreactor. *J. Food Eng.*, 75, 115–119, 2006. https://doi.org/10.1016/j.jfoodeng.2005.04.001.
39. Silva, S.S., Roberto, I.C., Felipe, M.G.A., Mancilha, I.M., Batch fermentation of xylose for xylitol production in stirred tank bioreactor. *Process Biochem.*, 31, 549–553, 1996. https://doi.org/10.1016/S0032-9592(96)00002-7.
40. Thilakavathi, M., Basak, T., Panda, T., Modeling of enzyme production kinetics. *Appl. Microbiol. Biotechnol.*, 73, 991–1007, 2006. https://doi.org/10.1007/s00253-006-0667-0.
41. Weiss, R.M. and Ollis, D.F., Extracellular microbial polysaccharides. I. Substrate, biomass, and product kinetic equations for batch xanthan gum fermentation. *Biotechnol. Bioeng.*, 22, 859–873, 1980. https://doi.org/10.1002/bit.260220410.
42. Amrane, A. and Prigent, Y., Mathematical model for lactic acid production from lactose in batch culture: Model development and simulation. *J. Chem. Technol. Biotechnol.*, 60, 241–246, 1994. https://doi.org/10.1002/jctb.280600303.
43. Aranda-Barradas, J.S., Garibay-Orijel, C., Badillo-Corona, J.A., Salgado-Manjarrez, E., A stoichiometric analysis of biological xylitol production. *Biochem. Eng. J.*, 50, 1–9, 2010. https://doi.org/10.1016/j.bej.2009.10.023.
44. Aranda-Barradas, J.S., Delia, M.L., Riba, J.P., Kinetic study and modelling of the xylitol production using *Candida parapsilosis* in oxygen-limited culture conditions. *Bioprocess Eng.*, 22, 0219–0225, 2000. https://doi.org/10.1007/s004490050723.
45. Tochampa, W., Sirisansaneeyakul, S., Vanichsriratana, W., Srinophakun, P., Bakker, H.H.C., Chisti, Y., A model of xylitol production by the yeast *Candida mogii*. *Bioprocess Biosyst. Eng.*, 28, 175–183, 2005. https://doi.org/10.1007/s00449-005-0025-0.
46. Aguiar, W.B., Faria, L.F.F., Couto, M.A.P.G., Araujo, O.Q.F., Pereira, N., Growth model and prediction of oxygen transfer rate for xylitol production from d-xylose by C. guilliermondii. *Biochem. Eng. J.*, 12, 49–59, 2002. https://doi.org/10.1016/S1369-703X(02)00040-2.
47. Nielsen, J. and Villadsen, J., Modelling of microbial kinetics. *Chem. Eng. Sci.*, 47, 4225–4270, 1992. https://doi.org/10.1016/0009-2509(92)85104-J.

48. Tsuchiya, H.M., Fredrickson, A.G., Aris, R., Dynamics of microbial cell populations, in: *Advances in Chemical Engineering*, pp. 125–206, Elsevier, Academic press, New York, 1966.
49. Zwietering, M.H., Jongenburger, I., Rombouts, F.M., vant Riet, K., Modeling of the bacterial growth curve. *Appl. Environ. Microbiol.*, 56, 1875–1881, 1990.
50. Andrews, J.F., A mathematical model for the continuous culture of microorganisms utilizing inhibitory substrates. *Biotechnol. Bioeng.*, 10, 707–723, 1968. https://doi.org/10.1002/bit.260100602.
51. Bentley, W.E. and Kompala, D.S., A novel structured kinetic modeling approach for the analysis of plasmid instability in recombinant bacterial cultures. *Biotechnol. Bioeng.*, 33, 49–61, 1989. https://doi.org/10.1002/bit.260330108.
52. Nielsen, J., Nikolajsen, K., Villadsen, J., Structured modeling of a microbial system: I. A theoretical study of lactic acid fermentation. *Biotechnol. Bioeng.*, 38, 1–10, 1991. https://doi.org/10.1002/bit.260380102.
53. Kim, B.G. and Shuler, M.L., A structured, segregated model for genetically modified Escherichia coli cells and its use for prediction of plasmid stability. *Biotechnol. Bioeng.*, 36, 581–592, 1990. https://doi.org/10.1002/bit.260360605.
54. Lee, S.B. and Bailey, J.E., Genetically structured models forlac promoter-operator function in the chromosome and in multicopy plasmids: Lac promoter function. *Biotechnol. Bioeng.*, 26, 1383–1389, 1984. https://doi.org/10.1002/bit.260261116.
55. García-Ochoa, F., Santos, V.E., Alcón, A., Simulation of xanthan gum production by a chemically structured kinetic model. *Math. Comput. Simul.*, 42, 187–195, 1996. https://doi.org/10.1016/0378-4754(95)00113-1.
56. Garcia-Ochoa, F., Santos, V.E., Alcón, A., Metabolic structured kinetic model for xanthan production. *Enzyme Microb. Technol.*, 23, 75–82, 1998. https://doi.org/10.1016/S0141-0229(98)00014-3.
57. Straight, J.V. and Ramkrishna, D., Complex growth dynamics in batch cultures: Experiments and cybernetic models. *Biotechnol. Bioeng.*, 37, 895–909, 1991. https://doi.org/10.1002/bit.260371002.
58. Young, J.D., Learning from the steersman: A natural history of cybernetic models. *Ind. Eng. Chem. Res.*, 54, 10162–10169, 2015. https://doi.org/10.1021/acs.iecr.5b01315.
59. Straight, J.V. and Ramkrishna, D., Cybernetic modeling and regulation of metabolic pathways. Growth on complementary nutrients. *Biotechnol. Prog.*, 10, 574–587, 1994. https://doi.org/10.1021/bp00030a002.
60. Turner, B.G., Ramkrishna, D., Jansen, N.B., Cybernetic modeling of bacteriol cultures at low growth rates: Single-substrate systems. *Biotechnol. Bioeng.*, 34, 252–261, 1989. https://doi.org/10.1002/bit.260340214.
61. Young, J., Henne, K., Morgan, J., Konopka, A., Ramkrishna, D., Cybernetic modeling of metabolism: Towards a framework for rational design of

recombinant organisms. *Chem. Eng. Sci.*, 59, 5041–5049, 2004. https://doi.org/10.1016/j.ces.2004.09.037.
62. Pappu, S.M.J. and Gummadi, S.N., Effect of cosubstrate on xylitol production by *Debaryomyces nepalensis* NCYC 3413: A cybernetic modelling approach, *Process Biochem.*, 69, 12–21, 2018. https://doi.org/10.1016/j.procbio.2018.03.023.
63. Kim, J.-S., Park, J.-B., Jang, S.-W., Ha, S.-J., Enhanced xylitol production by mutant *kluyveromyces marxianus* 36907-FMEL1 due to improved xylose reductase activity. *Appl. Biochem. Biotechnol.*, 176, 1975–1984, 2015. https://doi.org/10.1007/s12010-015-1694-z.
64. Zhang, Q., Li, Y., Xia, L., Liu, Z., Enhanced xylitol production from statistically optimized fermentation of cotton stalk hydrolysate by immobilized *candida tropicalis. Chem. Biochem. Eng.*, 28, 87–93, 2014.
65. Misra, S., Gupta, P., Raghuwanshi, S., Dutt, K., Saxena, R.K., Comparative study on different strategies involved for xylitol purification from culture media fermented by Candida tropicalis. *Sep. Purif. Technol.*, 78, 266–273, 2011. https://doi.org/10.1016/j.seppur.2011.02.018.
66. Park, S.H., Kim, H.J., Cho, J.-I., Optimal central composite designs for fitting second order response surface linear regression models, in: *Recent Advances in Linear Models and Related Areas*, pp. 323–339, Physica-Verlag HD, Heidelberg, 2008.
67. Silva, C.J.S.M. and Roberto, I.C., Optimization of xylitol production by *Candida guilliermondii* FTI 20037 using response surface methodology. *Process Biochem.*, 36, 1119–1124, 2001. https://doi.org/10.1016/S0032-9592(01)00153-4.
68. Silva, C.J.S.M. and Roberto, I.C., Statistical screening method for selection of important variables on xylitol biosynthesis from rice straw hydrolysate by *Candida guilliermondii* FTI 20037. *Biotechnol. Tech.*, 13, 743–747, 1999. https://doi.org/10.1023/A:1008964724812.
69. Faveri, D.D., Torre, P., Perego, P., Converti, A., Statistical investigation on the effects of starting xylose concentration and oxygen mass flowrate on xylitol production from rice straw hydrolyzate by response surface methodology. *J. Food Eng.*, 65, 383–389, 2004. https://doi.org/10.1016/j.jfoodeng.2004.01.037.
70. Mussatto, S.I. and Roberto, I.C., Optimal experimental condition for hemicellulosic hydrolyzate Treatment with activated charcoal for xylitol production. *Biotechnol. Prog.*, 20, 134–139, 2008. https://doi.org/10.1021/bp034207i.
71. Gurpilhares, D.B., Pessoa, A., Roberto, I.C., Glucose-6-phosphate dehydrogenase and xylitol production by *Candida guilliermondii* FTI 20037 using statistical experimental design. *Process Biochem.*, 41, 631–637, 2006. https://doi.org/10.1016/j.procbio.2005.08.008.
72. Mussatto, S.I. and Roberto, I.C., Establishment of the optimum initial xylose concentration and nutritional supplementation of brewer's spent grain hydrolysate for xylitol production by *Candida guilliermondii. Process Biochem.*, 43, 540–546, 2008. https://doi.org/10.1016/j.procbio.2008.01.013.

73. Cheng, K.-K., Zhang, J.-A., Ling, H.-Z., Ping, W.-X., Huang, W., Ge, J.-P., Xu, J.-M., Optimization of pH and acetic acid concentration for bioconversion of hemicellulose from corncobs to xylitol by *Candida tropicalis*. *Biochem. Eng. J.*, 43, 203–207, 2009. https://doi.org/10.1016/j.bej.2008.09.012.
74. Sarrouh, B.F. and da Silva, S.S., Application of response surface methodology for optimization of xylitol production from lignocellulosic hydrolysate in a fluidized bed reactor. *Chem. Eng. Technol.*, 33, 1481–1487, 2010. https://doi.org/10.1002/ceat.200900546.
75. Ling, H., Cheng, K., Ge, J., Ping, W., Statistical optimization of xylitol production from corncob hemicellulose hydrolysate by *Candida tropicalis* HDY-02. *N. Biotechnol.*, 28, 673–678, 2011. https://doi.org/10.1016/j.nbt.2010.05.004.
76. Ramesh, S., Muthuvelayudham, R., Rajesh Kannan, R., Viruthagiri, T., Enhanced production of xylitol from corncob by *Pachysolen tannophilus* using response surface methodology. *Int. J. Food Sci.*, 2013, 1–8, 2013. https://doi.org/10.1155/2013/514676.
77. Baishan, F., Hongwen, C., Xiaolan, X., Ning, W., Zongding, H., Using genetic algorithms coupling neural networks in a study of xylitol production: Medium optimisation. *Process Biochem.*, 38, 979–985, 2003. https://doi.org/10.1016/S0032-9592(02)00094-8.
78. Sampaio, F.C., De Faveri, D., Mantovani, H.C., Passos, F.M.L., Perego, P., Converti, A., Use of response surface methodology for optimization of xylitol production by the new yeast strain *Debaryomyces hansenii* UFV-170. *J. Food Eng.*, 76, 376–386, 2006. https://doi.org/10.1016/j.jfoodeng.2005.05.055.
79. Rao, R.S., Prakasham, R.S., Prasad, K.K., Rajesham, S., Sarma, P.N., Rao, L.V., Xylitol production by *Candida* sp.: Parameter optimization using Taguchi approach. *Process Biochem.*, 39, 951–956, 2004. https://doi.org/10.1016/S0032-9592(03)00207-3.
80. Kumdam, H.B., Murthy, S.N., Gummadi, S.N., A statistical approach to optimize xylitol production by *Debaryomyces nepalensis* NCYC 3413 *in vitro*. *Food Nutr. Sci.*, 03, 1027–1036, 2012. https://doi.org/10.4236/fns.2012.38136.
81. Vieira Cortez, D. and Conceicao Roberto, I., Improved xylitol production in media containing phenolic aldehydes: Application of response surface methodology for optimization and modeling of bioprocess. *J. Chem. Technol. Biotechnol.*, 85, 33–38, 2010. https://doi.org/10.1002/jctb.2265.
82. Pappu, J.S.M. and Gummadi, S.N., Multi response optimization for enhanced xylitol production by *Debaryomyces nepalensis* in bioreactor. *3 Biotech.*, 6, 151, 2016. https://doi.org/10.1007/s13205-016-0467-x.
83. Montgomery, D.C., Using fractional factorial designs for robust process development, *Qual. Eng.*, 3, 193–205, 1990. https://doi.org/10.1080/08982119008918849.

10

Recovery of Valuable Products from Vegetable Wastes

Pooja Aich[1], Balraj Sudha[2], Kanagaraj Suganya[2], Bishwambhar Mishra[3], Bapatla Sumithra[3], Sanjeeb Kumar Mandal[3*] and Sundaravadivelu Sumathi[2†]

[1]*Department of Medicinal and Applied Chemistry, Kaohsiung Medical University, Kaohsiung City, Taiwan*
[2]*Department of Biochemistry, Biotechnology, and Bioinformatics, Avinashilingam Institute for Home Science and Higher Education for Women, Coimbatore, Tamil Nadu, India*
[3]*Department of Biotechnology, Chaitanya Bharathi Institute of Technology, Hyderabad, India*

Abstract

Vegetables are essential for human health because they provide a significant quantity of vitamins, minerals, and fiber in our balanced meals. Vegetable production produces a large amount of liquid and solid waste, containing many valuable high-value components with significant economic benefits. If not used or disposed of properly, it may cause pollution. On the other hand, this waste material is high in valuable compounds and is thrown into the environment. These are unique, fresh, and cost-effective bases of flavoring, additives, dietary fiber, protein, antioxidants, and antimicrobials, which could be used in the beverage and food industries to produce beneficial products. As a result, new concepts concerning the practice of these by-products for more utilization in the manufacture of high-nutritive edible products or supplements have piqued interest as these are high-value products with the potential for economic recovery. This review aims to encourage vegetable production and processing by emphasizing the possibility of extracting active molecules from vegetable waste and their application in industries. These advantages would pave the way for the potential use of vegetable waste for therapeutic and nutraceutical applications. The retrieval of these bioactive substances from industrial by-products is currently generating much interest, owing to their

**Corresponding author*: sanjeeb.vit@gmail.com; ORCID: 0000-0002-3920-9456
†*Corresponding author*: sumathi_bc@avinuty.ac.in

Sanjay Kumar, Narendra Kumar and Shahid-ul-Islam (eds.) Role of Microbes in Industrial Products and Processes, (255–278) © 2023 Scrivener Publishing LLC

positive effects on people. This review concentrates on the vegetable by-products with bioactive properties and their potential applications in health sectors (nutraceuticals) and food production (functional foods).

***Keywords*:** Bioactive compounds, by-product, vegetables, waste utilization

10.1 Introduction

Vegetables are the most widely consumed horticultural crops, and they are consumed fresh, organically, and processed due to their health-promoting compounds and nutrients. The growth of the population and the changing dietary habits are contributing significantly to the production and dispensation of horticultural crops, particularly vegetables. Agricultural and processing industry losses and waste have become a primary dietary, economic, and environmental concern [1]. Vegetable processing operations produce substantial by-product waste, which makes up about 25–30% of the entire category of commodities. Seeds, skin, rind, and pomace comprise most of the waste. They are high in bioactive compounds that could also be useful, like polyphenols, carotenoids, dietary fibers, enzymes, vitamins, and oils. Such natural products would be used in several industries, including the food industry to make usable or enriched foods, the textile industry, the healthcare, and the pharmaceutical industries. Using waste to produce various predominant bioactive compounds is a significant step toward long-term expansion [2].

The processing of polyphenols, dietary fibers, and other bioactive compounds from vegetable waste has also been studied, and it is a major source of phytochemicals. Only flesh or pulp is eaten in most vegetables. However, research has shown that the seeds, peels, and other non-commonly consumed vegetable components contain significant amounts of phytochemicals and essential nutrients. Lemon peels, for example, comprise phenolic levels that are more than 15% higher than those found in pulp [3]. The biological activities are detected, quantified, extracted, and evaluated for fruit and vegetable with bioactive compounds. Over the last few centuries, which are an important field of medical and health research. In particular, absorption of vegetables and fruits has also been connected with decreased risk of cancer and cardiovascular diseases. Phytoconstituents in vegetables, particularly phenolics and dietary fibers that function as antioxidant carriers in our body, are recommended to be the main bioactive compound accountable for physical condition, primarily because of their antioxidant property [4].

Studies in the revalorization of vegetable by-products have grown enormously over the past 20 years. The immense possible advantages of biologically active compounds in these by-products have been the driving force behind an enormous number of significant scientific studies and considerable global funding sources. The derived compounds could improve food science (functional foods), the pharmaceutical sector, and biotechnological processing of other essential bioactive substances (e.g., enzymes). The impact of food waste is getting worse, and it's affecting every aspect of waste management, from managing waste. On per global scale, developed countries are estimated to waste more food than developing countries [5].

Recently, considerable focus has been placed on food waste recovery, recycling, and reconditioning. Studies have been conducted to switch wasted food into high-valuable products. Biofertilizers, bio-pigments and dyes, biofuels, emulsifiers, organic and essential oils, bio preservatives, and SCP (single-cell protein) are all converted into usable value-added commodities. Vegetable waste has the ability to be used to make value-added products and biofuels, making it an efficient way to manage wasted food. It is important to implement strategies for efficient waste management. Reducing waste at its source is the best solution to waste management [6]. In order to follow an effective strategy of integrated waste management, the socio-economic aspects of generated waste and disposal must also be considered. As part of human society, food waste is generated on a small domestic level as well as on a broad industrial level. The waste management activities practiced in developed nations are composting, incineration, sanitary landfills, etc. Waste is stored and mainly deposited in the open or burned in the open. This has an immediate impact on the environment as well as human health. Food waste is discarded in the open or in landfills. It decomposes due to the activity of various microorganisms. This releases a variety of gases (carbon dioxide and methane) that contribute to the greenhouse effect, which causes global warming [7]. It is necessary to resolve the food waste problem by converting it into various valuable products that are both environmentally friendly and efficient. Separate value-added materials from carrots, onions, peas, tomatoes, and sugar beet include biohydrogen, biomethane, biogas, single-cell protein, bioethanol, mushrooms, alpha-L-arabinofuranosidase, organic acids, vinegar, oligomers, fertilizers, animal feed, and glycoalkaloids [8].

The most challenging problem facing humankind worldwide is food waste. The food processing, household, and hospitality industries generate biodegradable waste, which is made up of carbohydrates, lipids, proteins, and inorganic compounds. Food waste management is very significant. In open areas, food waste is commonly incinerated or discarded, which can cause significant health and environmental problems [9]. It is possible to transform food waste into phytochemicals, bioactive substances, food additives, dietary fibers, animal feed, bio pigments and dyes, edible and essential oils, emulsifiers, biofertilizers, bio preservatives, biofuels, and single-cell proteins are some of the value-added materials used to treat food waste. Food waste with added value can be very eco-friendly [10].

Every year, one-third of the world's 1300 million tons of production of food is unused. Wastage occurs due to a variety of factors, including disease and destruction, strict structure and performance quality requirements, processes of packaging and marketing, and consumer behavior. A lack of cooperation between various players in the supply chain, infrastructure, off-cuts, and over-production often affects waste levels.

Valuable food waste assets, in particular horticultural waste (fruit and vegetables), can be retrieved and turned into new things, including packaged foods, nutraceuticals, complementary healthcare, and cosmeceuticals. To develop fruit and vegetable extracts and dried powder containing nutritious and bioactive materials, creative processing methods/technologies can be applied. The use of discarded streams from these crops increases the sector's viability and greatly improves food security.

10.2 Extraction of Valuable Products from Onion Waste

Onions are vegetables consumed worldwide. The processing of a large quantity of onion waste makes it necessary to consider its disposal and transformation into useful goods.

10.2.1 Disposal Through Dehydration of Wastes

Onions are common in Argentina's Cuyo region and play a significant role in universal vegetable production. Production levels still exceed immediate fresh selling markets, and excesses are discarded while still appropriate. In order to alleviate these difficult ecological concerns, this study explores the improvement of unused products through dehydration, thus ensuring their eventual food usage. As a result, *Crioularoxa* investigated the temperatures that would preserve the onions' key physicochemical characteristics while establishing optimum heat and mass transfer coefficients while considerably reducing energy utilization and CO_2 emissions at 60°C and 70°C [11].

10.2.2 Supercapacitors from Onion Peels

Onion husks (*Allium cepa L.*), plentiful biomass, have been reluctantly examined as a sustainable carbon source for preparing value-added absorbent carbon frameworks (OHCFs) through a direct carbonization and activation process. The synthetic system for automotive purposes is as simple as scaling up. The OHCF-2 sample-based two-electrode symmetrical supercapacitor (synthesized from 2 M K_2CO_3 solution) has outstanding electrochemical efficiency with a strong generic capacitor (up to 188 F/g at 1 A/g) [12].

10.2.3 Insulating Material from Onion Peel Dust

Agricultural raw material fibers represent a significant class of renewable materials used in construction sites. Many wooden product requirements list substitute sources for lignocellulosic fibers, which are the primary sources of functional lignocellulosic fibers in biomass. The fibers and particles of agricultural materials are similar to the fibers and particles of wood in several ways. The number of specific features (5) varies depending on the species and the degree of processing. Huge quantities of agricultural fibers are food production leftovers and are usually burned or used in landfilling as no other substitute to using them has been discovered. To achieve this, waste peanut shells and onion skins were ground to 1 mm in size, and samples were made using 4 × 4 × 16 cm of fly ash, perlite, pumice, cement, barite, and gypsum. These samples were kept at 1000°C for 24 hours in an oven. The samples' apparent specific gravities, ductility, water uptake rates, and impact resistance were used to determine their identities [13].

According to a review, onion peel dust is a cheap material for extracting nitrate from an aqueous medium, and it's simple to regenerate depleted adsorbent. Dust OPD was used to remove nitrate from the aqueous process after being treated with double distilled water and dried out in a hot air oven. The effect of pH, initial nitrate concentration, contact time, the temperature on nitrate ion elimination, and the adsorbent dose was evaluated. The OPD analyzed nitrate adsorption's adsorption, thermodynamics, and kinetics [14].

10.2.4 Effects of Onion Extracts

Gallic acid, quercetin, ferulic acid, kaempferol, and protocatechuic acid are also some of the phenolic compounds present in onion (*Allium cepa L.*) extract [15]. HPLC analysis was conducted to evaluate the concentrations of flavonoids and phenolic acids in OPE. In addition, the effect of OPE on inflammation and antioxidant-related gene expression in lipopolysaccharide (LPS)-promoted HT-29 cells were investigated [16].

The researcher investigated the capability of onion dry scale extract to enhance eggplant resistance to bacterial infection. This assessment was carried out by analyzing the improvement in the metabolism of polyamines and antioxidants in two groups of eggplant seedlings infected with *Ralstonia solanacearum* (the first one was offered a dry scale onion extract treatment, while another was kept alone). The extract of the dry onion scales increased the polyamine content. Polyamine oxidase activity in the treated infected plants enlarged mostly on the seventh day after the outbreak. The activity of PAO and H_2O_2 both declined significantly on the 14th day after infection [17].

10.2.5 Valorization of Onion Waste

Onions (*Allium cepa* L.) are abundant in flavanols that have been perceived to support human health. Flavonols such as quercetin and quercetin glycosides have been isolated and tested against enzymes of clinical significance in Alzheimer's disease and diabetes from onion solid waste (OSW). They have been shown to have cytotoxic and antioxidant effects [18].

10.2.6 Electrocatalyst

The utilization of onion skin waste as a raw material in producing non-noble metal electrocatalysts is a helpful complement to the recycling of excess biomass wastes like onion skin. Chemical treatment with HNO_3 activated the biocarbon to be generated at pyrolysis temperature. This heat has a specific impact on the formation and textural features of the onion skin waste-derived biocarbon, showing that it impacts the catalytic activity for the oxygen evolution reaction [19].

10.2.7 Extraction of Valuable Products from Capsicum Waste

In Australia, especially in Queensland, capsicum is a major horticultural crop. Australia produces approximately 76,000 tons of capsicums annually, of which Queensland accounts for 50–65% of the total output. About 30–40% of the total produce worth ~$300 million is lost or wasted and is presently dumped for different reasons [20].

10.3 Capsicum Seeds as the Source of a Bioactive Compound

Recent studies have concentrated on leftover by-products that include bioactive chemical substances. Though Capsicum seeds have some culinary and medicinal applications, there is still seed mass waste from pepper processing. In general, many of the world's leading pepper-producing countries lack the infrastructure and facilities needed for such processing technologies. Therefore pepper seeds are regularly discarded, buried, or used as animal feed which causes negative impacts on the climate and an unintended economic loss for farmers. There is a latent opportunity in the pepper processing industry to valorize pepper seeds to acquire added-value by-products and thereby minimize waste. Pepper seeds are rich in volatile compounds and flavonoids, phenolic acids, carotenoids, antioxidants, and vitamins A, C, and E. The family's distinctive alkaloids and capsaicinoids have been connected to a variety of beneficial pharmacological and biochemical effects. Other bioactive substances found in pepper seeds include lectins, saponins, and polyunsaturated fatty acids [21].

10.3.1 Generation of Single-Cell Proteins from Waste Capsicum Powder

Leftover capsicum powder was collected as a substrate to create a single cell protein (SCP) during capsanthin extraction. Capsanthin is a natural pigment derived from dry red capsicum and extracted with solvents [22]. Consumers prefer natural pigments because they are often present in natural foods and are easily metabolized. The use of natural dyes in food products is gaining popularity around the world. The red capsicum has a long history in China. Concerns about environmental pollution have sparked interest in turning waste materials into commercially useful products, especially SCP. The composition of CPM suggests that it may be a suitable nutrient medium for microbes. The total sugar, nitrogen, and phosphorus content of CPM were primarily influenced by the red capsicum variety used, planting conditions, and harvest time. Capsicum powder naturally contains organic acids such as tartaric and citric acids, making it acidic (pH 5) in nature.

10.3.2 Chili (*Capsicum Annum*) Spent Residue

Chili spent residue (CHSR) was analyzed as a potential source of protein, dietary fiber, polyphenols, and minerals. CHSR, a by-product of the spice oleoresin industry, is a potential source of 44.4% total dietary fiber, 34.9% insoluble dietary fiber, and 9.5% soluble dietary fiber, which is more than many vegetables and fruits, as well as protein content (19.7%). CHSR, a by-product of the spice oleoresin industry, is a potential source of 44.4% total dietary fiber, 34.9% insoluble dietary fiber, and 9.5% soluble dietary fiber, both of which are higher than in humans. The effect of CHSR incorporation on rheological, microstructural, and physical, sensory, and nutritional characteristics have been evaluated. In addition to that, use of CHSR in the preparation of bread as a source of the functional ingredient has also been evaluated [23]. The optimum CHSR concentration that could be applied to bread without affecting the texture was found to be 10% in this direction. At optimal dosage, the addition of CHSR to bread significantly increased the protein content (7.8–11%) and nearly doubled the bread's dietary fiber and mineral content. The addition of CHSR at 5–15% of the bread volume affected the properties of the bread, i.e., specific volume, crust color, and shape; crumb color, grain, and texture, resulting in sensory ratings of 60–74 compared to the control bread score of 84.

Additives such as fungal alpha-amylase enzyme, dry gluten sodium stearoyl-2-lactylate powder with 10% CHSR, however, improved the strength of the dough, decreased the crumb firmness value, specific volume, and increased the overall bread quality score. With the addition of CHSR above 10%, scanning electron microscopic studies showed a disturbance in the consistency of the protein matrix in bread dough.

10.3.3 Chili Seeds Extract as an Antimicrobial Agent

The most common microorganisms identified in foods and/or possible diseases have been evaluated with pepper seed extracts and chosen pigments, capsaicins, phenols, and capsiates. It has been found that the main antimicrobial action of capsaicinoids is against common pathogenic and opportunistic fungi such as *Aspergillus spp.*, *Fusarium spp.* But also, against such bacteria such as *Helicobacter pylori, Streptococcus spp., Pseudomonas aeruginosa*, or *Listeria monocytogenes* and yeasts such as *Saccharomyces cerevisiae* or *Candida albicans* [24–26].

Vibrio cholerae or *Vibrio monocytogenes* have been shown to decrease the virulence of pathogenic infections by inhibiting pro-inflammatory factors or the development of toxins [27–29]. Capsaicinoid antimicrobial tests have shown that it hinders the development of several strains of *Staphylococcus aureus, Listeria monocytogenes, Escherichia coli,* and *Bacillus subtilis* [24].

Although capsaicinoids have these specific antimicrobial activities, the positive relationship between antimicrobial activities, pungency, concentration, and potential irritation effects makes them less appropriate for use as a food preservative [30]. It is crucial to investigate the effects of alternative chemicals found in pepper seeds and fruit in this field.

Chrysoeriol is noted as an antimicrobial agent rich in a number of plants and peppers [31]. Several studies have shown that chrysoeriol inhibits the growth of *Klebsiella pneumoniae, Enterococcus faecalis, B. subtilis, S. aureus, L. monocytogenes,* and *Candida albicans* [32, 33].

It may be an essential nutrient preserver against a variety of pathogenic and opportunistic microbes due to the low concentrations at which it exerts its antimicrobial effect. Some lectins obtained from peppers are shown to have antifungal qualities preventing the growth of opportunistic fungi such as *Fusarium moniliforme* and *Aspergillus flavus* by limiting exterior adherence [34].

Furthermore, a recent study tested natural peptides from the *C. baccatum* var. pendulum for their antiadhesive properties against biofilms formed by Staphylococcus epidermidis [35]. Biofilms are a major source of pathogenicity; thus, any new research in this area is quite exciting. Finally, these findings may contribute to developing novel uses for natural antimicrobials, which is crucial given the worldwide worry about the spread of antibiotic-resistant microbes [32, 33].

10.4 Extraction of Valuable Products from Cauliflower Waste

Cauliflower is one of the cruciferous vegetables in the species. Cauliflower is a Brassicaceae family tropical plant grown in India and China [36].

10.4.1 Cauliflower Leaves Powder Waste to Utilize in Traditional Product

India is the second-largest vegetable producer in the world [37]. Cauliflower leaves are rich in iron, beta carotene, and calcium, but they have an upper waste index [38]. Cauliflower leaves are also considered waste products, but they can be used in value-added products to care for anemia and micronutrient deficiency [39]. The leaves contribute around 50% of the overall production of cauliflower. Various drying methods are tray drying, microwave oven drying, and sun drying for drying cauliflower leaves for specific periods at different temperatures. It can be processed after drying for use during the lean season [40].

These vegetables are also high in plant metabolites, including glucosinolates, anthocyanins, flavonoids, trepans, S-methyl cysteine sulfoxide, and other compounds that boost nutritional, physicochemical, and sensory efficiency, as well as being very good for your health.

10.4.2 Nutritional Evaluation of Dehydrated Cauliflower Stems Powder

The stalk and dense green leaves covering the head are usually discarded, leaving only the head to be eaten. All the dehydrated stems of cauliflower powder were incorporated in sev and mathri in various ratios. For the sensory evaluation of the integrated recipes, panels of 10 participants were chosen through triangle research. The assessment was carried out using the 9-point Hedonic Rating Scale and Composite scoring test. Using ANOVA, CD, and *t*-test, data collected during the investigation was statistically analyzed. The nutritional measurements showed a significant increase in calcium and iron when dehydrated cauliflower stems powder was added to the recipe.

Dehydrated cauliflower stem powder was concluded to be productively used in the product to increase its nutritional content [41].

10.5 Extraction of Valuable Products from Tomato Waste

The tomato (*Solanum lycopersicum* L.), is neither a vegetable nor a fruit, but a berry in botanical terms, is currently distributed worldwide and is a key ingredient in the cuisines of most cultures. The tomato originated in South America and was introduced into Mexico over there.

10.5.1 Tomato Wastes as a Source of Essential Raw Materials

The beginning of mechanical harvesting is thought to have twice the amount of this waste material [42]. Tomato waste is generally achieved in the form of seeds and skin residues during processing. Tomato seed flake was identified as a source of protein and the seeds as a source of edible oil [43].

The chemical composition of different tomato processing waste fractions was studied. Hexane extracted lipids were found to make up about 27% of the seeds. The most common fatty acid was oleic acid, which was followed by palmitic acid. The seed flake attained after hexane extraction pointed out a high protein content (-40%) with a high threonine and lysine content. Elemental analysis of the tomato seed flour extracted from hexane showed that the major elements were Na, K, Mg, and Ca, followed by Mn, Fe, Cu, and Zn. When acetone was extracted from wet tomato skins, a brightly colored concentrate contained mostly lycopene resulted (total yield of about 12 mg per 100 g of wet tomato skins). The pigment-free skin residue was discovered to contain about 11% protein, with the most prevalent essential amino acids being lysine, valine, and leucine. Elemental analysis indicated that the key elements in tomato skins were Ca, K, Na, and Mg.

10.6 Extraction of Valuable Products from Ginger Waste

Ginger, used both as a spice and medicine, is a valuable commercial crop grown for its aromatic rhizomes. The dried ginger spice and preserved, crystallized ginger is highly valued. Ginger is a perennial herb, but as a spice, it is commonly grown as an annual for harvesting.

10.6.1 Ginger (*Zingiber officinale* Roscoe) Spent Residue

Excessive gelatinization properties and poor *in vitro* starch digestibility (45%) of starch were isolated from discarded ginger and waste ginger which could be used for texture stabilization in preparing foods at high temperatures. Ginger spent is rich in carbohydrates, and *Saccharomyces cerevisiae's* carbohydrates fermentation has led to bioethanol development [44]. A compound with antioxidant properties was isolated from ginger spent. 3,5-diacetoxy-7-(3,4-dihydroxy phenyl) heptanes with important antioxidant properties that might be used in food composition were obtained from ginger spent methanolic extract when partitioning with butanol. The introduction of ginger into the rabbit diet resulted in enhanced enlargement efficiency, carcasses, serum cholesterol, and serum lipid reduction [45]. The ginger spent crush used in green polyurethane composites' fabric increased the composites' ductile potency and thermic balance [46]. Procurement of polyurethane/ginger composites with the integration of finely ground ginger spent crush at various levels of addition, *viz.* 0%, 2.5%, 5%, 7.5%, and 10% by weight, and tested for mechanical properties such as surface hardness and tensile behaviors. Thermogravimetric study of the green compound has tested thermic balance and their mode of thermic degeneration nature. Thermogravimetric research has observed a three-step thermal degeneration procedure in polyurethane-ginger composites. The thermal degradation kinetic parameter, activation energy (Ea), was measured for the green composites by Coats-Redfern and Broido's models. Wide-angle X-ray scattering data have calculated lattice strain and crystal size. The relationship between the structure-property of the composites and spent ginger was recognized based on the source of the above parameters.

10.7 Extraction of Valuable Products from Carrot Waste

Carrots (*Daucus carota* L.) are very cheap. These are highly nutritious because of richness in beta-carotene and contain a large amount of thiamine, riboflavin, pyridoxine, and cyanocobalamin [47]. The limiting threat of many cancers, particularly prostate cancer, is associated with high carotenoid intake.

10.7.1 Value-Added Product from Carrot Pomace

The most popular non-alcoholic beverages are carrot juices and mixes. Although substantial advancements in processing procedures such as mash heating, depolymerizing enzymes, and decanter technology. However, many beneficial chemicals such as uronic acids, carotenes, and natural sugars remain in the pomace. Only 60–70% of juice yield is registered, and up to 80% of the carotene with the pomace can be lost. Singh *et al.* [48] conducted a study to explore the possibility of utilizing waste residues (pomace) obtained during the extraction of carrot juice to prepare a value-added product based on carrot condensed milk manufactured goods. In two ways, the carrot pomace was osmotically treated: first, dropping in 65° Brix sucrose syrup; secondly, applying 35% sucrose (dry powder) to the pomace. Besides that, the product was convectively dehydrated to a moisture content of 4–5% (wet base) at 60°C and vacuum packed in an aluminum laminated box (100 gauge). The dehydrated product was kept at room temperature (28–42°C) for 6 months before making a carrot-based condensed milk product. Overall acceptability of the osmo-convectively dehydrated pomace-prepared product was moderate to excellent [48].

10.7.2 Processing and Stability of Carotenoid Powder from Carrot Pulp Waste

Carotenoids, such as â-carotene, have sparked a lot of interest in the last decade because of their potential to protect against certain types of cancer [49]. Beta-Carotene is a necessary biological compound because of its provitamin A activity [50]. Because only 60–70% of carrot juice is extracted, a large amount of carrot pulp waste is produced, posing significant removal trouble for the food industry. Conversely, it has been discovered that carrot pulp waste still retains a significant number of carotenoids [51].

10.8 Extraction of Valuable Products from Coriander Waste

Coriander is an evergreen shrub belonging to the Apiaceae family. Chinese parsley, dhania, and cilantro are some of the other names for it. All plant parts are safe to consume, but fresh leaves and dried seeds are the most commonly used in cooking.

10.8.1 Coriander (*Coriandrum sativum*) Spent

After oleoresin and oil removal, coriander was found to have a remarkable amount of total dietary fiber (51.7%) and high mineral levels, including potassium (4151 ppm), calcium (4013 ppm), iron (124 ppm), manganese (34.7 ppm) and zinc (19.7 ppm) [52]. Coriander, which is high in minerals and dietary fiber, was added to the flour at a rate of 10% to make bread. The substitution of spent coriander for bread flour had no effect on the loaf volume, taste, crumb grain, or shape (10%).

10.9 Extraction of Valuable Products from Potato Waste

After China and the Russian Federation, India is the third biggest producer of potatoes in the world. The primary agenda of the twenty-first century of global scientists was an environmental-friendly product. In India, potato production in 2010 was estimated at approximately 34391 thousand tons (FAO). It is estimated that the quantity of waste and by-products of potato production is about 12–20% of their total production.

10.9.1 Utilization of Potato Waste for Animal Feed

Potato peel is a commodity that is typically discarded due to microbial spoilage in the food processing industry, mainly in the manufacture of potato chips, posing an environmental concern. Potato peel waste has traditionally been used to make animal feed and a low-value fertilizer that results in the waste of abundant nutritional materials with antibacterial, antioxidant, chemopreventive, apoptotic, and anti-inflammatory activity.

The previous study examines the principles and technologies of several advanced potato peel waste creation methods used in the food processing, phytopharmaceutical, and biosynthesis industries and their benefits and drawbacks from the standpoints of popularization and economics and the potential for future mass production of these products. The production of biogas, phenolic acids, and lactic acids from potato peel waste has a better chance of contributing to China's food strategy, whereas ultrasonic steroidal alkaloid extraction may have great pharmaceutical potential, based on a comparison of their potential. Higher costs still exist in these sectors, so future efforts should be made to reduce the costs of these extraction methods and promote the development of potato waste biomass.

10.9.2 Bioplastics

In 2011, the global production of bioplastics was 1.2 million tons, while the global production of plastics was 250 million tons. As a result, it's fair to say that bioplastics have a limited market. In 2011, starch blends accounted for 11.3% of all bioplastics produced, but this number can only rise [53]. Potatoes are considered a rich source of starch, one of the most powerful polysaccharides that can be used in various ways. Starch is used to make amylose and amylopectin. Amylose causes starch crystallinity and entanglement, which increases the amylose/amylopectin ratio's value during processing and contributes to increased potency and reduced strain. Bioplastic made from starch can be used to replace existing plastics while causing no harm to the environment. It can be simply changed with current technology because the raw material is readily available and inexpensive. Compostable starch-based plastic can even be turned into a food item.

Thermoplastic, which can be processed using many methods, such as various forms of molding, typically used for synthetic polymers, and casting, is another alternative to using starch. By using waste materials from processing (e.g., potato peels) and even genetically modified species, the total usable starch supply can be increased. Bioplastics that are suitable for food processing, film production, and medicinal applications, like plastic products, may also be manufactured from plant proteins. Rice and potatoes are plants that are ideal for the production of gluten-free bioplastics, as wheat can be detrimental to health. Due to patatin content comparable to wheat protein plastic, this plastic had low water absorption values (9%) [54, 55].

10.9.3 Medicine and Pharmacy

Glycoalkaloids are mostly present in the outer layer of the potato tuber and reduce towards the middle, but their quantity rises during storage. These substances do but have benefits, such as lowering cholesterol, reducing allergies, pain, inflammation, lowering liver glycogen content, and lowering blood sugar levels. When used in pharmaceuticals, such as creams, it can reduce herpes damage and avoid new outbreaks for up to 9 months [56]. The majority of phenolic compounds (up to 90%) contain chlorogenic acid, which isn't very useful on its own but can be hydrolyzed into caffeic and quinic acids. Although quinic acid is used as a raw matter in many drug manufacturing processes, it has high antibacterial activity.

Glycoalkaloids could be present in the whole potato plant and can be used as anti-allergens, antibiotics, antipyretics, hyperglycemic and anti-inflammatory compounds in pharmaceuticals at exact doses and situations. These compounds must be extracted from potatoes because they can be harmful to humans when concentrations reach 200 mg per kilogram of potatoes, even though most of them are concentrated during processing, such as heating [56, 57].

Suberin and its monomers are another substance derived from potato peels. Cuticular wax is a waxy material that coats the external sections of plants and is derived from leaves. It contains hydrocarbons and alcohols, such as cholesterol, in a mixture with other compounds. Potato tubers have a suberized layer that protects them from damage, infections, and draught. The amount of suberin obtained is determined by how long the potato tubers have been stored. After six months of storage, the most important quantities of raw suberin fraction obtained were 20–29% of the peel [58].

Because of its healing properties, starch has also been investigated as a substance for the production of pharmaceutical products. In the pharmaceutical industry, gel-like starch is used as a binder-filler in the production of tablets. Capsules made with potato starch as a base material, whether soft or hard, are gelatin-free, fragrance-free, flavorless, and free of complications. Starch could be used in nanotechnology, and nanotubes are the preferred material for producing medical devices [59].

10.10 Extraction of Valuable Products from Beetroot Waste

Beetroot (*Beta vulgaris* L.) is a regular, multi-consumption vegetable with numerous nutritional benefits. It stands out among modestly processed vegetables due to the insistence on healthier provisions and its expediency. Its substantial fiber content has functional characteristics that provide the food industry with interesting technical properties, such as enriching pasta, cakes, and cookies.

10.10.1 Flour Extracted from Beetroot Waste

The minimally processed vegetable industry, as well as beetroot production, generates significant amounts of waste. The most effective method for extracting flour from minimally processed dried beetroot waste at various temperatures, creating a dye, and assessing its durability over 45 days. Beetroot waste dehydrated at 70°C generates flour with much higher antioxidant properties and betalain content than waste flour dried at 60°C or 80°C, although chlorination had no influence on the procedure because the microbiological findings supported its usage. Throughout the study, the colorant derived from beetroot waste showed 20 days of color stability and possible antioxidant activity, allowing it to be used as a purposeful preservative to develop nutritional properties and food products. These findings are encouraging because simply processed beetroot waste can be used as a substitute source of natural and functional constituents with betalain content and high antioxidant activity [60].

10.11 Extraction of Valuable Products from Bitter Gourd Waste

Momordica charantia, also known as bitter gourd, bitter melon, or karela, is a Cucurbitaceae plant that grows in moist and subtropical climates all over the world [61]. Bitter gourd vines are native to Asia, but due to the nutritional significance of both ripe and unripe fruits, they are now widely grown throughout the world, including in tropical countries, namely India, Brazil, Taiwan, and China [61, 62].

10.11.1 Bitter Gourd Seed Oil as a Nutraceutical Purpose

The bitter gourd seed oil has a high-level lipid content, mostly alpha-eleostearic acid, as well as a significant phytosterol level. In addition, it is of good quality. This finding boosts bitter gourd's farming perspective and might promote the marketing of its high-value components. For edible purposes, its seed oil may be used. On the other hand, its high alpha-eleostearic acid and phytosterol content support its therapeutic purpose. So, these products have demonstrated possible health-beneficial special properties [63].

10.12 Extraction of Valuable Products from Bottle Gourd Waste

Bottle gourd is one of the common edible plant sources considered a reservoir of essential nutrients. It is cultivated in India and is available all year round. A low-calorie diet, Bottle gourd is used in various Indian preparations and delicious dishes. As a veggie alternative, long thin strips of bottle gourd are used in sushi or can simply be consumed as a juice. For eating, the seeds are candied or roasted. Bottle gourd juice is healthy with a lot of medicinal properties for health.

10.13 Waste Utilization in Cucurbits

Cucurbits are an essential group of vegetables because of their nutritious value and their use for culinary and dessert purposes. But due to their thick rind and big seeds, they generate huge quantities of waste during consumption and processing. Such waste causes potential care and disposal issues for them. They are the type of enormous food that possesses medicinal values. It can be used to extract and refine oil into items such as pickles, preservatives, candy, edible foods, etc. Besides, such waste is possible raw stuff for the manufacture of biofuels. Cucurbit peel enzymes are used to identify and measure toxic metals in industrial waste and groundwater. Many studies have been conducted and concluded that the conversion of such waste into constructive goods and waste used as a raw matter for other industries [64].

10.14 Extraction of Useful Compounds from Pea Waste

Pea (*Pisum sativum*) is a cool-season crop and one of the most common legumes, grown in the temperate climate regions, either alone or in combination with small grains, and is widely consumed as a legume or vegetable throughout the world for the purpose of both human consumption and animal feeding. Various pea processing industries attempt to conserve and sell them with the advent of canning, freezing, and cold storage to make the seasonal crop available year-round.

10.14.1 Utilization of Pea Waste for Developing Biodegradable Product

The pea processing industry requires the preservation of green peas by freezing and selling them as a by-product for seasonal restriction and creating a very large amount of waste. Not only does insufficient management of such waste contribute to environmental degradation and contamination, but also to the depletion of important biomass resources. Researchers conducted a study that revealed that the nutritional value of waste pea peels was estimated and used effectively to produce bio-degradable products by adding value. The result showed that waste pea peels have a high crude protein (19.79%) nutritional value and have a decent amount of ash (7.87%), fat (2.27%), and fiber (19.79%). The biofilm obtained from this study has a good tensile strength (5.96 MPa), thickness (70 μm), and solubility in water (2.46%). Therefore, bio-film may be an alternative for synthetic plastic with the benefit of work generation, energy recovery, and protection of livelihoods, eventually contributing to sustainable growth [65].

10.14.2 Pea Peel Waste for Cellulase Production Using *Trichoderma reesei*

On our world, a broad range of waste bio-resources is available for processing into bioproducts. To synthesize biomass proteins and enzymes, waste is used as a source of energy with microorganisms in biological systems.

Due to the high amount and elimination charge, the substantial volumes of by-products created during the synthesis of plant-based products make an economic and environmental issue. Pea peel waste is used as an undervalued, unused energy source that can supply as a potential resource for the development of cellulase. Under the condition of solid-state, Batch

Table 10.1 Nature of vegetable wastes along with their typical loss percentage.

Commodity	Nature of waste	Typical losses and waste (%)	References
Onion	Outer leaves/peels	NA	[11–13]
Capsicum	Seeds, skin	30 to 40	[21]
Cauliflower	Leaves, stems	~50	[37, 41]
Tomato	Core, skin, and seeds	20	[42, 43]
Ginger	Ginger spent	NA	[44, 45]
Coriander	Spent	-	[52]
Carrot	Pomace, pulp	60 to 75	[48, 49]
Potato	Peel	15	[56]
Beetroot	Skin, leaves	NA	[60]
Bitter gourd	Seeds	40	[63]
Bottle gourd	Skin, seeds	20 to 45	[64]
Peas	Shell	40	[65]
Garlic	Skin	NA	[67, 68]

experiments were attained with the use of pea peel waste which is used as a carbon source for cellulose production by *Trichoderma reesei*. In that, 30°C temperature and pH 5.0 were set up to be the optimal conditions for the development of cellulase. Researchers demonstrated the effectiveness of pea peel waste, whey, and hydrolysate of wheat starch in the production of cellulase by *T. reesei*. The use of economic pea peel waste for cellulose production might be a fresh, commercial, and useful method for synthesizing cellulase and solid waste management [66].

10.15 Conclusion

To summarize, the vegetable derivative is a central resource of bioactive compounds and, as a result, should be best valued rather than discarded in landfills. The nature, preparation with manufacturing, and types of vegetable-derived waste are highlighted in this report. It also measures the bioactive target compounds present in the vegetable waste, namely dietary fibers, enzymes, phenolic compounds, flavors, and organic acids. This exhibits the enormous amount of deprivation and waste, not only due to the considerable amount of non-edible materials, but also due to insufficient field organization, harvesting, categorization, shipping, storage, and marketing (relative humidity and temperature) and industrial infrastructure, as well as waste produced due to dispersion. A substantial quantity of wasted and discarded vegetables and their components stand for the lack of edible food products and the loss of by-products, such as bioactive compounds, with major possible profit for various industries and uses. In order to attain greater recovery rates of bioactive substances from waste products, more advanced approaches must be used. Extracted substances can be employed in food, cosmetics, pharmaceutical, chemical companies, and food science and food production.

References

1. Ran, X.L., Zhang, M., Wang, Y., Adhikari, B., Novel technologies applied for recovery and value addition of high value compounds from plant by-products: A review. *Crit. Rev. Food Sci. Nutr.*, 59, 450, 2019.
2. Sagar, N.A., Pareek, S., Sharma, S., Yahia, E.M., Lobo, M.G., Fruit and vegetable waste: Bioactive compounds, their extraction, and possible utilization. *Compr. Rev. Food Sci. Food Saf.*, 17, 512, 2018.
3. Sharma, K., Mahato, N., Cho, M.H., Lee, Y.R., Converting citrus wastes into value-added products: Economic and environmentally friendly approaches. *Nutrition*, 34, 29, 2017.
4. Singh, B.K., Koley, T.K., Maurya, A., Singh, P.M., Singh, B., Phytochemical and antioxidative potential of orange, red, yellow, rainbow and black coloured tropical carrots (*Daucus carota* subsp. SativusSchubl. & Martens). *Physiol. Mol. Biol. Plants*, 24, 899, 2018.
5. Reynolds, A., Mann, J., Cummings, J., Winter, N., Mete, E., Te Morenga, L., Carbohydrate quality and human health: A series of systematic reviews and meta-analyses. *Lancet*, 393, 434, 2019.
6. Satari, B. and Karimi, K., Citrus processing wastes: Environmental impacts, recent advances, and future perspectives in total valorization. *Resour. Conserv. Recycl.*, 129, 153, 2018.
7. Vodnar, D.C., Calinoiu, L.F., Dulf, F.V., Stefanescu, B.E., Crisan, G., Socaciu, C., Identification of the bioactive compounds and antioxidant, antimutagenic and antimicrobial activities of thermally processed agro-industrial waste. *Food Chem.*, 231, 131, 2017.

8. Kumar, K., Yadav, A.N., Kumar, V., Vyas, P., Dhaliwal, H.S., Food waste: A potential bioresource for extraction of nutraceuticals and bioactive compounds. *Bioresour. Bioprocess.*, 4, 1, 18, 2017.
9. Szabo, K., Catoi, A.F., Vodnar, D.C., Bioactive compounds extracted from tomato processing by-products as a source of valuable nutrients. *Plant Foods. Hum. Nutr.*, 73, 268, 2018.
10. Veneziani, G., Novelli, E., Esposto, S., Taticchi, A., Servili, M., Applications of recovered bioactive compounds in food products, in: *Olive Mill Waste*, pp. 231–253, Academic Press, London, United Kingdom, 2017.
11. Fabani, M.P., Román, M.C., Rodriguez, R. and Mazza, G., Minimization of the adverse environmental effects of discarded onions by avoiding disposal through dehydration and food-use. *J. Environ. Manage.*, 271, 110947, 2020.
12. Wang, D., Liu, S., Fang, G., Geng, G., Ma, J., From trash to treasure: Direct transformation of onion husks into three-dimensional interconnected porous carbon frameworks for high-performance supercapacitors in organic electrolyte. *Electrochim. Acta*, 216, 405, 2016.
13. Binici, H. and Orhan, A., Insulation material production from onion skin and peanut shell fibres, fly ash, pumice, perlite, barite, cement and gypsum. *Mater. Today Commun.*, 10, 14, 2017.
14. Kumar, M., Dahuja, A., Sachdev, A., Kaur, C., Varghese, E., Saha, S., Sairam, K.V.S.S., Evaluation of enzyme and microwave-assisted conditions on extraction of anthocyanins and total phenolics from black soybean (Glycine max L.) seed coat. *Int. J. Biol. Macromol.*, 135, 1070, 2019.
15. Kim, J., Kim, J.S., Park, E., Cytotoxic and anti-inflammatory effects of onion peel extract on lipopolysaccharide-stimulated human colon carcinoma cells. *Food Chem. Toxicol.*, 62, 199, 2013.
16. Kim, S.W., Ko, M.J., Chung, M.S., Extraction of the flavonol quercetin from onion waste by combined treatment with intense pulsed light and subcritical water extraction. *J. Clean. Prod.*, 231, 1192, 2019.
17. Naguib, D.M. and Badawy, N.M., Phenol removal from wastewater using waste products. *J. Environ. Chem. Eng.*, 8, 103592, 2020.
18. Alonso-Lemus, I.L., Escobar-Morales, B., Lardizabal-Gutierrez, D., de la Torre-Saenz, L., Quintana-Owen, P., Rodriguez-Varela, F.J., Onion skin waste-derived biocarbon as alternative non-noble metal electrocatalyst towards ORR in alkaline media. *Int. J. Hydrog. Energy.*, 44, 12409, 2019.
19. Kothari, D., Lee, W.D. and Kim, S.K., Allium flavonols: Health benefits, molecular targets, and bioavailability. *Antioxidants*, 9, 888, 2020.
20. Echave, J., Pereira, G.A., Carpena, M., Prieto, A.M., Simal-Gandara, J., Capsicum seeds as a source of bioactive compounds: Biological properties, extraction systems, and industrial application, in: *Capsicum*, IntechOpen, London, UK, 2020.
21. Uquiche, E., Valle, J.M., Ortiz, J., Supercritical carbon dioxide extraction of red pepper (*Capsicum annuum* L.) oleoresin. *J. Food. Eng.*, 65, 55, 2004.
22. STATISTICS, A.B.O., Agricultural Commodities: Small Area Data, Australia, 2005-06 (Reissue), 2008.

23. Sowbhagya, H.B., Soumya, C., Indrani, D., Srinivas, P., Physico-chemical characteristics of chilli spent residue and its effect on the rheological, micro-structural and nutritional qualities of bread. *J. Food Sci. Technol.*, 52, 7218, 2015.
24. Ciulu-Costinescu, F., Chifiriuc, M.C., Popa, M., Bleotu, C., Neamtu, J., Marina, L., Screening of polyphenol content and *in vitro* studies of antioxidant, antibacterial and cytotoxic activities of *Capsicum annuum* extracts. *Rev. Chim.*, 66, 1261, 2015.
25. Diz, M.S.S., Carvalho, A.O., Rodrigues, R., Neves-Ferreira, A.G.C., da Cunha, M., Alves, E.W., Okorokova-Facanha, A.L., Oliveira, M.A., Perales, J., Machado, O.L., Gomes, V.M., Antimicrobial peptides from chilli pepper seeds causes yeast plasma membrane permeabilization and inhibits the acidification of the medium by yeast cells. *Biochim. Biophys. Acta*, 1760, 1323, 2006.
26. Bertao, M.R., Moraes, M.C., Palmieri, D.A., Silva, L.P., da Silva, R.M.G., Cytotoxicity, genotoxicity and antioxidant activity of extracts from *Capsicum* spp. *Res. J. Med. Plant*, 10, 265, 2016.
27. Marini, E., Magi, G., Mingoia, M., Pugnaloni, A., Facinelli, B., Antimicrobial and anti-virulence activity of capsaicin against erythromycin-resistant, cell-invasive group A streptococci. *Front. Microbiol.*, 6, 1, 2015.
28. Lee, I.O., Lee, K.H., Pyo, J.H., Kim, J.H., Choi, Y.J., Lee, Y.C., Anti-inflammatory effect of capsaicin in *Helicobacter pylori*-infected gastric epithelial cells. *Helicobacter*, 12, 510, 2007.
29. Yamasaki, S., Asakura, M., Neogi, S.B., Hinenoya, A., Iwaoka, E., Aoki, S., Inhibition of virulence potential of *Vibrio cholerae* by natural compounds. *Indian J. Med. Res.*, 133, 232, 2011.
30. Luo, X.J., Peng, J., Li, Y.J., Recent advances in the study on capsaicinoids and capsinoids. *Eur. J. Pharmacol.*, 650, 1, 2011.
31. Bashyal, P., Parajuli, P., Pandey, R.P., Sohng, J.K., Microbial biosynthesis of antibacterial chrysoeriol in recombinant *Escherichia coli* and bioactivity assessment. *Catalysts*, 9, 2, 1–15, 2019.
32. Nascimento, P.L.A., Nascimento, T.C.E.S., Ramos, N.S.M., Silva, G.R., Gomes, J.E.G., Falcao, R.E.A., Quantification, antioxidant and antimicrobial activity of phenolics isolated from different extracts of *Capsicum frutescens* (*Pigmenta malagueta*). *Molecules*, 19, 5434, 2014.
33. Salehi, B., Hernández-Álvarez, A.J., Contreras, M.D.M., Martorell, M., Ramírez-Alarcón, K., Melgar-Lalanne, G., Potential phytopharmacy and food applications of *Capsicum* spp.: A comprehensive review. *Nat. Prod. Commun.*, 13, 1543, 2018.
34. Kuku, A., Odekanyin, O., Adeniran, K., Adewusi, M., Olonade, T., Purification of a mannose/glucose-specific lectin with antifungal activity from pepper seeds (*Capsicum annum*). *Afr. J. Biochem. Res.*, 3, 272, 2009.
35. Von Borowski, R.G., Barros, M.P., da Silva, D.B., Lopes, N.P., Zimmer, K.R., Staats, C.C., Red pepper peptide coatings control *Staphylococcus epidermidis* adhesion and biofilm formation. *Int. J. Pharm.*, 574, 118872, 2020.

36. Ambrosone, C.B. and Tang, L., Cruciferous vegetable intake and cancer prevention: Role of nutrigenetics. *Cancer Prev. Res. (Phila Pa)*, 2, 298, 2009.
37. Buyaneswari, K.M. and Ramya, G.A., Study on overall acceptability of *Brassica oleracea* leaves (cauliflower leaves) incorporated food products and its impact on treating anemic college going girls. *Int. J. Curr. Res. Dev.*, 291, 38, 2014.
38. Kowsalya, S. and Sangheetha, M., Acceptability and nutrient profile of cauliflower leaves (*Brassica oleraceae var. botrytis*). *India J. Nutr. Diet.*, 36, 332, 1999.
39. Verhagen, H., Cancer prevention by natural food constituents. *Int. Food Ingred.*, 122, 22, 1993.
40. Singh, G., Asha, K., Sehgal, S., Development and nutritional evaluation of products prepared from dried powder of cauliflower leaves. *J. Food Sci. Technol.*, 42, 137, 2005.
41. Sharma, R. and Prasad, R., Nutritional evaluation of dehydrated stems powder of cauliflower incorporated in mathri and sev. *J. Nutr. Food. Sci.*, 8, 1, 2018.
42. Ben-Gera, I. and Kramer, A., The utilization of food industries wastes. *Adv. Food Res.*, 17, 77, 1969.
43. Boni, R. and Sodini, G., *The Quality of Foods and Beverages Chemistry and Technology*, G. Charalambous, and G. Inglett, (Eds.), pp. 153–169, Academic Press, London, New York, 1981.
44. Konar, E.M., Harde, S.M., Kagliwal, L.D., Singhal, R.S., Value-added bioethanol from spent ginger obtained after oleoresin extraction. *Ind. Crops Prod.*, 42, 299, 2013.
45. Omage, J., Onimisi, P., Adegbite, E., Agunbiade, M., The effect of ginger (*Zingiber officinale roscoe*) waste meal on growth performance, carcass characteristics, serum lipid and serum cholesterol profiles of rabbit. *Pak. J. Nutr.*, 6, 359, 2007.
46. Chethana, M., Madhukar, B.S., Somashekar, R., Hanta, S., The influence of ginger spent loading on mechanical, thermal and microstructural behaviours of polyurethane green composites. *J. Compos. Mater.*, 48, 2251, 2014.
47. Manjunatha, S.S., Mohan Kumar, B.L., Das Gupta, D.K., Development and evaluation of carrot kheer mix. *J. Food Sci. Technol.*, 40, 310, 2003.
48. Singh, B., Panesar, P.S., Nanda, V., Utilization of carrot pomace for the preparation of a value-added product. *World J. Dairy Food. Sci.*, 1, 22, 2006.
49. Bast, A., van den Berg, H., Van der Plas, R.M., Haenen, G.R.M., A-Carotene as antioxidant. *Eur. J. Clin. Nutr.*, 50, 554, 1996.
50. Olsen, J.A., Provitamin A function of carotenoids: The conversion of â-carotene to vitamin A. *J. Nutr.*, 119, 105, 1989.
51. Krishna, G., Carotene and tocopherol in agro-industrial by-products and wastes of the tropics. *Agric. Wastes*, 12, 235, 1985.
52. Chien, Y.L. and Potty, V., Studies on use of de-aromatised spices as a source of dietary fibre and minerals in bread. *J. Food. Technol.*, 39, 285, 1996.

53. Bruder, U., Bioplastics and biocomposites, in: *User's Guide to Plastic*, Carl Hanser Verlag, Munich, 2015.
54. Gomez-Heincke, D., Martínez, I., Stading, M., Gallegos, C., Partal, P., Improvement of mechanical and water absorption properties of plant protein based bioplastics. *Food Hydrocoll.*, 73, 21, 2017.
55. Mooney, B.P., The second green revolution? Production of plant-based biodegradable plastics. *Biochem. J.*, 418, 219, 2009.
56. Friedman, M., Potato glycoalkaloids and metabolites: Roles in the plant and in the diet. *J. Agric. Food Chem.*, 54, 8655, 2006.
57. Fewell, A.M. and Roddick, J.G., Potato glycoalkaloid impairment of fungal development. *Mycol. Res.*, 101, 597, 1997.
58. Jarvinen, R., Rauhala, H., Holopainen, U., Kallio, H., Differences in suberin content and composition between two varieties of potatoes (*Solanum tuberosum*) and effect of post-harvest storage to the composition. *LWT Food Sci. Technol.*, 44, 1355, 2011.
59. Outline, C., Starch and derivatives as pharmaceutical excipients, in: *Controlled Drug Delivery*, p. 21, 2015.
60. Dalla Costa, A.P., Hermes, V.S., de Oliveira Rios, A., Flores, S.H., Minimally processed beetroot waste as an alternative source to obtain functional ingredients. *J. Food Sci. Technol.*, 54, 2050, 2017.
61. Anjum, F., Shahid, M., Bukhari, S.A., Anwar, S., Latif, S., Study of quality characteristics and efficacy of extraction solvent/technique on the antioxidant activity of bitter gourd seed. *Food Process Technol.*, 4, 2, 2013.
62. Saha, S.S., Patra, M., Ghosh, M., *In vitro* antioxidant study of vegetable oils containing conjugated linolenic acid isomers. *LWT Food Sci. Technol.*, 46, 10, 2012.
63. Yoshime, L.T., de Melo, I.L.P., Sattler, J.A.G., de Carvalho, E.B.T., Mancini-Filho, J., Bitter gourd (*Momordica charantia* L.) seed oil as a naturally rich source of bioactive compounds for nutraceutical purposes. *Nutrire*, 41, 1, 2016.
64. Shetty, A.A., Rana, R., Buckseth, T., Preetham, S.P., Waste utilization in cucurbits: A review. *Waste Biomass Valorization*, 3, 363, 2012.
65. Upasana, D.V., Zaidi, M.G.H., Shukla, A.K., Utilization of pea industry waste for developing biodegradable product. *Pantnagar J. Res.*, 17, 262, 2019.
66. Verma, N., Bansal, M.C., Kumar, V., Pea peel waste: A lignocellulosic waste and its utility in cellulase production by *Trichoderma reesei* under solid-state cultivation. *Bioresource*, 6, 1505–1519, 2011.
67. Pandey, U.B., Garlic, in: *Handbook of Herbs and Spices*, Second, vol. 17, K.V. Peter (Ed.), pp. 299–318, Woodhead Publishing Limited, Elsevier, Cambridge, U.K, 2012.
68. de Albuquerque Landi, F.F., Fabiani, C., Castellani, B., Cotana, F., Pisello, A.L., Environmental assessment of four waste cooking oil valorization pathways. *Waste Manage.*, 138, 219, 2022.

11

Role of Microorganisms in Bioelectrochemical Systems for Hydrogen and Bioelectricity Production

Samsudeen Naina Mohamed[1], Boobalan Thulasinathan[2], Arun Alagarsamy[2], Sharon Mano Pappu J.[3], Bindhya K. Purushothaman[1] and Tamilmani Jayabalan[3]*

[1]*Department of Chemical Engineering, National Institute of Technology Tiruchirappalli, Tamil Nadu, India*
[2]*Department of Microbiology, Alagappa University, Karaikudi, Tamil Nadu, India*
[3]*Department of Biotechnology, National Institute of Technology Andhra Pradesh, Tadepalligudem, India*

Abstract

Over the years, technological progress has been witnessed from simple household fermentation to intricate large-scale industrial operations by deploying target-specific microorganisms to meet out sustainability. Recent research on various bioelectrochemical systems has evolved with the tremendous advancement in the production of bioelectricity, biohydrogen, and many value-added chemicals. The selection of microorganisms apt to the bioelectrochemical systems is crucial in influencing the performance as well as the yield of the bioprocess. Their key role and different mechanisms are highlighted in correspondence with Microbial Fuel cell, Microbial Electrolysis Cell, and Microbial desalination cell. In this chapter, various products of the individual bioelectrochemical systems are elaborated with parameters such as power density, current density, hydrogen yield, coulombic efficiency, and in a certain case, treatment efficiency. A variety of substrates in view of their use in different bioelectrochemical systems have been interrelated with the nature of the microorganisms to comprehend their role in the performance and output. This book chapter concludes with futuristic prospects and associated challenges with overall insights into the role of microorganisms in the modern-day production of bioelectricity and biohydrogen.

**Corresponding author*: tamilsiyaan@gmail.com

Sanjay Kumar, Narendra Kumar and Shahid-ul-Islam (eds.) Role of Microbes in Industrial Products and Processes, (279–304) © 2023 Scrivener Publishing LLC

***Keywords*:** Microbes, bioelectrochemical, MFC, MEC, MDC, bioelectricity, biohydrogen

11.1 Introduction

Microorganisms play a prominent role both in beneficial and detrimental activities associated with the living aspects of the human being. Besides the negative impacts, several microorganisms/microorganism-related products have been engineered and commercialized with the aid of modern-day technologies to industrial scale. Many commercial end products are being replaced with bio-based ones with less impact on the environment and more sustainability. Hydrogen is a significant industrial product used as one of the best energy candidates in the automotive and power sector. It could also be produced by the biological route avoiding the complex nature of the conventional energy-intensive production and high-temperature operation. Various green methods involved in sustainable hydrogen production are biophotolysis, photo-fermentation, dark fermentation and microbial electrolysis are the technologies involving the assistance of microorganisms for the conversion of chemicals into fuels [1, 2]. Bio photolysis is a novel biological technique involving photosynthetic microorganisms to split the water molecules into H_2 and O_2, requiring a large area for light collection, and the oxygen inhibition makes this a low-efficiency process [3]. Photo fermentation uses photosynthetic bacteria to produce hydrogen, which is catalyzed by nitrogenase under nitrogen-deficient conditions using light and employs anaerobic photoreactors with complicated designs [4]. Dark fermentation is a promising biohydrogen production process that involves moderate operating conditions, with a major drawback being the lower conversion of substrates due to the thermodynamic limitations [5]. Microbial Electrolysis Cell (MEC) is one of the significant Bioelectrochemical systems (BES), which is projected as a favorable sustainable green technology to tackle the ever-rising energy demands, specifically by utilizing wastewater as a substrate and simultaneously removing pollutants from the wastewater.

Primitive to MEC, Microbial Fuel Cell is the BESs initially developed to produce bioelectricity using specific types of microorganisms from a variety of substrates. Over the years, the technical progression in bioelectrochemical systems has significantly improved with robust and modern systems for high productivity. The microorganisms involved in BESs are specialized as their self-design or nature-design of decomposing organic matter or so-called "waste" or any wastewater into useful value-added products (either hydrogen or electricity) is a great boon to mankind. In this chapter, we see the role of these exclusive microorganisms with the history, mechanism involved, technology transition & advancements, and applications.

11.2 Evolution of BESs

In early of the 20th century, the astounding fact was discovered that 'some microorganisms could extracellularly transfer the electrons produced through degradation of organic matter. M. C. Potter was the first to research this and demonstrated his works as 'Electrical effects accompanying the decomposition of organic compounds [6]. However, it was the continuation of his paper published in 1910 describing the breakdown of organic compounds through microbes that occurred with simultaneous electrical energy liberation. The microorganisms were originally designated as Electro-Active Bacteria (EAB) and their biofilm formation could assist a competent organic removal, especially carbon sources, hence producing renewable energy in the form of bioelectricity in MFC and biohydrogen in MEC.

In 2005, based on the evolution of BES from primitive MFC, the idea of MEC technology was proposed [7]. Initially, this concept was started with the name of electrochemically assisted membrane bioreactor, later changed to bio-catalyzed electrolysis and then electrohydrogenesis, and lastly, as the present nomenclature-MEC. Many technological transformations had taken place in BES, and many advancements of systems like desalination, osmosis, dialysis, etc., were taken into integration and modification. Kim and Logan established a novel system named Microbial reverse electrodialysis Cell (MREC) by piling up five membrane pairs of reverse electrodialysis stack together into a MEC without the need of external power supply [8]. Mehanna and co-workers integrated the MEC concept with MDC and designated it as a microbial electrodialysis cell (MEDC) [9]. In recent innovations, combined MEC and MDC had been renamed as the microbial electrolysis desalination chemical production cell(MEDCC) for the dual purpose of water desalination of water and value-added chemical production [10]. Several other advancements have been taking place over the years, from simple MFCs to complex integrated robust systems.

11.3 Electrochemically Active Microbes and Extracellular Electron Transfer (EET) Mechanisms

Primarily, electro-active microorganisms are the main feature of bioelectrochemical systems having the main responsibility of electron transfer from the organic waste by degrading it effectively. Alternatively, some organisms are called exoelectrogens as they possess excellent electron transportability from the substrate to extracellular electron acceptors (EEA), short of any intermediary [11]. These organisms depend upon the external factors of MEC, namely pH, buffer strength, and temperature. The other names are electricigens and Anode Respiring Bacteria (ARB). These organisms' wide range of existence has been found in diverse environments like domestic and industrial wastewater, ocean and marine sediments, sewage sludge, etc. The known isolated diverse genetic group so far includes Acidobacteria, Actinobacteria, and Proteobacteria (α, β, γ, δ, and Epsilon types), and Firmicutes. These microorganisms are capable of transporting electrons to electrodes (conductive materials) through various surface contact mechanisms. Specifically, *sulfurreducens* [12], *metallireducens* [13] species of *Geobacter*, *oneidensis* strain 14063 [14], *oneidensis* [15] *oneidensis* MR-1 [16] species of *Shewanella*, *aeruginosa* [17], *fluorescence* [18] species of *Pseudomonas*, *pneumoniae* strain L17 [19], *variicola* [20] species of *Klebsiella pneumonia* and even *Escherichia coli* [21], *Rhodoferaxferrireducens* [22], etc. have been extensively used in MFCs.

There are certain differences between MFC and MEC organisms. Variety of microorganisms involved in various types of BESs are listed in Table 11.1. In MECs, the nature of the consortium has a significant hold in the quantity and quality of hydrogen produced using the system. Only limited studies are made on community analysis of anodes. Many organisms like *Shewanella*, *Geobacter*, *Pseudomonas*, *Desulfovibrio*, and *Rhodopseudomonas* are observed in these studies. The co-enrichment in MEC has been reported with EET microorganisms such as *Ochrobactrumanthropi Rhodoferaxferrireducens*, *Shewanellaputrefaciens*, *Rhodopseudomonas palustris*, and *Geobactersulfurreducens* [23] as these groups are used as mixed cultures in MFCs and other BESs. Also, methanogens are limiting factors in the HPR of reactors, and necessary precautions are to be followed to avoid methanogenesis. These methanogens severely impact both the yield and purity of H_2 production in the

Table 11.1 Variety of microorganisms used in BESs.

Organisms	Substrate	Bes design	Reference
Bacteroides members	Acetate	Single-chamber, 100 ml	[41]
Anaerobic sewage sludge	Acetate	Two-chamber, 326 ml	[42]
Thermincola species	Acetate	Two-chamber, 350 ml	[43]
Geobacter strains &mixed cultures	Acetate	Single-chamber, 18 ml	[44]
ARB biofilm (*Geobacter*-enriched)	Acetate	Single-chamber continuous, 140 mL (0.88 mL min^{-1})	[45]
Pelobacter sp./ Anaerobic digested sludge	Acetate	Two-chamber, 28 mL	[46]
Mixed cultures	Acetate	Two-chamber continuous, 200 mL (2.6 mL min^{-1})	[47]
Mixed cultures	Acetate	Two-chamber, 6.6 L	[48]
Mixed cultures	Acetate	Two-chamber, 200 mL	[49]
Anaerobic sludge	Acetate	Two-chamber, 120 mL	[50]
Mixed cultures	Domestic WW	Two-chamber continuous, 292 mL (0.368 g L^{-1})	[51]
Mixed cultures	Fermentation effluent	Single-chamber, 26 mL	[52]

(*Continued*)

Table 11.1 Variety of microorganisms used in BESs. (*Continued*)

Organisms	Substrate	Bes design	Reference
Mixed cultures	Food WW	Single-chamber, 28 mL	[53]
Ferroglobus and *Geoglobus* species	Hydrocarbons	Single-chamber, 5 ml	[54]
Geobacter strain biofilm	Potato WW	Single-chamber, 28 mL	[55]
Activated sludge biofilm	Rice straw	Two-chamber, 380 ml	[55]
Mixed cultures	Swine WW	Single-chamber, 28 mL	[57]
Clostridium strains enriched biofilm	Synthetic	Single-chamber, 28 ml	[58]

MECs [24]. On the other hand, pure culture facilitates controlled methane-free systems but sometimes results in low HPR [25]. Acetoclastic and hydrogenotrophic methanogens are involved in methane production in MECs from non-pure mixed cultures. *Methanosaetaceae* and *Methanosarcinaceae*are examples of the acetoclastic type, which can produce CH_4 by converting acetate substrate. *Methanomicrobiales* and *Methanobacteriales* are hydrogenotrophic types that convert methane from CO_2 and H_2 [26]. Another type, namely thermophilic microorganisms *Bacillus licheniformis, Therminicola* mediate the bio-electrocatalysis responsible as anode respiring bacteria advantageous comparatively to mesophilic microorganisms mainly by enhancing the rate of oxidation at anode [27].

In addition, there are various types of EET mechanisms involved in the transport of electrons to electrodes from the substrate [28] shown in Figure 11.1. The electron transfer rate directly depends on the type of mechanisms involved, and they can be listed as:

- Direct electron transfer
- Electron shuttle
- Conductive biofilm
- Conductive pili

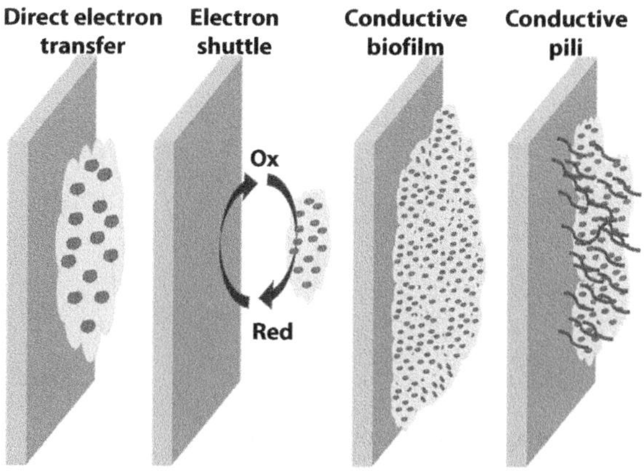

Figure 11.1 Various types of EET mechanisms.

11.4 Bioelectrochemical Systems

11.4.1 Microbial Fuel Cell (MFC)

MFCs are the primitive form of bioelectrochemical systems that can convert waste biological matter or bulk biomass into electrical current augmented by the catalysis of microorganisms [29]. The electrocatalytic microorganisms perform the main operation of oxidation of the substrates, which acts as fuel in this technology, after which they are responsible for the transport of electrons, thus enabling the significant difference from the conventional and commercial fuel cell systems. Last two decades, this technology has attracted more research as it is a cutting-edge technology involving simultaneous wastewater treatment with bioelectricity production.

In comparison to conventional bioenergy production, MFC technology is advantageous in several ways, such as (1) easy availability of raw material (substrate), as all organic matter could be used as a fuel in the microbial fuel cell; (2) the product is clean as the process is clean which directly converts the substrate or the biomass into electricity. Also, the undesired product is CO_2 gas only, which demand no further treatment; (3) reduced sludge generation; (4) normal operating condition generally milder conditions such as lower temperature and lower strength effluents can be used; (5) anaerobic operation and no aeration; (6) broad applications including removal of pollutants, hydrogen production and bioproduction [30]. The limitation of the technology is discussed in the challenges of the BES systems.

As MFCs are fueled by organic matter and bio catalyzed, the working principle can be explained with widely used dual-chambered MFC

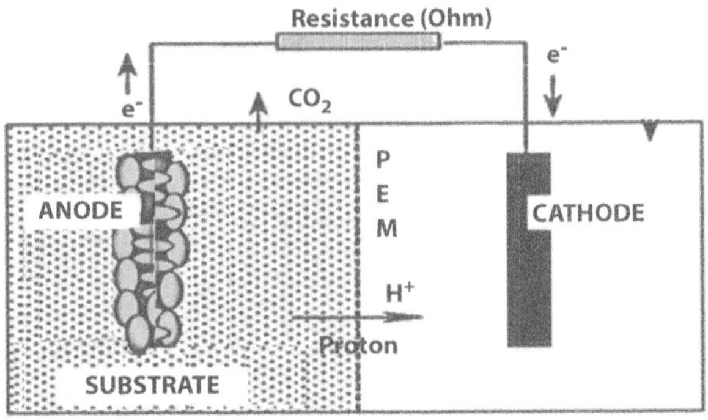

Figure 11.2 Schematic diagram of microbial fuel cell.

(Figure 11.2). It consists of an anaerobic anode compartment and an aerobic cathode compartment with a proton exchange membrane (PEM) placed in between them. Typically, the microorganisms are of a specific type that develops a biofilm over the anode in addition to the oxidation of the biological matter to protons and electrons. The H+ ions generated can only surpass the specially designed membrane and reach the other chamber, whereas the e-travel through the external circuit from the anode to the cathode, thus flow of e-responsible for bioelectricity production. The potential redox variant between the anode (low)and cathode (high) is the major driving force accredited to the electron transport, i.e., from the microorganism to the anode in the anaerobic compartment, then through the circuit with a current-carrying conductor (wire) connected to the cathode in the next compartment, in which the recombination of the electrons, protons, and oxygen (if available) take place. At the same electricity generation occurs, providing the suitable justification of the nomenclature of microbial fuel cells [29]. The wide variety of substrates comprising of numerous organic compounds in different waste and/or wastewater are used as fuels in MFCs for greener and cleaner biopower.

11.4.2 Microbial Electrolysis Cell (MEC)

MEC is one of the most promising technological developments in bioelectrochemical systems (BES). Similar to MFC, in MEC, the microorganisms disintegrate the organic content of the wastewater into electrons (e−) and protons (H^+), thus producing hydrogen with the addition of small voltage (<1.23V) less than water electrolysis [31].

This MEC technology is helpful in generating hydrogen from various wastewater and synthetic substrates. It has evolved from Microbial Fuel Cell

(MFC) that can produce electricity and simultaneously accomplish wastewater treatment [32]. Apart from hydrogen, a slight modification of the process produces other value-added products such as methane, acetate, hydrogen peroxide, ethanol, and formic acid [33]. Analogous to MFC, MEC also comprises a cathode and anode separated with a membrane and connected to an external power supply (DC), as shown in Figure 11.3. Recent research in this technology has focused majorly on the use of industrial and domestic effluents as a substrate rather than synthetic effluent [34].

In MEC, microbial interactions are responsible for producing hydrogen through the electrochemical conversion of the waste, which is the outcome of its metabolic action. The generated electron current moves to the anode, whereas the protons are discharged into solution directly, so they surpass the membrane and reach the cathode side for its recombination with electrons to form H_2. The main considerations on thermodynamics, electrolysis, and kinetics are essential to comprehend the MEC system better. For instance, Acetic acid is a substrate; the reactions involved are shown below:

At anode,

$$CH_3COOH + 2H_2O \rightarrow 2CO_2 + 8e^- + 8H^+ \tag{11.1}$$

At cathode,

$$8H^+ + 8e^- \rightarrow 4H^2 \tag{11.2}$$

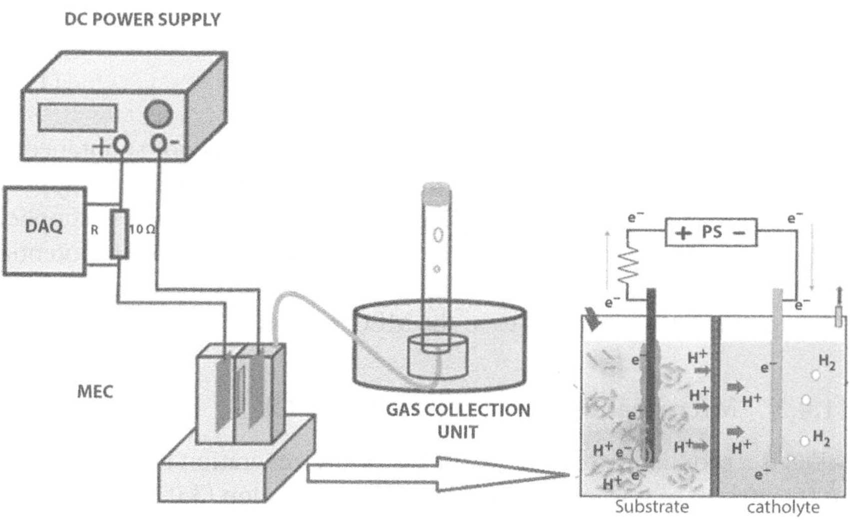

Figure 11.3 Schematic diagram of microbial electrolysis cell.

Overall reaction,

$$CH_3COOH + 2H_2O \rightarrow 2CO_2 + 4H^2 \tag{11.3}$$

It is better to calculate the reduction potential of cathode primarily so that each half-cell reaction can be used based on the Nernst equation as follows:

$$E_{cell} = E^0_{cell} - \frac{RT}{nF} \ln\left(\frac{Reactants}{Products}\right) \tag{11.4}$$

For cathode,

$$E_{cat} = E^0_{cat} - \frac{RT}{nF} \ln\left(\frac{pH_2}{\left[H^+\right]^8}\right) = -0.414\ V \tag{11.5}$$

For anode,

$$E_{an} = E^0_{an} - \frac{RT}{nF} \ln\left(\frac{\left[CH_3COOH\right]}{\left[H^+\right]^8}\right) = -0.300V \tag{11.6}$$

$$E_{cell} = E_{cat} - E_{an} = -0.114V \tag{11.7}$$

From these equations, it is clear that this reaction is not spontaneous, and to be favorable production, an external voltage must be supplied additionally of a value −0.114 V. Normally, an input voltage is supplied higher than this theoretical value but less than water electrolysis potential (>1.23V). Practically, this voltage is supplied to overcome ohmic, activation, and mass transport losses in the MEC system.

11.4.3 Microbial Desalination Cell (MDC)

MDC is a promising technology for the future generation for desalination and treatment of wastewater and electricity generation [35]. MDC consists of wastewater in an anode chamber and does not need superfluous bacteria because the effluent consists of organic matter, which is electro-energetic. The bacteria proliferate their count, forming bio-film, and generating electricity. The cling bio-film over the anode terminal surface oxidizes, the bio-pollutants present in emanating discharge electrons and protons [36]. A recently developed MDC consists of an anode, cathode,and salt chamber. The middle chamber is embedded with cation exchange membrane (CEM) and anion exchange membrane (AEM) on respective chambers. In the anode chamber, the substrate degradation promoting bioelectricity generation occurs, and the cathode chamber completes the electric loop with the mass transport of ions through respective membranes resulting in desalination in the middle chamber [37]. Over the years, different types of designs have been proposed for MDC. Microbial osmosis desalination cell (MODS) has evolved from MDC by supplanting AEM with an osmotic membrane. In this technology, ions move across the membrane while water can likewise to the saline water from the anode chamber because of higher osmotic pressure in the salt chamber. This can become disadvantageous as instead of diluting the salt, undesirable ions transport happens, slowing down the overall performance [38]. Stacked microbial desalination cell (SMDC) is a new type of design of CEM and AEM coupled that converts one saltwater chamber into a series of simpler units. SMDC can remove salt of 98% from saline water [39]. Bipolar membrane (coated selective layers of CEM and AEM) placed next to the anode AEM and CEM in Bipolar MDC novel design [36]. Anode chamber receives Cl^{-1} from the middle chamber; consequently, accumulation of Cl^{-1} results in acid production and affects the performance of MDC [40]. Other BES mentioned in the technical progression over time are at the initial development stages. Many efforts have been put over those systems to achieve better-optimized results.

11.5 Value-Added Products from BES

11.5.1 Bioelectricity Production

Bioelectrocatalysis in MFCs is the mechanism promoting electrochemical reactions where the biotic component (microorganisms) enhances the electron transfer kinetics through interaction with abiotic components (electrode). Similar to other BES, MFC performance is enhanced by the addition of electrocatalytic materials to the template electrode. Through the years, the advancement in intense research with various biocatalysts and electrocatalysts has progressed effectively for the overall MFC output improvement. In the experimentation of synthetic substrates, various sugars such as glucose, lactose, sucrose, galactose, maltose, and mannose have been employed in *Proteus vulgaris* enriched MFCs [59]. Liu *et al.* [60] demonstrated a single-chamber MFC operated with organic acids such as acetate and butyrate with power densities (PD) of 661 mW/m^2 and 349 mW/m^2 respectively. The effluent from the Protein food industry in dual-chambered MFC, achieved PD of 81 mW/m^2 with COD reduction of 95% [61]. While D-glucuronic acid fuelled MFC achieved 2770 $mW\ m^{-2}$ as the maximum PD with COD removal and Coulombic Efficiency (CE) of 89% and 24%, respectively. In the case of D-mannose substrate, it was not utilized effectively, and the reported PD was 1240 $mW\ m^{-2}$ with COD removal80% and CE ranging between 21–37% [62]. A PD of 1410.2 $mW\ m^{-2}$ was attained in Molasses wastewater fed MFCs [63]. A variety of volatile fatty acids (VFAs) were employed in MFC, yielding a PD of 240 $mW\ m^{-2}$ [64]. In Dairy effluents, the CE, maximum PD and COD removal were 37.16%, 621.13 mW/m^2, and 90.46%, respectively [65]. In nitrate substrate fueled MFC system (Single chamber), the microorganism inoculum was 10% aerobic denitrifying mixture culture exhibited an excellent PD of 1250 mW/m^2 with 90% COD removal [66]. Ullah and Zeshan demonstrated acetate as the effective substrate with 114 mW/m^2 as maximum PD and 79% COD removal [67]. In a study of testing rice husk charcoal as an electrode, poultry wastewater was used as a substrate. The MFC attained 6.9 Wm^{-3} PD with 40%COD removal efficiency [68]. A two-stage hybrid constructed wetland-MFC was studied with swine wastewater which showed 33.3 mW/m^3 PD and 72% COD removal [69]. The higher PD of 31 $W\ m^{-2}$ was achieved using sulfate-reducing bacteria in molasses wastewater substrate [70]. Recently, the natural microflora of Septic tank wastewater biodegraded the polycyclic aromatic hydrocarbons (PAHs) along with power generation with the help of mushroom and accounted for a PD $110 \pm 10\ mW\ m^{-2}$, CE 40% and COD removal 61% [71].

11.5.2 Biohydrogen Production

The variety of substrates employed in MECs directly influences the MEC performance of MEC, because of its nutrients or organic source and sometimes the *in-situ* microorganisms responsible for the quality and the quantity of hydrogen produced. MEC performance parameters such as columbic efficiency, hydrogen recoveries, and COD removal directly depend upon the nature of the microorganisms and the substrate as well. Synthetic substrates are investigated to acquire the basic knowledge and mechanism involved in MEC. For instance, Liu *et al.* [7] used acetate as substrate in dual-chamber MEC which achieved substrate degradation 90% CE 78%

and overall hydrogen production efficiency 60–73%. Many fermentation end products and simpler substrates are generally used in MECs. Cheng and Logan [72] have used substrates like glucose, cellulose, acetic acid, butyric acid, lactic acid, propionic acid, and valeric acid with a production rate of 1.23, 0.11, 1.10, 0.45, 1.04, 0.72, and 0.14 $m^3/m^3/D$, respectively [72]. Lu *et al.* [73] experimented with MEC with glucose as substrate with a better yield of 6 mol H_2/mol glucose and HPR of 25 ± 0.03–0.37 ± 0.04 m^3 $H_2/m^3/D$. Even with trehalose as a substrate for M.E.C., Xu *et al.* [74] attained HPR of 0.25 m^3 H_2m^{-3} D^{-1}, 46.4% Coulomb efficiency, and 203% energy recovery efficiency at an optimal concentration of 50 mmol L^{-1}.

Proteins have also been studied as substrates in M.E.C. Lu *et al.* [75] studied bioelectrohydrogenesis using bovine serum albumin (BSA) and peptone in MEC. The simple protein attained HPR value 0.42 ± 0.07 $m^3/m^3/D$ while the complex protein (peptone) under the same conditions, achieved 0.05 ± 0.01 $m^3/m^3/D$. Complex substrates such as lignocellulose and cellulose were also studied in MEC by Lalaurette *et al.* and Wang *et al.* [56, 58, 76]. Predominantly, Domestic wastewater (DWW) has extensively experimented in MEC as substrate. Ditzig *et al.* [51] investigated DWW in a 300 mL double-chambered reactor operated in a batch mode. Escapa *et al.* [77] reported the peak output of HPR at 0.3 L L_r^{-1}/D by varying hydraulic retention time in continuous flow mode: 3–6 h. A pilot-scale MEC was designed and operated with DWW with 120 L capacity operated over 3-month time period [78].

Industrial effluents with organic constituents have been used as fuel in MEC for biohydrogen production. Ren *et al.* [79] demonstrated the de-oiled refinery wastewater sample in MEC with industry effluent as substrate yielding 2.1 ± 0.2 A/m^2 current density, 79% COD removal, and 82% BOD removal. Tenca *et al.* [53] examined two different types of wastewaters, namely, methanol-rich industrial (IN) wastewater and food processing (FP)wastewater. The main finding was that IN wastewater displayed greater gas generation (0.8–1.8 m^3 m^{-3} D^{-1}), and COD reduction (1.8–2.8 kg COD m^{-3} D^{-1}) in comparison with FP wastewater. IN wastewater had overall energy recovery ranging between 3.1–3.8 (kWh kg^{-1} COD removed), and for FP wastewater, this was (−0.7 to 1.2 kWh kg^{-1} COD). Kiely *et al.* [55] demonstrated two different wastewaters, namely potato wastewater, dairy wastewater in MFC, and microbial communities' analysis for hydrogen production. A pilot-scale MEC setup (continuous flow) of 1000 L capacity was fabricated and investigated for the current generation and COD reduction employing winery wastewater by Cusick *et al.* [80]. Post enrichment, a steady SCOD removal of 62% ± 20% at hydraulic retention period of 24 h (0.9 V applied voltage), and after 100 days of operation,

Table 11.2 Biohydrogen production in MEC.

S. no	Design	Capacity(L)	Applied voltage (V)	Substrate	HPR (L/L/D)	Ref.
1.	DCC	0.042	0.6	Acetic acid	1.10	[72]
				Glucose	1.23	
				Butyric acid	0.45	
				Lactic acid	1.04	
				Propionic acid	0.72	
				Valeric acid	0.14	
2.	SCC	0.028	0.6	Acetate	1.99	[82]
3.	DCB	6.6	0.5	Acetate	0.02	[48]
4.	DCB	0.028	0.8	Acetate	0.052	[46]
5.	SCC	0.5	0.6	Acetate	0.53	[25]
6.	SCC	0.018	0.5	Acetate	2.0	[83]
7.	SCB	0.028	0.6	Acetate	1.7	[84]
		0.048	1.06	Acetate	0.76	
8.	DCB	0.24	0.8	Acetate	0.008–0.018	[50]
9.	SCC	0.40	1.0	Acetate	1.58	[85]
10.	SCC	0.026	0.6–0.9	Glucose	1.99–3.12	[73]
11.	SCC	0.028	0.5	Synthetic effluent Lignocellulose	0.96–1.11	[58]
12.	SCC	0.028	0.6	Proteins	0.42	[75]
13.	DCB	0.584	0.5	DWW	0.154	[51]
14.	SCC	0.028	0.5	Swine WW	0.9–1.1	[57]
15.	SCC	0.026	0.6	Fermentation effluent	1.41	[52]

(Continued)

Table 11.2 Biohydrogen production in MEC. (*Continued*)

S. no	Design	Capacity(L)	Applied voltage (V)	Substrate	HPR (L/L/D)	Ref.
16.	SCC	0.028	0.7	Industrial & FP WW	0.08–0.17	[53]
17.	SCC	0.028	0.9	Potato WW	0.74	[55]
18.	CFPS	1.0	0.9	Winery WW	0.19	[80]

*DCC, Dual Chamber Cylindrical/tubular; SCC, Single Chamber Cylindrical/tubular; DCB, Dual Chamber Box/Cube; DCF, Dual Chamber Flow cell; SCB, Single Chamber Box/Cube; CFPS, Continuous Flow Pilot Scale.

a maximum current density of 7.4 Am^{-3} was achieved with an HPR of 0.19 ± 0.04 LL^{-1} D^{-1} with about 86% methane conversion [80]. In the study of Wang *et al.* [81], the feasibility of operating the MEC at low temperatures was evaluated for molasses wastewater in a temperature range lower than 10 °C. Table 11.2 summarizes the various substrates/wastewater-fed MECs along with their HPR and other details.

11.6 Other Products and Chemical Production in Various BESs

As MFC mainly aims at bioenergy production, the valuable chemical production in larger scope is not feasible. However, nitrogen, phosphorus, and heavy metal removal can be easily accomplished by microbial operation. Coupling with other techniques, for example, integrating membrane separation with MFC had achieved hygienic large-scale water production [86]. In recent advancements, the photosynthetic MFC or the algal-based MFCs have the advantage of algal biomass harvesting, which in turn could lead to various valuable chemicals such as lutein, chlorophyll, xanthin, and carotene pigments [87]. In contrast, M.E.C. MEC is helpful in the production of many value-added chemicals other than pure hydrogen production. MECs' cathodic compartment has been pioneered largely with various applications, especially Microbial Electro Synthesis (MES) arises as a new concept of effective production of value-added products, based on the type of mechanism involved either the reduction of H_2O or Hydrogen.

Methane, Hydrogen peroxide, ethanol, formic acid are widely synthesized chemicals in MECs.

Methane gas is a potential candidate among the fuels to fulfill the global energy needs with the main benefit of higher calorific value. As discussed in the variety of microorganisms, if methanogenesis occurs naturally in the substrates (or due to mixed culture), methane production is substantial in MEC with no concerns about the quality as the hydrogen gas synthesized as the specific methanogens would convert it. The large scale of this methane production can be facilitated in a single chamber MEC completed eliminating the cost burden due to the presence of a costlier PEM membrane. This is a biorefinery approach where the hydrogen purity is compromised with a huge amount of methane gas production. At an applied voltage <0.7 V, a single-chambered MEC operated with *Methanobacterium palustre* had a current capture efficiency of 96% and production rate of 0.9 to 656 mmol-CH_4 d^{-1} m^{-2} of the biocathode [88]. The methane production rate varies with parameters such as culture, nature, and composition of wastewater or substrate and a prime factor -reactor architecture [89]. Instead of working on energy-intensive inhibition of methanogens, methane production in MEC is cost-effective and advantageous in several ways.

Hydrogen peroxide is a widely used industrial as well as medicinal chemistry, which could be synthesized in MEC with a slight difference in the cathode side mechanism. The production of $H_2O_2$was first demonstrated by the principle of organic matter bioelectrochemical degradation at anode integrated with the reduction of oxygen to H_2O_2 at the cathode. At an externally applied additional voltage of 0.5 V, the performance of H_2O_2 production rate was ranging between 1.7–2.1 kg m^{-3} D^{-1} and 83.1% overall efficiency in the simpler substrate of acetate [89]. Not only in MECs but also MFCs, the production of H_2O_2 simultaneous with bioelectricity generation has been achieved without any external power supply [90]. H_2O_2 produced in MECs serves as a good source for the Fenton reaction that can be carried out in BESs [91].

Simpler chemicals such as Formic acid can be synthesized in MECs as it has significant applications in the tannery, pharma field, and as laboratory chemicals. A special mechanism in MECs has to happen to favor the production of formic acid by the decomposition of the organics in anode along with the reduction of CO_2 occurring in the cathode. Recent advancements of this operation coupled with salt removal led to the concept of "microbial electrolysis desalination and chemical production cell' (MEDCC). It is a cutting edge production of formic acid than that of the electrodialysis bipolar membrane owing to less electric input [92]. Stacking of MFC with self-supporting electricity generation for electrochemical conversion of CO_2 into formic acid (101.8 mg L^{-1} D^{-1}) was attained with 64.8% CE [93].

11.7 Prospects and Challenges in BES Technology

The foremost important challenge in the construction of BESs is a selection of electrodes, the development of suitable materials and catalysts, membrane, and various reactor designs. The reactor design is one of the most important factors that affect the current density or hydrogen production by influencing the internal resistance of BES. Several types of reactor configurations (single, dual, up-flow, etc.) have been experimented with in lab-scale studies and are being promoted to semi-pilot and pilot-scale operation. They are operated under different conditions for enhancing the BES performance and optimized to reduce the overall cost of MECs. The selection characteristics of anode material for BESs are non-corrosive, highly conductive (electrical), highly biocompatible to support the adherence and proliferation of microorganisms, high surface to volume ratio, less overpotential, easy fabrication with low cost and scalability [94]. Based on the characteristics, many anode materials have been developed over the years and have been broadly categorized as carbon or non-carbon-based materials. Another route is the development of suitable pretreatment methods to enhance the properties of surfaces and to facilitate the adhesion of microorganisms and electron transport.

In general, carbon-based materials used for anodes are also suitable as cathode materials, but some modifications are essentially required for increasing the overall efficiency of BESs. There is a significant need for a suitable catalyst to drive the electron transfer kinetics at low overpotential. In the case of MECs, cathodic hydrogen production on carbon materials is linked to the problem of high overpotential. Noble metals such as platinum, palladium, etc., can be used because of their high stability and better catalytic activity in HER. Platinum (Pt) is a widely used noble metal for catalysts in MECs [94]. It is reported that the use of noble metal-based cathodic catalysts contributed to 47% of the total capital cost of MEC [95]. Various emerging catalysts are pioneered, such as phosphides, sulphides, and oxides of transition metals, focusing on hydrogen recovery and improving the power density of the system on applications in bioelectrochemical systems [96]. Apart from these factors, other operational parameters like electrolyte conditions (pH, concentration), temperature, and applied voltage (in the case of MEC) also play a vital role in BES performance.

Apart from the substrate and microorganisms, suitable material development against real-time effluents is a crucial challenge for scalability and sustainability. Traditionally, carbonaceous materials are used as both anode and cathode, owing to several advantages. In an anode, carbon-based materials are directly employed due to their biocompatibility and high conductivity. However, in the case of the cathode, precious metals (e.g., Pt, Au, etc.) have to be used to improve the hydrogen evolution kinetics. Pt-based cathodes have been reported with higher HPR, but the main setback is the incurred catalyst cost that contributes one-third of the total cost. So, it is essential to develop suitable cathode material with enhanced properties compared to the conventional benchmark catalysts. And commercial membrane cost is also high. Thus, the main critical point of BES is its high economics incurred by the materials of construction. The choice of electrode, membrane, and reactor architecture directly impact the overall cost. The biocompatibility of microorganisms involved in the splitting of proton and electron from the organics and their transport mechanism must be considered and supported with suitable choice. It is very tedious to develop a replacement for commercial membrane for practical applications without compromising the quality. Poisoning effects of precious metal catalysts are also a prime reason for research of alternative options for the long run and extending to a large-scale operation. Other than that, operational problems and culture maintenance are all major concerns related to technology transfer from bench-scale to pilot and then to industrial scale. The focus on value-added chemicals and simultaneous wastewater treatment can positively affect this technology. The coupling or integration of one or more BESsor with conventional treatment methods can improve the overall economy.

The variety of substrates (synthetic, domestic and industrial) and microorganisms is crucial in developing various specifications of BESs. Over the years more than two decades, the technological sustainability of the BESs technologies has been realized into the commercial market success. The high performance of BESs can be impacted due to the complexity of the substrate. The treatment efficiency or COD removal efficiency is directly affected due to the strength of wastewater considered for the BES operation. Other factors like useless biomass and metabolites, less substrate degradation, methanogenesis are responsible for the poor performance of the BESs. Scaling up strategies must be improved for the longer sustainability goals for cleaner and greener energy.

11.8 Conclusion

Considering various perspectives and challenges involved in BESs technology, the research, and development with the multi-disciplinary approach towards electrochemistry, microbiology, biotechnology, material engineering, etc., with a decent pace will prove its caliber to tackle many energy and environmental issues for sustenance. Purely, biological factors, i.e., the role of the microorganisms in the BESs, are highly significant in the technological advancement towards various future sustainable goals. BESs research with intense care and strenuous efforts in material development, identification of individual strains, simultaneous treatment will promote and positively influence the large-scale applications. Commercialization with techno-economic feasibility studies could authenticate the right direction in employing the microorganisms to benefit mankind through their self-design.

References

1. Acar, C. and Dincer, I., Comparative assessment of hydrogen production methods from renewable and non-renewable sources. *Int. J. Hydrogen Energy*, 39, 1, 1–12, 2014.
2. Manish, S. and Banerjee, R., Comparison of biohydrogen production processes. *Int. J. Hydrogen Energy*, 33, 1, 279–286, 2008.
3. Ding, C., Yang, K.-L., He, J., Biological and fermentative production of hydrogen, in: *Handbook of Biofuels Production*, Second, R. Luque, C.S.K. Lin, K. Wilson, J. Clark, (Eds.), pp. 303–333, Woodhead Publishing, Cambridge, England, 2016.
4. Zhang, Q. and Zhang, Z., Biological hydrogen production from renewable resources by photofermentation, vol. 3, pp. 137–160, Elsevier, Cambridge, Massachusetts, 2018.
5. Singh, V. and Das, D., Potential of hydrogen production from biomass, in: *Science and Engineering of Hydrogen-Based Energy Technologies*, P.E.V. De Miranda, (Ed.), pp. 123–164, Academic Press, Cambridge, Massachusetts, 2019.
6. Potter, M.C., Electrical effects accompanying the decomposition of organic compounds. *Proc. R. Soc. Lond. B Biol. Sci.*, 84, 571, 260–276, 1911.
7. Liu, H., Grot, S., Logan, B.E., Electrochemically assisted microbial production of hydrogen from acetate. *Environ. Sci. Technol.*, 39, 11, 4317–4320, 2005.
8. Kim, Y. and Logan, B.E., Hydrogen production from inexhaustible supplies of fresh and salt water using microbial reverse-electrodialysis electrolysis cells. *Proc. Natl. Acad. Sci.*, 108, 39, 16176–16181, 2011.

9. Mehanna, M., Kiely, P.D., Call, D.F., Logan, B.E., Microbial electrodialysis cell for simultaneous water desalination and hydrogen gas production. *Environ. Sci. Technol.*, 44, 24, 9578–9583, 2010.
10. Chen, S., Liu, G., Zhang, R., Qin, B., Luo, Y., Development of the microbial electrolysis desalination and chemical-production cell for desalination as well as acid and alkali productions. *Environ. Sci. Technol.*, 46, 4, 2467–2472, 2012.
11. Angelaalincy, M.J., Navanietha Krishnaraj, R., Shakambari, G., Ashokkumar, B., Kathiresan, S., Varalakshmi, P., Biofilm engineering approaches for improving the performance of microbial fuel cells and bioelectrochemical systems. *Front. Energy Res.*, 6, 1–12, July 2018.
12. Ishii, S., Watanabe, K., Yabuki, S., Logan, B.E., Sekiguchi, Y., Comparison of electrode reduction activities of geobacter sulfurreducens and an enriched consortium in an air-cathode microbial fuel cell. *Appl. Environ. Microbiol.*, 74, 23, 7348–7355, 2008.
13. Min, B., Cheng, S., Logan, B.E., Electricity generation using membrane and salt bridge microbial fuel cells. *Water Res.*, 39, 9, 1675–1686, 2005.
14. Fernando, E., Keshavarz, T., Kyazze, G., Enhanced bio-decolourisation of acid orange 7 by Shewanella oneidensis through co-metabolism in a microbial fuel cell. *Int. Biodeterior. Biodegradation*, 72, 1–9, 2012.
15. Ringeisen, B.R., Henderson, E., Wu, P.K., Pietron, J., Ray, R., Little, B., Biffinger, J.C., Jones-Meehan, J.M., High power density from a miniature microbial fuel cell using shewanella oneidensis DSP10. *Environ. Sci. Technol.*, 40, 8, 2629–2634, 2006.
16. Baron, D., LaBelle, E., Coursolle, D., Gralnick, J.A., Bond, D.R., Electrochemical measurement of electron transfer kinetics by shewanella oneidensis MR-1. *J. Biol. Chem.*, 284, 42, 28865–28873, 2009.
17. Zhang, M., Ma, Z., Zhao, N., Zhang, K., Song, H., Increased power generation from cylindrical microbial fuel cell inoculated with P. aeruginosa. *Biosens. Bioelectron.*, 141, 111394, 2019.
18. Kaushik, A. and Jadhav, S.K., Conversion of waste to electricity in a microbial fuel cell using newly identified bacteria: *Pseudomonas fluorescens*. *Int. J. Environ. Sci. Technol.*, 14, 1771–1780, 2017.
19. Liu, L., Li, F., Feng, C., Li, X., Microbial fuel cell with an azo-dye-feeding cathode. *Appl. Microbiol. Biotechnol.*, 85, 1, 175–183, 2009.
20. Islam, M.A., Ethiraj, B., Cheng, C.K., Yousuf, A., Khan, M.M.R., An insight of synergy between *Pseudomonas Aeruginosa and Klebsiella Variicola* in a microbial fuel cell. *ACS Sustain. Chem. Eng.*, 6, 3, 4130–4137, 2018.
21. Cournet, A., Délia, M.-L., Bergel, A., Roques, C., Bergé, M., Electrochemical reduction of oxygen catalyzed by a wide range of bacteria including gram-positive. *Electrochem. Commun.*, 12, 4, 505–508, 2010.
22. Liu, Z., Lian, J., Du, Z., Li, H., Construction of sugar-based microbial fuel cells by dissimilatory metal reduction bacteria. *Chin. J. Biotechnol.*, 22, 1, 131–137, 2006.

23. Fedorovich, V., Knighton, M.C., Pagaling, E., Ward, F.B., Free, A., Goryanin, I., Novel electrochemically active bacterium phylogenetically related to Arcobacter butzleri, isolated from a microbial fuel cell. *Appl. Environ. Microbiol.*, 75, 23, 7326–7334, 2009.
24. Tice, R.C. and Kim, Y., Methanogenesis control by electrolytic oxygen production in microbial electrolysis cells. *Int. J. Hydrogen Energy*, 39, 7, 3079–3086, 2014.
25. Hu, H., Fan, Y., Liu, H., Hydrogen production using single-chamber membrane-free microbial electrolysis cells. *Water Res.*, 42, 15, 4172–4178, 2008.
26. Karthikeyan, R., Cheng, K.Y., Selvam, A., Bose, A., Wong, J.W.C., Bioelectrohydrogenesis and inhibition of methanogenic activity in microbial electrolysis cells-A review. *Biotechnol. Adv.*, 35, 6, 758–771, 2017.
27. Rathinam, N.K., Bibra, M., Salem, D.R., Sani, R.K., Thermophiles for biohydrogen production in microbial electrolytic cells. *Bioresour. Technol.*, 277, 171–178, January 2019.
28. Reguera, G., McCarthy, K.D., Mehta, T., Nicoll, J.S., Tuominen, M.T., Lovley, D.R., Extracellular electron transfer via microbial nanowires. *Nature*, 435, 7045, 1098–1101, 2005.
29. Cheng, S. and Logan, B.E., Increasing power generation for scaling up single-chamber air cathode microbial fuel cells. *Bioresour. Technol.*, 102, 6, 4468–4473, 2011.
30. Rabaey, K. and Verstraete, W., Microbial fuel cells: Novel biotechnology for energy generation. *Trends Biotechnol.*, 23, 6, 291–298, 2005.
31. Liu, H., Grot, S., Logan, B.E., Electrochemically assisted microbial production of hydrogen from acetate. *Environ. Sci. Technol.*, 39, 11, 4317–4320, 2005.
32. Sleutels, T.H.J.A., Ter Heijne, A., Buisman, C.J.N., Hamelers, H.V.M., Bioelectrochemical systems: An outlook for practical applications. *ChemSusChem*, 5, 6, 1012–1019, 2012.
33. Kadier, A., Kalil, M.S., Abdeshahian, P., Chandrasekhar, K., Mohamed, A., Azman, N.F., Logroño, W., Simayi, Y., Hamid, A.A., Recent advances and emerging challenges in microbial electrolysis cells (MECs) for microbial production of hydrogen and value-added chemicals. *Renew. Sust. Energ. Rev.*, 61, 501–525, 2016.
34. Kadier, A., Simayi, Y., Kalil, M.S., Abdeshahian, P., Hamid, A.A., A review of the substrates used in microbial electrolysis cells (MECs) for producing sustainable and clean hydrogen gas. *Renew. Energy*, 71, 466–472, 2014.
35. Ragab, M., Elawwad, A., Abdel-Halim, H., Evaluating the performance of microbial desalination cells subjected to different operating temperatures. *Desalination*, 462, 56–66, April 2019.
36. Carmalin Sophia, A., Bhalambaal, V.M., Lima, E.C., Thirunavoukkarasu, M., Microbial desalination cell technology: Contribution to sustainable waste

water treatment process, current status and future applications. *J. Environ. Chem. Eng.*, 4, 3, 3468–3478, 2016.

37. Saeed, H.M., Husseini, G.A., Yousef, S., Saif, J., Al-Asheh, S., Abu Fara, A., Azzam, S., Khawaga, R., Aidan, A., Microbial desalination cell technology: A review and a case study. *Desalination*, 359, 1–13, 2015.
38. Zhang, B. and He, Z., Integrated salinity reduction and water recovery in an osmotic microbial desalination cell. *RSC Adv.*, 2, 8, 3265–3269, 2012.
39. Chen, X., Xia, X., Liang, P., Cao, X., Sun, H., Huang, X., Stacked microbial desalination cells to enhance water desalination efficiency. *Environ. Sci. Technol.*, 45, 6, 2465–2470, 2011.
40. Chen, S., Liu, G., Zhang, R., Qin, B., Luo, Y., Development of the microbial electrolysis desalination and chemical-production cell for desalination as well as acid and alkali productions. *Environ. Sci. Technol.*, 46, 4, 2467–2472, 2012.
41. Shehab, N.A., Ortiz-Medina, J.F., Katuri, K.P., Hari, A.R., Amy, G., Logan, B.E., Saikaly, P.E., Enrichment of extremophilic exoelectrogens in microbial electrolysis cells using Red Sea brine pools as inocula. *Bioresour. Technol.*, 239, 82–86, 2017.
42. Kyazze, G., Popov, A., Dinsdale, R., Esteves, S., Hawkes, F., Premier, G., Guwy, A., Influence of catholyte pH and temperature on hydrogen production from acetate using a two chamber concentric tubular microbial electrolysis cell. *Int. J. Hydrogen Energy*, 35, 15, 7716–7722, 2010.
43. Lusk, B.G., Parameswaran, P., Popat, S.C., Rittmann, B.E., Torres, C.I., The effect of pH and buffer concentration on anode biofilms of Thermincola ferriacetica. *Bioelectrochemistry*, 112, 47–52, 2016.
44. Call, D.F., Wagner, R.C., Logan, B.E., Hydrogen production by geobacter species and a mixed consortium in a microbial electrolysis cell. *Appl. Environ. Microbiol.*, 75, 24, 7579–7587, 2009.
45. Lee, H.-S., Torres, C.I., Parameswaran, P., Rittmann, B.E., Fate of H2 in an upflow single-chamber microbial electrolysis cell using a metal-catalyst-free cathode. *Environ. Sci. Technol.*, 43, 20, 7971–7976, 2009.
46. Chae, K.J., Choi, M.J., Lee, J., Ajayi, F.F., Kim, I.S., Biohydrogen production via biocatalyzed electrolysis in acetate-fed bioelectrochemical cells and microbial community analysis. *Int. J. Hydrogen Energy*, 33, 19, 5184–5192, 2008.
47. Jeremiasse, A.W., Bergsma, J., Kleijn, J.M., Saakes, M., Buisman, C.J.N., Stuart, M.C., Hamelers, H.V.M., Performance of metal alloys as hydrogen evolution reaction catalysts in a microbial electrolysis cell. *Int. J. Hydrogen Energy*, 36, 17, 10482–10489, 2011.
48. Rozendal, R.A., Hamelers, H.V.M., Euverink, G.J.W., Metz, S.J., Buisman, C.J.N., Principle and perspectives of hydrogen production through biocatalyzed electrolysis. *Int. J. Hydrogen Energy*, 31, 12, 1632–1640, 2006.

49. Lee, M.Y., Kim, K.Y., Yang, E., Kim, I.S., Evaluation of hydrogen production and internal resistance in forward osmosis membrane integrated microbial electrolysis cells. *Bioresour. Technol.*, 187, 106–112, 2015.
50. Xiao, L., Wen, Z., Ci, S., Chen, J., He, Z., Carbon/iron-based nanorod catalysts for hydrogen production in microbial electrolysis cells. *Nano Energy*, 1, 5, 751–756, 2012.
51. Ditzig, J., Liu, H., Logan, B.E., Production of hydrogen from domestic wastewater using a bioelectrochemically assisted microbial reactor (BEAMR). *Int. J. Hydrogen Energy*, 32, 13, 2296–2304, 2007.
52. Lu, L., Ren, N., Xing, D., Logan, B.E., Hydrogen production with effluent from an ethanol–H2-coproducing fermentation reactor using a single-chamber microbial electrolysis cell. *Biosens. Bioelectron.*, 24, 10, 3055–3060, 2009.
53. Tenca, A., Cusick, R.D., Schievano, A., Oberti, R., Logan, B.E., Evaluation of low cost cathode materials for treatment of industrial and food processing wastewater using microbial electrolysis cells. *Int. J. Hydrogen Energy*, 38, 4, 1859–1865, 2013.
54. Yilmazel, Y.D., Zhu, X., Kim, K.-Y., Holmes, D.E., Logan, B.E., Electrical current generation in microbial electrolysis cells by hyperthermophilic archaea Ferroglobus placidus and geoglobus ahangari. *Bioelectrochemistry*, 119, 142–149, 2018.
55. Kiely, P.D., Cusick, R., Call, D.F., Selembo, P.A., Regan, J.M., Logan, B.E., Anode microbial communities produced by changing from microbial fuel cell to microbial electrolysis cell operation using two different wastewaters. *Bioresour. Technol.*, 102, 1, 388–394, 2011.
56. Wang, Y.-Z., Zhang, L., Xu, T., Ding, K., Influence of initial anolyte pH and temperature on hydrogen production through simultaneous saccharification and fermentation of lignocellulose in microbial electrolysis cell. *Int. J. Hydrogen Energy*, 42, 36, 22663–22670, 2017.
57. Wagner, R.C., Regan, J.M., Oh, S.-E., Zuo, Y., Logan, B.E., Hydrogen and methane production from swine wastewater using microbial electrolysis cells. *Water Res.*, 43, 5, 1480–1488, 2009.
58. Lalaurette, E., Thammannagowda, S., Mohagheghi, A., Maness, P.-C., Logan, B.E., Hydrogen production from cellulose in a two-stage process combining fermentation and electrohydrogenesis. *Int. J. Hydrogen Energy*, 34, 15, 6201–6210, 2009.
59. Kim, N., Choi, Y., Jung, S., Kim, S., Effect of initial carbon sources on the performance of microbial fuel cells containing Proteus vulgaris. *Biotechnol. Bioeng.*, 70, 1, 109–114, 2000.
60. Liu, H., Cheng, S., Logan, B.E., Production of electricity from acetate or butyrate using a single-chamber microbial fuel cell. *Environ. Sci. Technol.*, 39, 2, 658–662, 2005.

61. Oh, S. and Logan, B.E., Hydrogen and electricity production from a food processing wastewater using fermentation and microbial fuel cell technologies. *Water Res.*, 39, 19, 4673–4682, 2005.
62. Catal, T., Li, K., Bermek, H., Liu, H., Electricity production from twelve monosaccharides using microbial fuel cells. *J. Power Sources*, 175, 1, 196–200, 2008.
63. Zhang, B., Zhao, H., Zhou, S., Shi, C., Wang, C., Ni, J., A novel UASB–MFC–BAF integrated system for high strength molasses wastewater treatment and bioelectricity generation. *Bioresour. Technol.*, 100, 23, 5687–5693, 2009.
64. Choi, J., Chang, H.N., Han, J.-I., Performance of microbial fuel cell with volatile fatty acids from food wastes. *Biotechnol. Lett.*, 33, 4, 705–714, 2011.
65. Mansoorian, H.J., Mahvi, A.H., Jafari, A.J., Khanjani, N., Evaluation of dairy industry wastewater treatment and simultaneous bioelectricity generation in a catalyst-less and mediator-less membrane microbial fuel cell. *J. Saudi Chem. Soc.*, 20, 1, 88–100, 2016.
66. Yang, N., Zhan, G., Li, D., Wang, X., He, X., Liu, H., Complete nitrogen removal and electricity production in Thauera-dominated air-cathode single chambered microbial fuel cell. *Chem. Eng. J.*, 356, 506–515, 2019.
67. Ullah, Z. and Zeshan, S., Effect of substrate type and concentration on the performance of a double chamber microbial fuel cell. *Water Sci. Technol.*, 81, 7, 1336–1344, 2020.
68. Oyiwona, G.E., Ogbonna, J.C., Anyanwu, C.U., Okabe, S., Electricity generation potential of poultry droppings wastewater in microbial fuel cell using rice husk charcoal electrodes. *Bioresour. Bioprocess*, 5, 1, 13, 2018.
69. Ren, B., Wang, T., Zhao, Y., Two-stage hybrid constructed wetland-microbial fuel cells for swine wastewater treatment and bioenergy generation. *Chemosphere*, 268, 128803, 2021.
70. Akgul, V., Cirik, K., Duyar, A., Koroglu, E.O., The effect of olive oil mill and molasses wastewater as a co-substrate during simultaneous textile wastewater treatment and energy generation. *Int. J. Glob. Warm.*, 23, 1, 30, 2021.
71. Thulasinathan, B., Jayabalan, T., Sethupathi, M., Kim, W., Muniyasamy, S., Sengottuvelan, N., Nainamohamed, S., Ponnuchamy, K., Alagarsamy, A., Bioelectricity generation by natural microflora of septic tank wastewater (STWW) and biodegradation of persistent petrogenic pollutants by basidiomycetes fungi: An integrated microbial fuel cell system. *J. Hazard. Mater.*, 412, 125228, 2021.
72. Cheng, S. and Logan, B.E., Sustainable and efficient biohydrogen production via electrohydrogenesis. *Proc. Natl. Acad. Sci.*, 104, 47, 18871–18873, 2007.
73. Lu, L., Xing, D., Ren, N., Logan, B.E., Syntrophic interactions drive the hydrogen production from glucose at low temperature in microbial electrolysis cells. *Bioresour. Technol.*, 124, 68–76, 2012.
74. Xu, L., Liu, W., Wu, Y., Lee, P., Wang, A., Li, S., Trehalose enhancing microbial electrolysis cell for hydrogen generation in low temperature (0°C). *Bioresour. Technol.*, 166, 458–463, 2014.

75. Lu, L., Xing, D., Xie, T., Ren, N., Logan, B.E., Hydrogen production from proteins via electrohydrogenesis in microbial electrolysis cells. *Biosens. Bioelectron.*, 25, 12, 2690–2695, 2010.
76. Wang, A., Sun, D., Cao, G., Wang, H., Ren, N., Wu, W.M., Logan, B.E., Integrated hydrogen production process from cellulose by combining dark fermentation, microbial fuel cells, and a microbial electrolysis cell. *Bioresour. Technol.*, 102, 5, 4137–4143, 2011.
77. Escapa, A., Gil-Carrera, L., García, V., Morán, A., Performance of a continuous flow microbial electrolysis cell (MEC) fed with domestic wastewater. *Bioresour. Technol.*, 117, 55–62, 2012.
78. Heidrich, E.S., Edwards, S.R., Dolfing, J., Cotterill, S.E., Curtis, T.P., Performance of a pilot scale microbial electrolysis cell fed on domestic wastewater at ambient temperatures for a 12month period. *Bioresour. Technol.*, 173, 87–95, 2014.
79. Ren, L., Siegert, M., Ivanov, I., Pisciotta, J.M., Logan, B.E., Treatability studies on different refinery wastewater samples using high-throughput microbial electrolysis cells (MECs). *Bioresour. Technol.*, 136, 322–328, 2013.
80. Cusick, R.D., Bryan, B., Parker, D.S., Merrill, M.D., Mehanna, M., Kiely, P.D., Liu, G., Logan, B.E., Performance of a pilot-scale continuous flow microbial electrolysis cell fed winery wastewater. *Appl. Microbiol. Biotechnol.*, 89, 6, 2053–2063, 2011.
81. Wang, Y., Guo, W.-Q., Xing, D.-F., Chang, J.-S., Ren, N.-Q., Hydrogen production using biocathode single-chamber microbial electrolysis cells fed by molasses wastewater at low temperature. *Int. J. Hydrogen Energy*, 39, 33, 19369–19375, 2014.
82. Call, D. and Logan, B.E., Hydrogen production in a single chamber microbial electrolysis cell lacking a membrane. *Environ. Sci. Technol.*, 42, 9, 3401–3406, 2008.
83. Hu, H., Fan, Y., Liu, H., Hydrogen production in single-chamber tubular microbial electrolysis cells using non-precious-metal catalysts. *Int. J. Hydrogen Energy*, 34, 20, 8535–8542, 2009.
84. Call, D.F., Merrill, M.D., Logan, B.E., High surface area stainless steel brushes as cathodes in microbial electrolysis cells. *Environ. Sci. Technol.*, 43, 6, 2179–2183, 2009.
85. Guo, K., Tang, X., Du, Z., Li, H., Hydrogen production from acetate in a cathode-on-top single-chamber microbial electrolysis cell with a mipor cathode. *Biochem. Eng. J.*, 51, 1–2, 48–52, 2010.
86. Zinadini, S., Zinatizadeh, A.A., Rahimi, M., Vatanpour, V., Bahrami, K., Energy recovery and hygienic water production from wastewater using an innovative integrated microbial fuel cell–membrane separation process. *Energy*, 141, 1350–1362, 2017.
87. Arun, S., Sinharoy, A., Pakshirajan, K., Lens, P.N.L., Algae based microbial fuel cells for wastewater treatment and recovery of value-added products. *Renew. Sust. Energ. Rev.*, 132, 110041, June 2020.

88. Cheng, S., Xing, D., Call, D.F., Logan, B.E., Direct biological conversion of electrical current into methane by electromethanogenesis. *Environ. Sci. Technol.*, 43, 10, 3953–3958, 2009.
89. Chae, K.-J., Choi, M.-J., Kim, K.-Y., Ajayi, F.F., Chang, I.-S., Kim, I.S., Selective inhibition of methanogens for the improvement of biohydrogen production in microbial electrolysis cells. *Int. J. Hydrogen Energy*, 35, 24, 13379–13386, 2010.
90. Shi-Jie, Y., Jing-Yuan, W., Nan-Qi, R., Xiu-Heng, W., Jin-Na, Z., Sustainable conversion of glucose into hydrogen peroxide in a solid polymer electrolyte microbial fuel cell. *ChemSusChem*, 3, 3, 334–338, 2010.
91. Feng, C.-H., Li, F.-B., Mai, H.-J., Li, X.-Z., Bio-electro-fenton process driven by microbial fuel cell for wastewater treatment. *Environ. Sci. Technol.*, 44, 5, 1875–1880, 2010.
92. Lu, Y., Luo, H., Yang, K., Liu, G., Zhang, R., Li, X., Ye, B., Formic acid production using a microbial electrolysis desalination and chemical-production cell. *Bioresour. Technol.*, 243, 118–125, 2017.
93. Zhao, H.-Z., Zhang, Y., Chang, Y.-Y., Li, Z.-S., Conversion of a substrate carbon source to formic acid for carbon dioxide emission reduction utilizing series-stacked microbial fuel cells. *J. Power Sources*, 217, 59–64, 2012.
94. Logan, B.E., Call, D., Cheng, S., Hamelers, H.V.M., Sleutels, T.H.J.A., Jeremiasse, A.W., Rozendal, R.A., Microbial electrolysis cells for high yield hydrogen gas production from organic matter. *Environ. Sci. Technol.*, 42, 23, 8630–8640, 2008.
95. Rozendal, R.A., Hamelers, H.V.M., Rabaey, K., Keller, J., Buisman, C.J.N., Towards practical implementation of bioelectrochemical wastewater treatment. *Trends Biotechnol.*, 26, 8, 450–459, 2008.
96. Ahmed, J., Yuan, Y., Zhou, L., Kim, S., Carbon supported cobalt oxide nanoparticles-iron phthalocyanine as alternative cathode catalyst for oxygen reduction in microbial fuel cells. *J. Power Sources*, 208, 170–175, 2012.

12

Biogas from Microalgae: Production Approaches and Strategies for Process Optimization

Jorge Alberto Vieira Costa[1*], Priscilla Quenia Muniz Bezerra[1], Juliana Botelho Moreira[2], Alex Nunes Molina[1] and Michele Greque de Morais[2]

[1]Laboratory of Biochemical Engineering, College of Chemistry and Food Engineering, Federal University of Rio Grande, Rio Grande, RS, Brazil
[2]Laboratory of Microbiology and Biochemistry, College of Chemistry and Food Engineering, Federal University of Rio Grande, Rio Grande, RS, Brazil

Abstract

The production of biogas from microalgae biomass is an attractive alternative in the biofuel industry. The cultivation of these microorganisms can be carried out with wastewater and industrial effluents, which contributes to reducing costs and negative impacts on the environment. Microalgal biomass is an ideal substrate for the nutrient demands of bacteria that act in anaerobic digestion. With microalgae technology, both raw biomass and residual biomass can be used for biogas production. Thus, it is possible to maximize the use of biomass by extracting various compounds and generating appropriate residual biomass for obtaining biogas. Moreover, the microalgal biorefinery concept encourages the production of this biofuel by suggesting the application of biomass in different industrial sectors. In this context, this chapter presents the biotechnological strategies, recovery, and pretreatments of microalgal biomass to maximize the production of biogas. Future trends inserted in the biorefinery context are also addressed for the sustainable production of biogas from microalgae.

Keywords: Anaerobic digestion, biogas upgrading, biomass microalgal, biorefinery, carbon dioxide, methane

**Corresponding author*: jorgealbertovc@terra.com.br

Sanjay Kumar, Narendra Kumar and Shahid-ul-Islam (eds.) Role of Microbes in Industrial Products and Processes, (305–330) © 2023 Scrivener Publishing LLC

12.1 Introduction

Biofuels obtained from microalgae have gained prominence due to their ecologically correct and renewable nature [1–3]. With the microalgae technology, it is possible to produce hydrogen (H_2) through a photobiological process, biodiesel from biomass lipids, bioethanol by the fermentation of carbohydrates, and biogas through anaerobic digestion of crude and/or residual microalgal biomass [3].

Anaerobic digestion has a low operating cost without the need to dry the organic matrix. Besides, the possibility of co-digestion provides an advantage in waste management [4, 5]. This biochemical process consists of the transformation of organic matter into biogas by bacterial species under oxygen-free conditions. Hydrolysis, acidogenesis, acetogenesis, and methanogenesis are steps observed during the production of biogas. Depending on the chemical composition of the digested organic matrix, biogas is composed mainly of methane (CH_4) (55–70%, v v^{-1}) and carbon dioxide (CO_2) (30–45%, v v^{-1}), and the effluent from anaerobic digestion is rich in nitrogen [3, 6].

Biogas can be produced from several sources of biomass, including animal manure, food, agricultural, forest residues, and those obtained after extraction of microalgal compounds. Microalgal biomass acts as an ideal substrate for the nutrient demands of bacteria that act in anaerobic digestion [7]. Moreover, other advantages include suitability for cultivation on non-arable land using wastewater and seawater as a growth medium, viability to recycle nutrients, minimum sludge production, lower operating costs, and lower energy consumption [6]. Anaerobic digestion based on microalgae was first reported using *Chlorella* sp. and *Scenedesmus* sp. with biogas production of 0.17–0.32 L CH_4 g VS^{-1} (volatile solids) [8]. Recently, a wide range of biogas yield (0.05 to 0.8 L CH_4 g VS^{-1}) has been reported for microalgae biomass, representing about 40–72% of the theoretical maximum production [3]. Furthermore, the potential for biogas production from microalgae biomass is related to the cell characteristics and cell wall of the strains, carbon, and nitrogen (C: N) ratio of the biomass and operational conditions of anaerobic digestion (pH, temperature, and configuration of the reactor) [5].

Although lignocellulosic waste is commonly used in the production of biogas, its biodegradability is slow. Lignin, an essential component of these residues, is resistant to hydrolysis and toxic to some microorganisms active in anaerobic digestion [6]. On the other hand, microalgae, which do not have lignin, tend to present greater biodegradability front the chemical reactions of anaerobic digestion [4]. The rupture of the cell wall of these microorganisms is one of the main actions to maximize the production of biogas. In this sense, pretreatment methods (physical, thermal, biological, or chemical) in biomass must be performed [9].

Once purified to CH_4, biogas can be used as a vehicle fuel or to supply the natural gas grid [10]. In this way, physical-chemical and biological approaches have been evaluated to remove or transform the CO_2 contained in the raw biogas to increase the CH_4 content. This process is denominated in biogas upgrading. In this context, the use of microalgae has been proposed as a promising alternative for this purpose [11]. Moreover, the production of biogas from microalgae can be integrated into wastewater treatment processes. Effluents from anaerobic digestion can be recycled for the cultivation of microalgae [5].

The residual biomass obtained from lipid and carbohydrate extraction processes in microalgae promotes the insertion of microalgal biotechnology in biorefinery systems to produce biodiesel, bioethanol, and biogas. Besides, microalgae can grow in the slurry (a byproduct of biogas production). Thus, the biogas upgrading is processed more cleanly since CO_2 and nutrients can be used directly by microalgae [6]. In this perspective, this chapter discusses biotechnological strategies, biomass recovery, and pretreatment to optimize biogas production. Moreover, future trends are mentioned regarding the use of microalgal biomass in the sustainable production of biogas within the concept of biorefinery.

12.2 Biogas Production from Anaerobic Digestion

Anaerobic digestion is commonly performed in four stages: hydrolysis, acidogenesis, acetogenesis, and methanogenesis [12] (Figure 12.1). The hydrolysis stage is responsible for the biodegradation of polymeric compounds into monomeric or oligomeric components. There is the conversion of carbohydrates into glucose and fructose, proteins into amino acids, and lipids into long-chain fatty acids. In the acidogenesis stage, short-chain volatile fatty acids, CO_2, H_2, and acetic acid are obtained from the decomposition of the components produced in hydrolysis [6]. In acetogenesis, short-chain volatile fatty acids are converted into acetic acid, and H_2 by acetogenic bacteria. In methanogenesis, the remaining acetic acid is transformed to methane by acetogenic and non-acetogenic bacteria. On the other hand, methanogenesis is performed by archaeal species [6, 13].

Organic loading rate, hydraulic retention time, solids retention time, temperature, pH, C: N ratio, and substrate particle size are influential

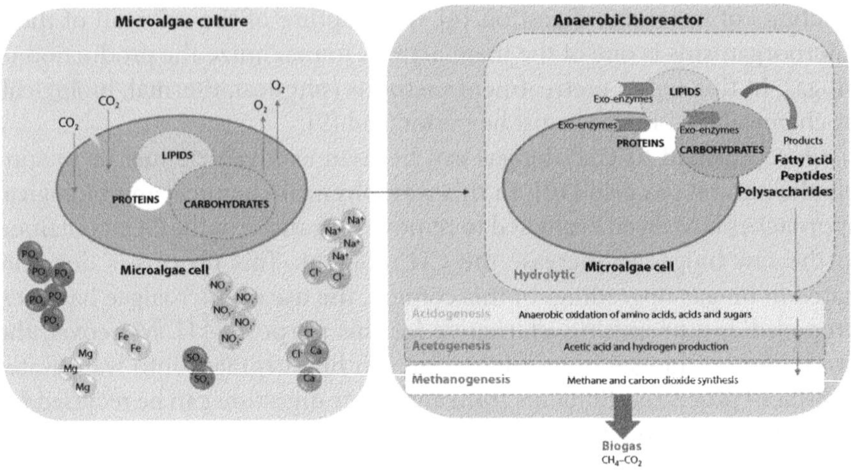

Figure 12.1 Schematic illustration of anaerobic digestion for the production of biogas from microalgae.

parameters of anaerobic digestion [12]. Jankowska *et al.* [14] reported that the high rate of the organic load increases the CH_4 yield, similarly to the solids retention time. In a conventional biogas plant, the hydraulic retention time is 30–50 days, while the anaerobic digestion of microalgae on a laboratory scale showed a hydraulic retention time between 15 and 30 days. Large-scale biogas plants are generally operated under mesophilic (~35–45°C) or thermophilic (~45–60°C) temperatures. In these conditions, CH_4 generation is the greatest [6, 15].

The production of CH_4 from microalgae biomass can be improved by increasing the temperature of anaerobic digestion. By using high temperatures, the photosynthetic activity of microalgae is reduced [14]. Therefore, the process favors the destruction of pathogens that act in the mesophilic temperature range [16]. The thermophilic approach contributes to the stabilization of residues, in a reduction of volatile organic compounds and less risk of foaming, which allows a higher rate of organic load and maximizes biogas production yields [6].

Maintaining an adequate pH is necessary for anaerobic digestion. Hydrolysis and acidogenesis are more efficient in the pH range between 4.5–7.0, while acetogenesis is at pH 6.8–8.2 [12]. The high concentration of ammonia dissolved during protein degradation decreases the rate of methanogenesis due to the low C: N ratio. Thus, carbon-rich substrates must be added to the digesters to increase the C: N ratio [14]. According to Weiland (2010) [17], the C: N ratio should preferably be in the range of 15–30.

Besides, the particle size of the substrate has influenced the efficiency of anaerobic digestion. Size reduction can benefit hydrolysis rates, especially in the case of substrates with a rigid structure, resistant to mechanical stress and enzymatic action [18].

The anaerobic digestion can be in a liquid or solid, according to the physical state of biomass. In solid anaerobic digestion, the productivity of biogas is approximately twice higher than that supplied in a liquid medium. With solid anaerobic digestion, relatively less volume in the digester and less energy are required. Moreover, residues remaining after anaerobic digestion in solid media are rich in phosphorus and nitrogenous organic compounds. Therefore, this effluent has been exploited for agricultural applications, such as the production of fertilizer [6, 19].

The two-stage concept was introduced to separate the stages of anaerobic digestion. Thus, each stage can be conducted at different pH, providing better control of the process [20]. In the two-stage process, acidogenesis and methanogenesis are followed distinctly to produce H_2 and biogas, respectively. Two-stage anaerobic digestion has an advantage when compared to the single-stage process. This advantage is related to the optimization of acidogenesis and methanogenesis. These processes are performed individually, offering greater yield and process stability [6].

12.3 Biogas Composition and Technologies for Biogas Upgrading

Biogas is mainly composed of CH_4 and CO_2, and lower concentrations of other components such as water vapor, hydrogen sulfide (H_2S), oxygen, nitrogen, and siloxanes. From this composition, only CH_4 is desired. Cooling, compression, absorption, or adsorption techniques can be applied to remove water vapor. To reduce the H_2S content, the processes of precipitation in the liquid or physical/chemical absorption of the gas can be carried out. For the release of oxygen and nitrogen, molecular sieves or membranes, as well as adsorption processes can be employed. The removal of siloxanes can be achieved according to the temperature applied in the anaerobic digestion process [10].

Biogas can be used directly to generate electricity, however, the high concentration of CO_2 (30–40%) [21] significantly affects its efficiency. As a result, the heating value of biogas is reduced, increasing compression and transportation costs [6, 16]. Therefore, the biogas needs to be purified to obtain CH_4, so it can be used as a vehicle fuel or for grid injection.

When upgraded to 98% CH_4, biogas has properties very similar to natural gas [10]. The upgrading process is performed through physical, chemical, or biological technologies to reduce the CO_2 content and increase the calorific value of biogas. Thus, this process allows the conversion of biogas to a higher fuel standard [11].

Several biogas upgrading processes are applied to purify the CH_4 content. They can be physical, chemical, physical-chemical, or biological methods. The technologies used to mitigate CO_2 from biogas can be carried out by physical or chemical absorption, adsorption by pressure oscillation, adsorption on solid surfaces, membrane separation, cryogenic separation, chemical hydrogenation process, among others. The removal of CO_2 from biogas can be achieved on a large scale, such as through adsorption and membrane separation [10]. Upgrading by chemical or physical techniques can purify CH_4 by up to 88–98%, removing about 99% of CO_2, H_2S, halocarbons, and siloxanes. Despite their efficiency, physicochemical approaches consume a lot of energy and are not friendly to the environment. For this reason, a high number of researches have been observed on the biological biogas upgrading [6].

Biological technologies can remove 80–100% CO_2 and up to 99% H_2S [22]. The use of CO_2 by microalgae offers a new method for upgrading biogas. The microalgal biomass produced after CO_2 absorption can be used as a biological matrix in the production of biodiesel, bioethanol, and biogas [10]. It is also possible to integrate the process of updating biogas with the cultivation of microalgae in wastewater, reinforcing the concept of biorefinery. As a result, removing pollutants from wastewater and producing macromolecules with various industrial applications makes microalgae a promissory system for improving the economy of the biogas upgrade system [10, 16]. Moreover, microalgae are tolerant to high concentrations of CO_2, which indicates another advantage of using microalgae in the upgrading process.

12.4 Microalgal Cultivation

Microalgae cultivation, intended for the production of biogas, is more favorable than other energy matrixes in the production of biofuels [23]. Microalgal biomass acts as an ideal substrate for the nutrient demands of bacteria that act in anaerobic digestion [7]. The microalgal cultivation system offers the possibility of using non-cultivable areas, using wastewater

[24], sea, and brackish groundwater. Furthermore, microalgae cultivation can occur simultaneously with the bioremediation of nutrients present in these water resources [25, 26], with lower operating costs and energy consumption [27].

Depending on the cultivation conditions, the microalgae can have greater photosynthetic efficiency, high biomass productivity, CO_2 fixation, and CH_4 yield. Moreover, microalgae can release less H_2S due to their reduced sulfur amino acid profile compared to other energetic substrates [4, 6, 28]. As reviewed by Bose *et al.* [29], the quality of microalgae biogas is also associated with optimized cultivation operational criteria that include: type of reactor used and the microorganism's high tolerance to pH and CO_2.

The C: N ratio below 20 can promote the release of ammonia and the total solids content above 30% in microalgal biomass. As a result, anaerobic digestion can compromise CH_4 yield [6, 30]. The high protein concentration (30–60%) in the biomass can reduce the C: N ratio (15:30), resulting in a high concentration of ammonia (due to the degradation of amino acids). Besides, this protein content can inhibit the action of anaerobic microorganisms (especially for methanogenic archeobacteria) [31, 32]. In this way, research has been stimulating microalgae cultivation strategies to reduce the protein content in the biomass of these microorganisms. Most of these cultivation systems limit the concentration of nitrogen sources. From this, the chemical characteristics of the biomass are modified, favoring the accumulation of lipids or carbohydrates [30].

Nutritional limitation can increase the lipid content in *Nannochloropsis* [33] and enhance the carbohydrate content in *Chlorella vulgaris* [34]. In this perspective, Mahdy *et al.* [35] investigated the cultivation of *Chlorella vulgaris* in a nitrogen-limited culture medium. The authors observed an increase in the carbohydrate content (1.8 times) in the microalgae biomass, in comparison to the biomass obtained in a culture medium with higher nitrogen content. After anaerobic digestion, the biomass of the carbohydrate-rich microalgae increased 1.5 times the CH_4 yield.

The highest results of theoretical yields of microalgae CH_4 can be derived from biomass rich in lipids (1.014 L CH_4 g VS^{-1}) compared to biomass with high protein content (0.851 L CH_4 g VS^{-1}) and carbohydrates (0.415 L CH_4 g VS^{-1}) [36]. Thus, the mechanisms for the accumulation of carbohydrates or lipids in microalgae biomass have been investigated and associated, mainly with the viability of using wastewater as water and a nutritional source in cultivation [37, 38].

12.5 Microalgal Biomass Recovery and Pre-Treatment

12.5.1 Recovery Techniques

The recovery of microalgal biomass allows the concentration of this biological matrix, enabling subsequent biochemical processes in the production of biofuels such as biogas [24]. Harvesting is one of the main limitations for obtaining microalgae biogas at the commercial level. The optimization in the microalgae harvesting processes aims to minimize production costs (3–15%) [24, 39]. In this way, the recovery methods used must have the potential to process the high production of microalgae. Besides, the pH of the medium, size, and cell structure of microorganisms and chemical components that may be added should also be considered [24, 40].

Microalgal biomass recovery methods can include one or more physical, chemical, or biological methods to allow adequate solid-liquid separation [40]. Coagulation, flocculation, flotation, gravity sedimentation, and centrifugation are used to recover large volumes of biomass [24, 41, 42]. The combined methods comprise sedimentation-flocculation-coagulation, flocculation-flotation, filtration-centrifugation and flotation-aggregation [24, 41–46].

Small-scale microalgae cultivations mostly use centrifugation to recover biomass [47, 48]. Sedimentation followed by centrifugation [5], or sedimentation-flocculation-coagulation can also be used [24]. Larger-scale cultivations (reactors with working ranges >150 L) use gravity sedimentation methods [49, 50] or centrifugation [35]. However, few studies direct the optimization of biomass recovery techniques aiming at the production of biogas [24]. Therefore, this makes it difficult to verify the influence of these methods on the biochemical processes of anaerobic digestion and CH_4 reduction.

Microalgae cultivation systems for biogas production perform biomass recovery in two stages. The first stage requires microalgal biomass varying between 2% and 7% of solid matter. In this case, biomass can be processed by coagulation, flocculation, flotation, sedimentation by gravity, or a combination of these. Subsequently, biomass is concentrated by centrifugation [5, 51]. Chemical agents that provide coagulation or flocculation should not be toxic or inhibit biochemical processes in anaerobic digestion [24].

The recovery of microalgae biomass by flocculation under an increase in the pH of the medium (pH >10) has received attention in small-scale research. This process has favored the reduction of operating costs and feasibility in obtaining biomass without toxic aggregates. García-pérez *et al.* [52] obtained recovery efficiency greater than 95% of *Chlorella*

Vulgaris biomass at pH 10.5. The biomass of freshwater microalgae (*Chlorella vulgaris*, *Scenedesmus* sp., and *Chlorococcum*) and marine species (*Nannochloropsis Oculata*, and *Phaeodactylum tricornutum*) were also favorable to recovery (> 90%) by flocculation at pH 10.6 [53].

According to Gutiérrez *et al.* [24], the flocculation method can be optimized by adding natural flocculants based on tannin and cationic action (Ecotan and Tanfloc). These flocculants are an alternative to replace chemicals based on aluminum and iron, commonly used in flocculation. These metallic salts, in high concentrations, can be adsorbed to biomass and limit the sequential biochemical processes of anaerobic digestion. The authors reported that the addition of 10 and 50 mg L^{-1} of Ecotan and Tanfloc, respectively, promoted the recovery (> 90%) of microalgal biomass of *Monoraphidium* sp., *Scenedesmus* sp., *Stigeoclorium* sp., *Nitzchia* sp., *Navicula* sp., and *Amphora* sp. in a short period (10–80 min) compared to traditional methods. Furthermore, the natural flocculants do not compromise the pH of the medium and the anaerobic processes, enabling the production of CH_4 from the flocculated biomass (162–166 mL CH_4 g VS^{-1}).

The proper choice of the microalgae species contributes to promoting the effectiveness of biomass recovery. Single-celled microalgae such as *Chlorella* sp., *Chlorella vulgaris*, or *Scenedesmus obliquus* tend to make harvesting difficult. For these cases, the combination of flocculation, followed by centrifugation, may be favorable [29]. However, strains of *Spirulina* are easily harvested by sedimentation since their spiral cell structure allows aggregation of cells, facilitating the harvesting [54]. Besides, high biomass production can also favor flocculation due to cell-cell interaction [51, 55].

Gravitational sedimentation is among the most used microalgae biomass recovery methods in studies aimed at the production of biogas [5, 24, 50]. Its practice promotes the separation of solid liquid by a difference in density. A sedimentation tank or Imhoff cone can be used to improve harvest efficiency [49].

Centrifugation can be an alternative to recover larger volumes of biomass. However, this technique demands high energy and equipment maintenance costs [56]. Its application is encouraged when using the recovered biomass for the production of high-value compounds for application in the chemical, pharmaceutical, or cosmetic industry. Among these compounds, there are unsaturated fatty acids, pigments, and polymeric materials such as polyhydroxyalkanoates [45, 57]. Mahdy *et al.* [35] used centrifugation to recover *Chlorella vulgaris* from 150 L of culture medium to investigate the potential of this microalga in the production of biogas.

Generally, filtration is a sequential method of coagulation, flocculation, flotation, or sedimentation steps. This process promotes the thickening of microalgal biomass by physical barriers (filters), separating the microalgae cells from the liquid medium. Filters can consist of materials, sometimes of biological or mineral origin, with different porosities [58]. This technique can compromise the cellular integrity of the microalgae and promote the clogging of the filtering pores. Thus, the use of filtration in high-density crops such as *Chlorella* sp. [58, 59] is limited. The flotation recovery method uses the injection of air microbubbles into the culture medium. In this case, there is no addition of chemicals to promote the transport of biomass to the surface and facilitate its harvest [51]. Both filtration and flotation are methods of little reported in the recovery of biomass destined for the production of biogas.

12.5.2 Pre-Treatment Methods

The production of biogas from the anaerobic digestion of microalgal biomass requires, in its majority, the pretreatment of the microalgae cellular matrix [5, 60]. This treatment facilitates the disruption of the cell wall, and the intracellular components are more accessible for anaerobic bacteria in the first stage of anaerobic digestion. Consequently, pretreatment positively interferes with CH_4 yield [5, 61, 62]. Therefore, it is essential to consider the structure and rigidity of the cell wall of the selected microalgae, as well as the pretreatment methods with low energy demand and operational costs, viable on an industrial scale [3].

The pretreatment of microalgal biomass can be carried out by biological, physical, chemical, or mechanical action (Figure 12.2). These methods can be used alone or in combination [3, 5, 61, 62]. Biological pretreatment is the most used method in the treatment of microalgal biomass for biogas production, mainly because it does not generate toxic and inhibitory compounds in anaerobic digestion. This method consists of using enzymes (cellulase, proteases, lipases, esterases, and pectinase) that break down polymeric compounds and convert them into low molecular weight substances [3, 61].

Çakmak and Ugurlu [5] found that cellulases and proteases showed greater activity in Porphyridium cruentum when they used temperatures between 25–55°C. However, when the microalgal biomass was pre-treated with protease only (0.5 mL g^{-1} of dry biomass), an increase in cell wall solubilization of up to 32% was identified. Moreover, improvements in CH_4 production were observed, both in mesophilic conditions (77%) and thermophilic conditions (100%). However, the researchers stressed that this method becomes costly given the high cost of commercial enzymes. Hom-Diaz *et al.* [62] observed that the action of the enzyme laccase combined

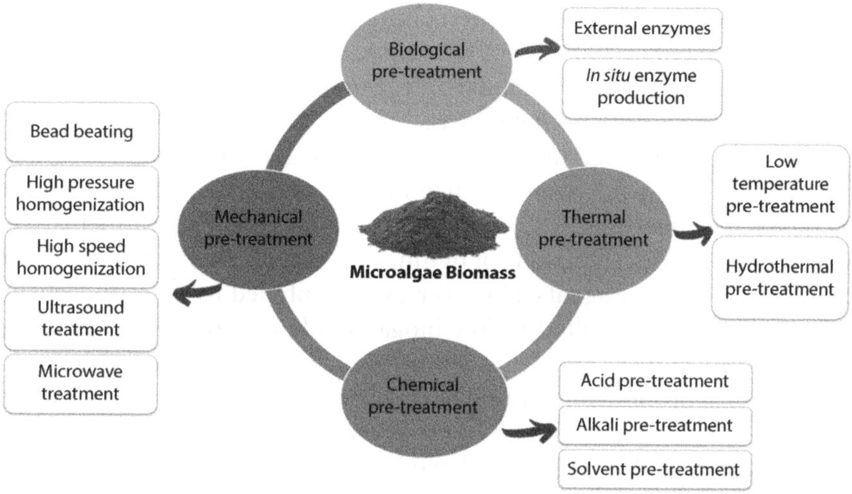

Figure 12.2 Pre-treatment methods used in microalgae biomass for biogas production.

with a fungal broth of *Trametes Versicolor* allowed increased solubility in the cell wall of *Oocystis* sp. and CH_4 yield by up to 74% (144 mL CH_4 g VS^{-1}). The authors concluded that this synergistic effect contributed to CH_4 production, considering that the result of the pretreatment under the isolated use of the enzyme promoted an increase of only 20%.

The thermal pretreatment includes heating operations, as well as freezing/thawing. However, the application of heat has favored the production of CH_4 when microalgal biomass is used as an organic matrix [63, 64]. Besides, this technique allows the use of temperatures <100 °C under atmospheric pressure and temperatures >100°C under high pressure (hydrothermal pretreatment) [32, 63–65]. These techniques are widely used to increase biogas production and are associated with biological and chemical methods [5, 63]. However, the use of these techniques may result in the production of recalcitrant compounds, resulting from the Maillard reaction, which can impair anaerobic digestion [64, 66].

Mendez *et al.* [64] found that the pretreatment of *Chlorella vulgaris* under heating to 180 °C promoted efficient solubilization of carbohydrates and proteins present in the microalgae biomass. However, the authors found a reduction in CH_4 production, possibly due to the formation of recalcitrant compounds. To minimize this problem, Xiao *et al.* [60] verified the efficiency of the hydrothermal pretreatment (35°C) powered by solar energy in the solubilization of the cell wall of *Chlorella* sp. This hydrothermal pretreatment promoted a higher yield in biogas production (40.8%), compared to biomass without pretreatment (26.2%) and with hydrothermal pretreatment at 160°C (35.9%). The researchers also pointed out that these results benefited the cellular integrity of the microalgae by exhaustion during the production of biogas.

The chemical pretreatment uses acid, alkaline, or oxidizing substances to solubilize the cellular appearance of microalgae. Moreover, chemical methods are often used in combination with thermal pretreatments. The absence or low concentrations of lignin, as in the case of microalgae, favor the action of alkaline agents. However, even combined with heating, chemical methods have resulted in low biogas yields (up to 20%) in microalgae such as *Chlorella* and *Scenedesmus* [63]. Studies attribute this to the presence of sporopollenin and algaenane present in these microorganisms, which give high rigidity and hinder the penetration of alkaline solvents [3, 67]. However, acidic substances have been acting more efficiently in the solubilization of these cellular structures. The disadvantage of using these substances is associated with the toxicity of the generated by-products and the formation of inhibitory compounds in anaerobic digestion [32].

Among the most used mechanical methods, microwave and ultrasonication pretreatments work to reduce the size of the particles, breaking the crystalline structure of the microalgae. According to Ometto *et al.* [65], in the context of biogas production, the application of ultrasound (35 MJ kg^{-1}) was efficient in *Chlorella vulgaris.* Besides, there was an increased CH_4 production by more than 44% (196 mL CH_4 g VS^{-1} of crude biomass). However, the intensity of energy applied depends on the microalgae species [68]. The use of microwave irradiation of 21.8, 43.6, and 65.4 kJ kg TS^{-1} (total solids) in the pretreatment of biomass was evaluated by Passos *et al.* [69]. This treatment contributed to the disintegration and digestibility of the microalgal membrane. In this way, the microwave treatment promoted a higher yield (78%) in the production of biogas.

12.6 Biotechnological Strategies for Biogas Production

Biotechnological strategies to optimize the production of biogas from microalgae biomass include optimization of cultivation parameters [29], choice of strain and biomass pretreatment method [5], as well as co-digestion (microalgal biomass combined with other substrates [50]. The cultivation of microalgae has stimulated the productivity of biomass and the simultaneous obtaining of bio compounds by these microorganisms.

Srinuanpan *et al.* [47] reported that in this cultivation system, microalgae act in the biofixation of CO_2, and when grown in wastewater, they can present a potential for bioremediation. Moreover, the chemical constituents present in the biomass can be modified by limiting/supplementing nutrients in the culture medium [35]. From this, the researchers have been observing viability in biomass productivity with high potential in the production of CH_4.

The use of microalgal biomass combined with other organic matrices (co-digestion) can improve the physical-chemical characteristics of the associated substrate, benefiting the processes of bioconversion in the production of biogas [70]. Besides, the increase in CH_4 yield was observed in a circular cultivation system, where the microalga was grown in effluents from the anaerobic digestion of animal manure (chicken manure)/ domestic sewage. The biomass resulting from this process is mixed with animal waste/domestic sewage and subjected to anaerobic digestion for the production of biogas [50, 71].

Moungmoon *et al.* [37] aimed to maximize the productivity of biogas from biomass of *Acutodesmus* sp. AARL G023 and *Chlorella* sp. AARL G049 cultivated these microalgae in wastewater from a chicken manure biogas digester. From this cultivation strategy, a reduction of ammonium-nitrogen (NH_4^+–N) (43.5%) and soluble reactive phosphorus (49.6%) levels were observed, as well as the biogas productivity of 1.44 mL g^{-1} of 58.3% (v v^{-1}) of CH_4. Mahdy *et al.* [35] analyzed the potential of CH_4 production from the mixture of *Chlorella vulgaris* and manure (cattle) (80/20 VS). This condition resulted in the highest value of biochemical CH_4 potential (431 mL CH_4 g VS^{-1}) compared to anaerobic digestion of microalgae alone (100/0) (415 mL CH_4 g VS^{-1}). The authors observed that despite high levels of ammonium (3.7 to 4.2 g NH_4^+–N L^{-1}), CH_4 yields reached up to 84%.

Fozer *et al.* [72] cultivated *Chlorella vulgaris* optimizing light intensity (243.5 and 96.8 μmol photons m^{-2} s^{-1}) promoted by light-emitting diode (LEDs) to stimulate the production of biomass for the production of biogas. The authors obtained biomass productivity of up to 74.62 mg L^{-1} d^{-1}. Srinuanpan *et al.* [47] cultivated *Scenedesmus* sp. with CO_2 (gas flow rate of 0.3 L h^{-1} per 1 L–microalgal culture) to improve the CH_4 content. The researchers observed a 30% increase in biogas production, lipid productivity of 88.57 mg L^{-1} d, and a CO_2 removal rate of 5.097 g-CO^2 d^{-1} per 1 L–microalgal culture. In another study, this same microalga was cultivated under light intensity (5.5 Klux), nitrogen source (0.8 g L^{-1} KNO_3), and CO_2 supply (0.3 L h^{-1} per 1 L microalgae culture), where was observed a high efficiency of CO_2 removal (> 96%) from biogas and CH_4 purification in biogas

(>98%), biomass production of 4.40 g L^{-1}, the lipid content of 34.10% and CO_2 removal rate of 6.50 g–CO_2 d L^{-1} of microalgal culture [48].

12.6.1 Screening of Potential Microalgae Species

The selection of the microalgae species is one of the essential factors to maximize the production of biogas and CH4 yield [6, 23, 29]. The low cellulose content and absence of lignin in the cellulose content and absence of lignin in the cellular structure of these microorganisms make microalgal biomass attractive for biogas production via anaerobic digestion [7, 73]. Its potential as a substrate in anaerobic digestion is associated with the growth capacity and structure of the cell wall [6].

The efficiency of biochemical reactions by biogas-producing bacteria depends on the type of microalgae used [30]. *Chlorella* sp. and *Scenedesmus* sp. are the most studied microalgae in obtaining biogas [2, 7, 74–76] (Table 12.1). Researches point out that these microorganisms stand out for their greater capacity for fixing CO_2, growth, and productivity of biomass in a mixotrophic environment [77, 78]. Most microalgae cultivation for biogas

Table 12.1 Microalgae used for CH_4 production.

Microalga	CH_4 production (mL CH_4 g VS^{-1})	Reference
Chlorella sp.	193–242	[2, 75, 79]
Chlorella vulgaris	431	[35]
Chlorella pyrenoidosa	264	[80]
Chlorella minutissima	166	[80]
Chlorella sorokiniana	298–388	[81]
Chlorella kessleri	335	[82]
Scenedesmus obliquus	177–199	[7]
Scenedesmus dimorphus	397	[7]
Nannochloropsis gaditana	130–204	[83, 84]
Dunaliella salina	505	[82]
Chlamydomonas sp.	333	[85]

production has been carried out with wastewater, which makes it possible to develop circular economy systems [47, 77, 78].

The composition of the microalgae cell wall differs with each species and may form extra or intracellular surfaces. Generally, the microalgae classes Cryptophyceae, Dinophyceae, and Euglenophyceae classes have only a single intracellular cell membrane (composed mainly of polysaccharides) [86]. Microalgae species without a cell wall (*Dunaliella* sp.) or that do not have cellulose or hemicellulose (*Chlamydomonas* sp.) may show greater efficiency in CH_4 production compared to microalgae that have a carbohydrate-based cell wall (*Chlorella* sp. or *Scenedesmus* sp.) [82].

The presence of sporopollenin and algaenane (polymers located in the insoluble extracellular portion) are common to *Chlorella* sp., *Chlorella vulgaris, Nannochloropsis* sp., *Pediastrum boryanum, Scenedesmus obliquus* [67] and attribute rigidity to the cell wall of these microalgae [3]. The rigid cell wall hinders the action of anaerobic bacteria in the first stage of anaerobic digestion (hydrolysis), reducing its accessibility to organic matter and, consequently, the efficiency of conversion of substrates to biogas [3, 67]. Besides, some studies highlight the presence of cellulosic microfibril, pectins, glycoproteins, glucosamine, calcium carbonate, or silicon in some cell structures of microalgae such as *Chlorella* sp. However, the concentration of these components is dependent on the rate and growth phase of the microalgae [87]. In this sense, the selection of pretreatment methods for microalgal biomass, to allow the rupture of cellular appearance, are essential. Moreover, it should be highlighted that most studies do not analyze the composition of the microalgae cell wall before anaerobic digestion and concentrate their analysis on intracellular components, such as carbohydrates, fatty acids, and proteins [75].

12.6.2 Production of Biogas by Microalgae Integrating a Biorefinery System

The cultivation of microalgae aiming at the production of biogas has shown a sustainable character since the microalgae cells have a high potential in the removal of CO_2 from the production of biogas (CH_4 + CO_2) and of nutrients (present in the effluents released after anaerobic digestion) [48]. As a result, microalgae use the nutrients in the effluent for their growth. The microalgal biomass produced feeds the anaerobic bioreactor again, and the CH_4 content is increased, thus characterizing a circular cultivation system. In this segment, the energy cost is reduced, without the generation of waste, favoring biological treatment in the purification of biogas and simultaneous generation of renewable energy [88]. The residual biomass can be used as a biological matrix in the production of biofuels, such as bioethanol and biodiesel. Furthermore, the pigments, fatty acids, essential amino acids, and polymers produced can be applied in the pharmaceutical, chemical, and food industries, supporting the biorefinery approach (Figure 12.3).

The reuse of nutrients and CO_2 from the anaerobic digestion of microalgal biomass is a promising alternative, in addition to contributing to the reduction of the nutritional costs required in the cultivation systems [89]. As reviewed by Cavinato *et al.* [89], the effluents from the microalgal anaerobic digester can be considered a source of carbon and nutrients such as inorganic phosphorus and nitrogen. However, an additional supply of other nutrients may be required to increase biomass production [90, 91].

Zhang *et al.* [92] reported that *Scenedesmus dimorphus* cells used nitrogen and phosphorus, which comes from the effluent of anaerobic digestion of microalgae biomass with a bioremediation efficiency of up to 74% and 35%, respectively. González-González *et al.* [7] confirmed the potential of this same microalga in the anaerobic digestion of biomass with an average CH_4 yield of 199 mL g VS^{-1}. The authors observed that the concentrations of nutrients in the liquid phase of the digester effluent were sufficient to allow the growth of the microalga. Other studies report the potential of

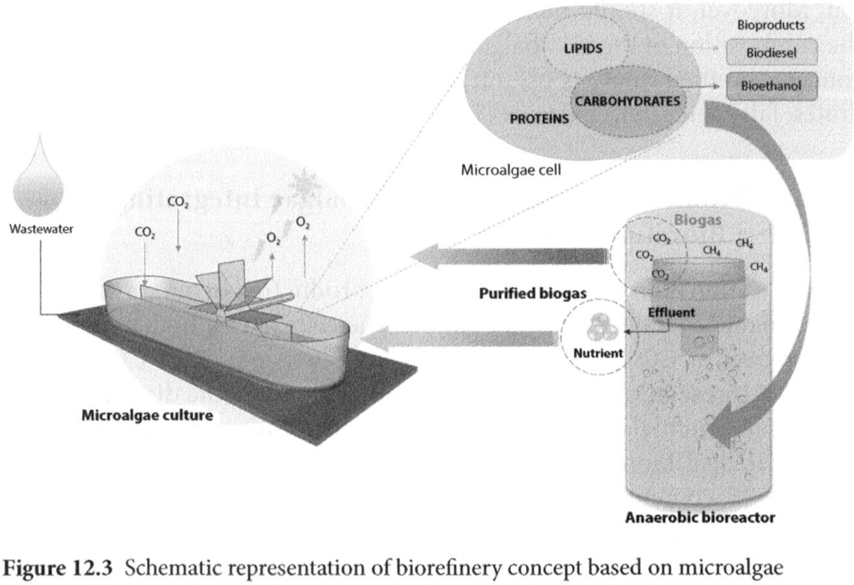

Figure 12.3 Schematic representation of biorefinery concept based on microalgae cultures. Biogas production from anaerobic digestion of microalgae biomass.

Scenedesmus obliquus, *Nannochloropsis* sp., *Chlorella* sp., and *Neochloris oleoabundans* in the fixation of CO_2 from biogas production >80% [11, 88, 93, 94].

Xinyi *et al.* [95] demonstrated the potential of anaerobic digestion effluent from the biomass of *Scenedesmus* sp. as a nutritional source in the growth of this microalgal culture. Shen *et al.* [77] used a similar strategy, cultivating *Chlorella vulgaris* in the effluent of anaerobic digestion from its biomass. The authors analyzed the biofixation capacity of CO_2 from the same system of anaerobic digestion. The study promoted biomass productivity up to 0.097 g L^{-1} d^{-1}, as well as efficiency in removing chemical demand for oxygen, total nitrogen, and CO_2 in 78.6%, 76.4%, and 64.0%, respectively. Moreover, the microalgae showed greater efficiency in the removal of total phosphorus by up to 80.2%. In this segment, the co-production of bio compounds produced by biomass has economic advantages over the isolated generation of biofuel [96].

12.7 Perspectives and Conclusions

The feasibility of the future application of CH_4 from microalgae is biased in the biogas industry. In this perspective, the cultivation of these microorganisms should focus on effluent bioremediation efficiency and high biomass productivity. The effectiveness in the production of this biofuel from microalgae is encouraged to realize the systems of circular cultivation, using nutrients and residual CO_2 for the microalgae cells. The extraction of compounds with high added value from microalgal biomass can improve the economic viability of cultivation and reinforce biorefinery strategies. Thus, the co-products produced by microalgae (pigments, macromolecules, fatty acids, essential amino acids, and polymers) can be applied in the food, pharmaceutical, and chemical industry. Besides, residual biomass can feed anaerobic digesters for biogas production. This energy source can be used as an alternative to other fossil fuels requiring less production cost, without negative impacts on the environment.

Acknowledgments

This research was developed within the scope of the Capes-PrInt Program (Process # 88887.310848/2018-00). This study was financed in part by the Coordenação de Aperfeiçoamento de Pessoal de Nível Superior - Brasil (CAPES) - Finance Code 001.

References

1. Unpaprom, Y., Tipnee, S., Ramaraj, R., Biodiesel from green alga *Scenedesmus acuminatus*. *Int. J. Sustain. Energy*, 4, 1–6, 2015.
2. Jehlee, M., Khongkliang, P., O-Thong, S., Biogas production from *Chlorella* sp. tistr 8411 biomass cultivated on biogas effluent of seafood processing wastewater. *Energy Proc.*, 138, 853–857, 2017.
3. Córdova, O., Santis, J., Ruiz-Fillipi, G., Zuñiga, M.E., Fermoso, F.G., Chamy, R., Microalgae digestive pretreatment for increasing biogas production. *Renew. Sust. Energ. Rev.*, 82, 2806–2813, 2018.
4. Saratale, R.G., Kumar, G., Banu, R., Xia, A., Periyasamy, S., Saratale, G.D., A critical review on anaerobic digestion of microalgae and macroalgae and co-digestion of biomass for enhanced methane generation. *Bioresour. Technol.*, 262, 319–332, 2018.
5. Çakmak, E.K. and Ugurlu, A., Enhanced biogas production of red microalgae via enzymatic pretreatment and preliminary economic assessment. *Algal Res.*, 50, 101979, 2020.
6. Zabed, H.M., Akter, S., Yun, J., Zhang, G., Zhang, Y., Qi, X., Biogas from microalgae: Technologies, challenges and opportunities. *Renew. Sust. Energ. Rev.*, 117, 109503, 2020.
7. González-González, L.M., Correa, D.F., Ryana, S., Jensen, P.D., Pratt, S., Schenk, P.M., Integrated biodiesel and biogas production from microalgae: Towards a sustainable closed loop through nutrient recycling. *Renew. Sust. Energ. Rev.*, 82, 1137–1148, 2018.
8. Golueke, C.G., Oswald, W.J., Gotaas, H.B., Anaerobic digestion of algae. *Appl. Microbiol.*, 5, 47–55, 1957.
9. Kendir, E. and Ugurlu, A., A comprehensive review on pretreatment of microalgae for biogas production. *Int. J. Energy Res.*, 42, 3711–3731, 2018.
10. Xie, L., Xu, J., Zhang, Y., He, Y., Biogas upgrading, in: *Advances in Bioenergy*, Y. Li and S.K. Khanal (Eds.), pp. 309–344, Elsevier, USA, 2020.
11. Meier, L., Barros, P., Torres, A., Vilchez, C., Jeison, D., Photosynthetic biogas upgrading using microalgae: Effect of light/dark photoperiod. *Renew. Energy*, 106, 17–23, 2017.
12. Klassen, V., Blifernez-Klassen, O., Wobbe, L., Schlüter, A., Kruse, O., Mussgnug, J.H., Efficiency and biotechnological aspects of biogas production from microalgal substrates. *J. Biotechnol.*, 234, 7–26, 2016.
13. Li, Y., Park, S.Y., Zhu, J., Solid-state anaerobic digestion for methane production from organic waste. *Renew. Sust. Energ. Rev.*, 15, 821–826, 2011.
14. Jankowska, E., Ashish, K., Oleskowicz-Popiel, S.P., Biogas from microalgae: Review on microalgae's cultivation, harvesting and pretreatment for anaerobic digestion. *Renew. Sust. Energ. Rev.*, 75, 692–709, 2017.
15. Bohutskyi, P., Betenbaugh, M.J., Bouwer, E.J., The effects of alternative pretreatment strategies on anaerobic digestion and methane production from different algal strains. *Bioresour. Technol.*, 155, 366–372, 2014.

16. Brigagão, G.V., Wiesberg, I.L., Pinto, J.L., Araujo, O.Q.F., Medeiros, J.L., Upstream and downstream processing of microalgal biogas: Emissions, energy and economic performances under carbon taxation. *Renew. Sust. Energ. Rev.*, 112, 508–520, 2019.
17. Weiland, P., Biogas production: current state and perspectives. *Appl. Microbiol. Biotechnol.*, 85, 849–60, 2010.
18. Sharma, S.K., Mishra, I.M., Sharma, M.P., Saini, J.S., Effect of particle size on biogas generation from biomass residues. *Biomass*, 17, 251–263, 1988.
19. Uggetti, E., Sialve, B., Trably, E., Steyer, J.P., Integrating microalgae production with anaerobic digestion: A biorefinery approach. *Biofuel Bioprod. Biorefin.*, 8, 516–529, 2014.
20. Parawira, W., Read, J.S., Mattiasson, B., Bjornsson, L., Energy production fromagricultural residues: High methane yields in pilot-scale two-stage anaerobic digestion. *Biomass Bioenergy*, 32, 44–50, 2008.
21. Zhou, K., Chaemchuen, S., Verpoort, F., Alternative materials in technologies for biogas upgrading via CO_2 capture. *Renew. Sust. Energ. Rev.*, 79, 1414–1441, 2017.
22. Rodero, M.R., Ángeles, R., Marín, D., Díaz, I., Colzi, A., Posadas, E., Lebrero, R., Muñoz, R., Biogas purification and upgrading technologies, in: *Biogas: Fundamentals, Process, and Operation*, M. Tabatabaei and H. Ghanavati (Eds.), pp. 239–276, Springer International Publishing, Cham, 2018.
23. Córdova, O. and Chamy, R., Microalgae to biogas: Microbiological communities involved, in: *Microalgae Cultivation for Biofuels Production*, A. Yousuf (Ed.), pp. 227–249, Academic Press, Sylhet, Bangladesh, 2020.
24. Gutiérrez, R., Passos, F., Ferrer, I., Uggetti, E., García, J., Harvesting microalgae from wastewater treatment systems with natural flocculants: Effect on biomass settling and biogas production. *Algal Res.*, 9, 204–211, 2015.
25. Bezerra, P.Q.M., Moraes, L., Cardoso, L.G., Druzian, J.I., Morais, M.G., Nunes, I.L., Costa, J.A.V., *Spirulina* sp. LEB 18 cultivation in seawater and reduced nutrients: Bioprocess strategy for increasing carbohydrates in biomass. *Bioresour. Technol.*, 316, 123883, 2020.
26. Duarte, J.H., Cardoso, L.G., de Souza, C.O., Nunes, I.L., Druzian, J.I., Morais, M.G., Costa, J.A.V., Brackish groundwater from Brazilian backlands in *Spirulina* cultures: Potential of carbohydrate and polyunsaturated fatty acid production. *Appl. Biochem. Biotechnol.*, 190, 907–917, 2020.
27. Costa, J.A.V., Freitas, B.C.B., Rosa, G.M., Moraes, L., Morais, M.G., Mitchell, B.G., Operational and economic aspects of *Spirulina*-based biorefinery. *Bioresour. Technol.*, 292, 121946, 2019.
28. Adarme, O.F.H., Baeta, B.E.L., Lima, D.R.S., Gurgel, L.V.A., Aquino, S.F., Methane and hydrogen production from anaerobic digestion of soluble fraction obtained by sugarcane bagasse ozonation. *Ind. Crops Prod.*, 109, 288–299, 2017.

29. Bose, A., Lina, R., Rajendran, K., O'Shea, R., Xia, A., Murphy, J.D., How to optimise photosynthetic biogas upgrading: A perspective on system design and microalgae selection. *Biotechnol. Adv.*, 37, 107444, 2019.
30. Klin, M., Pniewski, F., Latała, A., Growth phase-dependent biochemical composition of green microalgae: Theoretical considerations for biogas production. *Bioresour. Technol.*, 303, 122875, 2020.
31. Kwietniewska, E. and Tys, J., Process characteristics, inhibition factors and methane yields of anaerobic digestion process, with particular focus on microalgal biomass fermentation. *Renew. Sust. Energ. Rev.*, 34, 491–500, 2014.
32. González-Fernández, C., Sialve, B., Molinuevo, B., Anaerobic digestion of microalgal biomass: Challenges, opportunities and research needs. *Bioresour. Technol.*, 198, 896–906, 2015.
33. Fakhry, E.M. and El Maghraby, D.M., Lipid accumulation in response to nitrogen limitation and variation of temperature in *Nannochloropsis salina*. *Bot. Stud.*, 56, 6, 2015.
34. Dragone, G., Fernandes, B.D., Abreu, A.A.P., Vicente, A.A., Teixeria, J.A., Nutrient limitation as a strategy for increasing starch accumulation in microalgaae. *Appl. Energy*, 88, 3331–3335, 2011.
35. Mahdy, A., Fotidis, I.A., Mancini, E., Ballesteros, M., González-Fernández, C., Angelidaki, I., Ammonia tolerant inocula provide a good base for anaerobic digestion of microalgae in third generation biogas process. *Bioresour. Technol.*, 225, 272–278, 2016.
36. Sialve, B., Bernet, N., Bernard, O., Anaerobic digestion of microalgae as a necessary step to make microalgal biodiesel sustainable. *Biotechnol. Adv.*, 27, 409–416, 2009.
37. Moungmoon, T., Chaichana, C., Pumas, C., Pathom-aree, W., Ruangrit, K., Pekkoh, J., Quantitative analysis of methane and glycolate production from microalgae using undiluted wastewater obtained from chicken-manure biogas digester. *Sci. Total Environ.*, 714, 136577, 2020.
38. Prandini, J.M., Silva, M.L.B., Mezzari, M.P., Pirolli, M., Michelon, W., Soares, H.M., Enhancement of nutrient removal from swine wastewater digestate coupled to biogas purification by microalgae *Scenedesmus* sp. *Bioresour. Technol.*, 202, 67–75, 2015.
39. Fasaei, F., Bitter, J.H., Slegers, P.M., Boxtel, A.J.B., Techno-economic evaluation of microalgae harvesting and dewatering systems. *Algal Res.*, 31, 347–362, 2018.
40. Mata, T.M., Martins, A.A., Caetano, N.S., Microalgae for biodiesel production and other applications: A review. *Renew. Sust. Energ. Rev.*, 14, 217–232, 2010.
41. Godos, I., Guzman, H.O., Soto, R., García-Encina, P.A., Bécares, E., Muñoz, R., Vargas, V.A., Coagulation flocculation-based removal of algal-bacterial biomass from piggery wastewater treatment. *Bioresour. Technol.*, 102, 923–927, 2011.

42. Salim, S., Bosma, R., Vermue, M.H., Wijffels, R.H., Harvesting of microalgae by bioflocculation. *J. Appl. Phycol.*, 23, 849–855, 2011.
43. Danquah, M.K., Ang, L., Uduman, N., Moheimani, N., Forde, G.M., Dewatering of microalgal culture for biodiesel production: Exploring polymer flocculation and tangential flow filtration. *J. Chem. Technol. Biotechnol.*, 84, 1078–1083, 2009.
44. Vandamme, D., Foubert, I., Meesschaert, B., Muylaert, K., Flocculation of microalgae using cationic starch. *J. Appl. Phycol.*, 22, 525–530, 2009.
45. Christenson, L. and Sims, R., Production and harvesting of microalgae for wastewater treatment, biofuels, and bioproducts. *Biotechnol. Adv.*, 29, 686–702, 2011.
46. Granados, M.R., Acién, F.G., Gomez, C., Fernández-Sevilla, J.M., Molina-Grima, E., Evaluation of flocculants for the recovery of freshwater microalgae. *Bioresour. Technol.*, 118, 102–110, 2012.
47. Srinuanpan, S., Cheirsilp, B., Kitcha, W., Prasertsan, P., Strategies to improve methane content in biogas by cultivation of Oleaginous microalgae and the evaluation of fuel properties of the microalgal lipids. *Renew. Energy*, 113, 1229–1241, 2017.
48. Srinuanpan, S., Cheirsilp, B., Prasertsan, P., Effective biogas upgrading and production of biodiesel feedstocks by strategic cultivation of oleaginous microalgae. *Energy*, 148, 766–774, 2018.
49. Solé-Bundó, M., Garfí, M., Matamoros, V., Ferrer, I., Co-digestion of microalgae and primary sludge: Effect on biogas production and microcontaminants removal. *Sci. Total Environ.*, 660, 974–981, 2019.
50. Vassalle, L., Díez-Montero, R., Machado, A.T.R., Moreira, C., Ferrer, I., Mota, C.R., Passos, F., Upflow anaerobic sludge blanket in microalgae-based sewage treatment: Codigestion for improving biogas production. *Bioresour. Technol.*, 300, 122677, 2020.
51. Brennan, L. and Owende, P., Biofuels from microalgae — A review of technologies for production, processing, and extractions of biofuels and co-products. *Renew. Sust. Energ. Rev.*, 14, 557–577, 2020.
52. García-pérez, J.S., Beuckels, A., Vandamme, D., Depraetere, O., Foubert, I., Parra, R., Koenraad, M., Influence of magnesium concentration, biomass concentration and pH on flocculation of *Chlorella vulgaris*. *Algal Res.*, 3, 24–29, 2014.
53. Wu, Z., Zhu, Y., Huang, W., Zhang, C., Li, T., Zhang, Y., Li, A., Evaluation of flocculation induced by pH increase for harvesting microalgae and reuse of flocculated medium. *Bioresour. Technol.*, 110, 496–502, 2012.
54. Komárek, J. and Johansen, J.R., Filamentous cyanobacteria, in: *Freshwater Algae of North America: Ecology and Classification*, J.D. Wehr, R.G. Sheath, J.P. Kociolek (Eds.), pp. 135–235, Academic Press, London, UK, 2015.
55. Chen, C.Y., Yeh, K.L., Aisyah, R., Lee, D.J., Chang, J.S., Cultivation, photobioreactor design and harvesting of microalgae for biodiesel production: A critical review. *Bioresour. Technol.*, 102, 71–81, 2011.

56. Molina-Grima, E., Belarbi, E.H., Acien Fernandez, F.G., Robles Medina, A., Chisti, Y., Recovery of microalgal biomass and metabolites: Process options and economics. *Biotechnol. Adv.*, 20, 491–515, 2003.
57. Rawat, I., Kumar, R.R., Mutanda, T., Bux, F., Dual role of microalgae: Phycoremediation of domestic wastewater and biomass production for sustainable biofuels production. *Appl. Energy*, 88, 3411–3424, 2011.
58. Barros, A.I., Gonçalves, A.L., Simões, M., Pires, J.C.M., Harvesting techniques applied to microalgae: A review. *Renew. Sust. Energ. Rev.*, 41, 1489–1500, 2015.
59. Singh, G. and Patidar, S.K., Microalgae harvesting techniques: A review. *J. Environ. Manage.*, 217, 499–508, 2018.
60. Xiao, C., Liao, Q., Fu, Q., Huang, Y., Xia, A., Shen, W., Chen, H., Zhu, X., Exergy analyses of biogas production from microalgae biomass via anaerobic digestion. *Bioresour. Technol.*, 289, 121709, 2019.
61. Passos, F., Hom-Diaz, A., Blanquez, P., Vicent, T., Ferrer, I., Improving biogas production from microalgae by enzymatic pretreatment. *Bioresour. Technol.*, 199, 347–351, 2015.
62. Hom-Diaz, A., Passos, F., Ferrer, I., Vicent, T., Blánquez, P., Enzymatic pretreatment of microalgae using fungal broth from *Trametes* versicolor and commercial laccase for improved biogas production. *Algal Res.*, 19, 184–188, 2016.
63. Mahdy, A., Mendez, L., Ballesteros, M., González-Fernández, C., Autohydrolysis and alkaline pretreatment effect on *Chlorella vulgaris* and *Scenedesmus* sp. methane production. *Energy*, 78, 48–52, 2014.
64. Mendez, L., Mahdy, A., Demuez, M., Ballesteros, M., González-Fernández, C., Effect of high pressure thermal pretreatment on *Chlorella vulgaris* biomass: Organic matter solubilisation and biochemical methane potential. *Fuel*, 117A, 674–679, 2014.
65. Ometto, F., Quiroga, G., Psenicka, P., Whitton, R., Jefferson, B., Villa, R., Impacts of microalgae pretreatments for improved anaerobic digestion: Thermal treatment, thermal hydrolysis, ultrasound and enzymatic hydrolysis. *Water Res.*, 65, 350–361, 2014.
66. Mendez, L., Mahdy, A., Ballesteros, M., González-Fernández, C., Biomethane production using fresh and thermally pretreated *Chlorella vulgaris* biomass: A comparison of batch and semi-continuous feeding mode. *Ecol. Eng.*, 84, 273–277, 2015.
67. Biller, P., Ross, A.B., Skil, S.C., Investigation of the presence of an aliphatic biopolymer in cyanobacteria: Implications for kerogen formation. *Org. Geochem.*, 81, 64–69, 2015.
68. Passos, F., Uggetti, E., Carrère, H., Ferrer, I., Pretreatment of microalgae to improve bioga production: A review. *Bioresour. Technol.*, 172, 403–412, 2014.

69. Passos, F., Solé, M., García, J., Ferrer, I., Biogas production from microalgae grown in wastewater: Effect of microwave pretreatment. *Appl. Energy*, 108, 168–175, 2013.
70. Gruber-Brunhumer, M.R., Montgomery, L.F.R., Nussbaumer, M., Schoepp, T., Zohar, E., Muccio, M., Ludwig, I., Bochmann, G., Drosg, B., Effects of partial maize silage substitution with microalgae on viscosity and biogas yields in continuous AD trials. *J. Biotechnol.*, 295, 80–89, 2019.
71. Li, R., Duan, N., Zhang, Y., Liu, Z., Li, B., Zhang, D., Lu, H., Dong, T., Co-digestion of chicken manure and microalgae *Chlorella* 1067 grown in the recycled digestate: Nutrients reuse and biogas enhancement. *Waste Manage.*, 70, 247–254, 2017.
72. Fozer, D., Kiss, B., Lorincz, L., Szekely, E., Mizsey, P., Nemeth, A., Improvement of microalgae biomass productivity and subsequent biogas yield of hydrothermal gasification via optimization of illumination. *Renew. Energy*, 1262–1272, 2019.
73. Liew, Y.X., Chan, Y.J., Manickam, S., Chong, M.F., Chong, S., Tiong, T.J., Pan, G.T., Enzymatic pretreatment to enhance anaerobic bioconversion of high strength wastewater to biogas: A review. *Sci. Total Environ.*, 713, 136373, 2020.
74. Lin, C., Nguyen, M.T., Lay, C., Starch-containing textile wastewater treatment for biogas and microalgae biomass production. *J. Clean. Prod.*, 168, 331–337, 2017.
75. Perendeci, N.A., Yılmaz, V., Taştan, B.E., Gökgöla, S., Fardinpoora, M., Namlı, A., Steyer, J.-P., Correlations between biochemical composition and biogas production during anaerobic digestion of microalgae and cyanobacteria isolated from different sources of Turkey. *Bioresour. Technol.*, 281, 209–216, 2019.
76. Llamas, M., Magdalena, J.A., Tomás-Pejó, E., González-Fernández, C., Microalgae-based anaerobic fermentation as a promising technology for producing biogas and microbial oils. *Energy*, 206, 118184, 2020.
77. Shen, X., Xue, Z., Sun, L., Zhao, C., Sun, S., Liu, J., Zhao, Y., Liu, J., Effect of GR24 concentrations on biogas upgrade and nutrient removal by microalgae-based technology. *Bioresour. Technol.*, 312, 123563, 2020.
78. Xiao, C., Fu, Q., Liao, Q., Huang, Y., Xia, A., Chen, H., Zu, X., Life cycle and economic assessments of biogas production from microalgae biomass with hydrothermal pretreatment via anaerobic digestion. *Renew. Energy*, 151, 70–78, 2019.
79. Wu, H., Li, J., Wang, C., Liao, Q., Fu, Q., Liu, Z., Sequent production of proteins and biogas from *Chlorella* sp. via CO_2 assisted hydrothermal treatment and anaerobic digestion. *J. Clean. Prod.*, 277, 123563, 2020.
80. Prajapati, S.K., Malik, A., Vijay, V.K., Comparative evaluation of biomass production and bioenergy generation potential of *Chlorella* sp. through anaerobic digestion. *Appl. Energy*, 114, 790–797, 2014.

81. Ayala-Parra, P., Liu, Y., Field, J.A., Sierra-Alvarez, R., Nutrient recovery and biogas generation from the anaerobic digestion of waste biomass from algal biofuel production. *Renew. Energy*, 108, 410–416, 2017.
82. Mussgnug, J.H., Klassen, V., Schlüter, A., Kruse, O., Microalgae as substrates for fermentative biogas production in a combined biorefinery concept. *J. Biotechnol.*, 150, 51–56, 2010.
83. Buxy, S., Diltz, R., Pullammanappallil, P., Biogasification of marine algae Nannochloropsis oculata, in: *Materials Challenges in Alternative and Renewable Energy II: Ceramic Transactions*, G. Wicks, J. Simon, R. Zidan, R. Brigmon, G. Fischman, S. Arepalli, A. Norris, M. McCluer (Eds.), pp. 59–67, The American Ceramic Society, Inc., Hoboken, NJ, EUA, 2012.
84. Park, S. and Li, Y., Evaluation of methane production and macronutrient degradation in the anaerobic co-digestion of algae biomass residue and lipid waste. *Bioresour. Technol.*, 111, 42–48, 2012.
85. Frigon, J.C., Matteau-Lebrun, F., Abdou, R.H., McGinn, P.J., O'Leary, S.J., Guiot, S.R., Screening microalgae strains for their productivity in methane following anaerobic digestion. *Appl. Energy*, 108, 100–107, 2013.
86. Cavalier-Smith, T., Symbiogenesis: Mechanisms, evolutionary consequenses, and systematic implications. *Annu. Rev. Ecol. Evol. Syst.*, 44, 145–172, 2013.
87. Nemcova, Y. and Kalina, T., Cell wall development, microfibril and pyrenoid structure in type strains of *Chlorella vulgaris, C. kessleri, C. sorokiniana* compared with *C. luteoviridis* (*Trebouxiophyceae, Chlorophyta*). *Algol. Stud.*, 10, 95–105, 2000.
88. Nagarajan, D., Lee, D., Chang, J., Integration of anaerobic digestion and microalgal cultivation for digestate bioremediation and biogas upgrading. *Bioresour. Technol.*, 290, 121804, 2019.
89. Cavinato, C., Ugurlu, A., Godos, I., Kendir, E., Gonzalez-Fernandez, C., Biogas production from microalgae, in: *Microalgae-Based Biofuels and Bioproducts*, C. Gonzalez-Fernandez and R. Muñoz (Eds.), pp. 185–182, Woodhead Publishing, Sawston, UK, 2017.
90. Cheng, J., Yang, Z., Huang, Y., Huang, L., Hu, L., Xu, D., Zhou, J., Cen, K., Improving growth rate of microalgae in a 1191 m^2 raceway pond to fix CO_2 from flue gas in a coal-fired power plant. *Bioresour. Technol.*, 190, 235–241, 2015.
91. Cheng, J., Xu, J., Huang, Y.L., Yuyou, Z., Junhu, C.K., Growth optimisation of microalga mutant at high CO_2 concentration to purify undiluted anaerobic digestion effluent of swine manure. *Bioresour. Technol.*, 177, 240–246, 2015.
92. Zhang, Y., Kendall, A., Yuan, J., A comparison of on-site nutrient and energy recycling technologies in algal oil production. *Resour. Conserv. Recycl.*, 88, 13–20, 2014.
93. Meier, L., Martínez, C., Vílchez, C., Bernard, O., Jeison, D., Evaluation of the feasibility of photosynthetic biogas upgrading: Simulation of a large-scale system. *Energy*, 189, 116313, 2019.

94. Srinuanpan, S., Cheirsilp, B., Boonsawang, P., Prasertsan, P., Immobilized oleaginous microalgae as effective two-phase purify unit for biogas and anaerobic digester effluent coupling with lipid production. *Bioresour. Technol.*, 281, 149–157, 2019.
95. Xinyi, E., Czarena, C., Mark, C., Application of recycled media and algae-based anaerobic digestate in *Scenedesmus* cultivation. *J. Renew. Sustain. Energy*, 8, 013116, 2016.
96. Jonker, J.G.G. and Faaij, A.P.C., Techno-economic assessment of micro-algae as feedstock for renewable bio-energy production. *Appl. Energy*, 102, 461—475, 2013.

13

Microalgae-Based Wastewater Treatment and Resource Recovery in the Form of Bioenergy and Other Bioproducts

Arindam Sinharoy[1]* and Khwairakpam Sanayaima Singh[2]

[1]Department of Microbiology, National University of Ireland Galway, Galway, University Road, Galway, Ireland
[2]Department of Chemical Engineering, National Institute of Technology Karnataka, Surathkal, Mangalore, Karnataka, India

Abstract
Treatment of industrial wastewater is one of the major challenges in the modern world. This is necessary for protecting our environment, and not taking proactive steps towards this can cause further environmental degradation. Biological treatment technologies received much interest in the past few decades due to their inexpensive character, and this method is considered sustainable and more environmentally friendly. Moreover, public acceptance of biological treatment technologies is higher than other conventional treatment technologies. In the past few decades, increased stress has been given to treating wastewater and recovering valuable resources, further incentivizing the industry to actively develop such waste mitigation processes. Such systems can improve the economics of wastewater treatment. Microalgae-based treatment systems can achieve such targets as microalgae can utilize a wide range of such wastewaters for their growth. The recovered algal biomass can be further processed to produce bioenergy and other valuable products. This chapter intends to discuss the feasibility of algal wastewater treatment from different industrial sources.

Keywords: Microalgae, resource recovery, biological wastewater treatment, bioenergy, bioproducts

**Corresponding author*: arindam.sinharoy004@gmail.com

Sanjay Kumar, Narendra Kumar and Shahid-ul-Islam (eds.) Role of Microbes in Industrial Products and Processes, (331–362) © 2023 Scrivener Publishing LLC

13.1 Introduction

One of the most significant global issues in the 21st century causing suffering to humankind is related to the energy crisis and environmental degradation [1]. Over past centuries, a massive increase in human population and industrial revolution has resulted in a fuel crisis, worldwide global warming, a massive increase in pollution, and associated human health issues [2]. Therefore, the researchers' greatest challenge is finding alternative energy sources and means to prevent and reduce environmental degradation. Due to this, biotechnological products and processes using renewable resources have gained attraction during recent times [3].

Biological systems can be useful for waste remediation and bioenergy production in this context. Among the different biological organisms, algae are more interesting. Due to their photosynthetic properties can convert solar energy to valuable biomass and simultaneously remove nitrogen, phosphorous, and other contaminants [4]. In the natural environment, due to the release of untreated wastewater into the surface water bodies, sometimes algal blooms can be seen that thrive on the nutrients available in such waste streams. This capability of algae can be harnessed to treat different wastewater from industries. Moreover, algal biomass produced from such treatment systems can be harvested and converted to energy and valuable products [5]. An overview of which is provided in Figure 13.1.

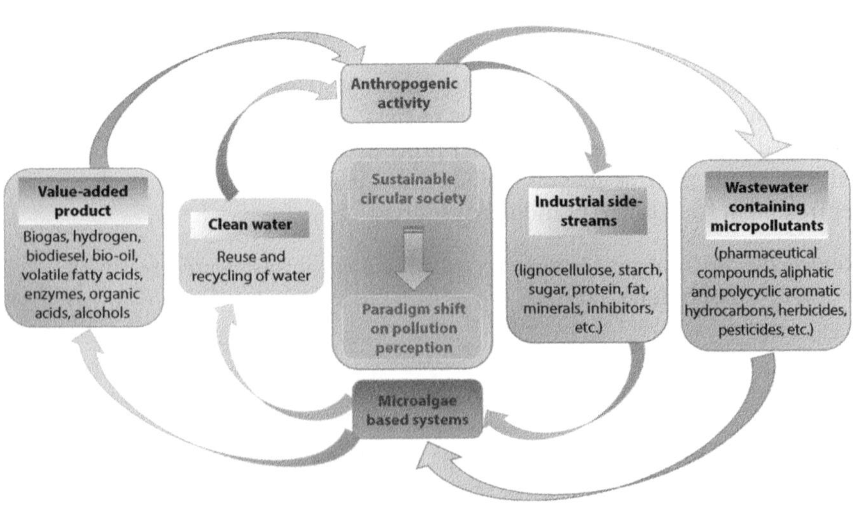

Figure 13.1 Overview of algae-based wastewater treatment and resource recovery.

Historically the research and development in algal wastewater treatment systems are quite old. For example, mass-scale production of common algal strains, namely *Chlorella* and *Dunaliella* for wastewater treatment, is reported even eight-decade before [6]. However, this has gained much interest and application potential in the past few years, even in developing countries such as India, China, etc. In terms of wastewater treatment capabilities, algae are versatile. They can be applied for a variety of purposes such as nutrient (nitrogen, phosphorous) removal [7], removal of chemical oxygen demand (COD), and biochemical oxygen demand (BOD) [8], heavy metal removal [9], polyaromatic hydrocarbons (PAH) removal [10], pharmaceuticals, micropollutants removal [11], and so on. The algal treatment system can achieve very low pollutant concentration in the wastewater to meet the regulatory standards. Another advantage of the algal wastewater treatment system is that this is a low-cost process that requires less energy and nutrients for its cultivation and maintenance. Furthermore, as already mentioned earlier, from algal treatment systems, high-value algal biomass can be obtained at the end of the treatment, which can be utilized for resource recovery. This chapter, therefore, focused on algal-based methods for treating industrial wastewaters and recovery of valuable resources like energy and other byproducts.

Algal cultivation systems are one of the major research areas essential for developing an algal wastewater treatment system. Broadly this can be classified into open systems such as algal ponds and closed systems such as different photobioreactors (Figure 13.2). Open algal pond systems such as high-rate open ponds or aerated algal ponds have been applied on the industrial scale due to their simplicity in design and operation. However, control in this type of system is difficult due to its open nature. In this context, closed bioreactor systems such as airlift reactor, bubble column reactor, flat-panel reactor, and tubular reactors have been reported for higher treatment efficiencies and biomass yield due to superior control on the operational parameters [12]. Following the treatment of wastewater in algal systems, biomass needs to be harvested, for which several technologies have been reported in the literature [13]. As described in Figure 13.2, these algal biomass harvesting techniques can be broadly categorized into physical, chemical, and biological methods. Under physical techniques, centrifugation, filtration, ultrafiltration, sedimentation have been used for algal biomass harvesting. Although most of these methods are very efficient in separating algal biomass from treated wastewater, many disadvantages hinder their application on a large scale. For example, centrifugation, filtration, and ultrafiltration involve substantial initial costs. Furthermore, additional costs for skilled labor and high maintenance were also required.

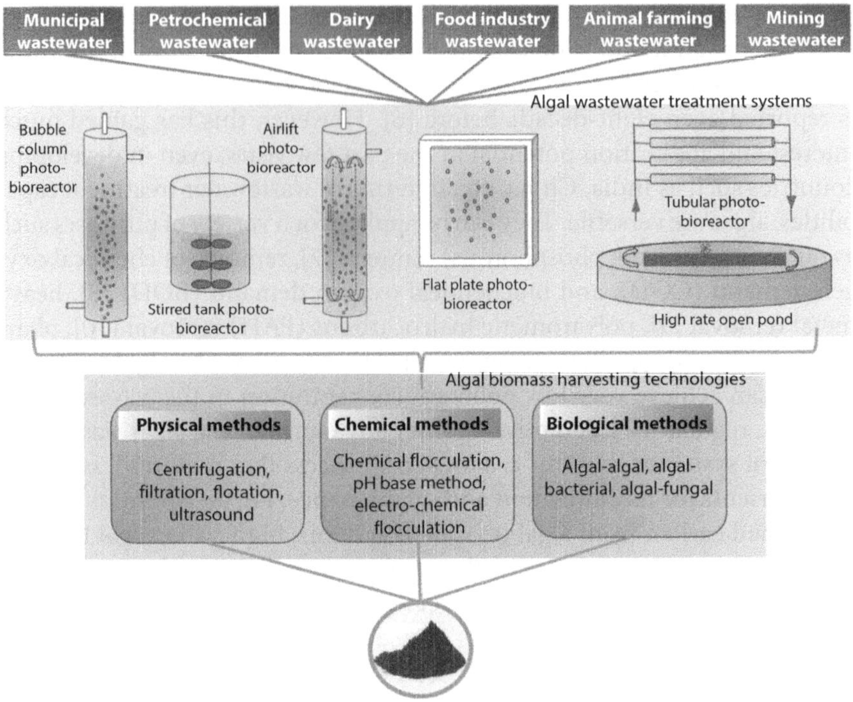

Figure 13.2 Algal wastewater treatment systems along with different algal biomass harvesting technologies are available.

Sedimentation, though low cost, cannot be applied to all types of algal species [14]. A chemical method such as flocculation using different chemicals can cause a problem in further biomass processing due to a chemical contaminant. Electro-flocculation needs additional power input, and their dependency on pH and biomass composition can be disadvantageous [13]. Biological methods are essentially flocculation assisted by either bacteria or algae. Both are considered more sustainable, does not involve any chemical and processes involving less cost and little energy requirement [15].

13.2 Algal Wastewater Treatment

13.2.1 Municipal Wastewater Treatment

Municipal or urban wastewater is one of the most common and most significant constituents of all the different wastewater generated worldwide. This type of wastewater can vary widely in quality and quantity depending on the location and sources from which these are received [16]. Typical constitutes of urban wastewater contain organic compounds such as carbohydrates, fats, proteins, amino acids, volatile acids, and inorganic compounds Na, Ca, P, Mg, Cl, S, PO_4, bicarbonate, NH_4 salts and heavy metals, etc. [16]. Conventional municipal wastewater treatment comprises an arrangement of physicochemical and biological treatment where biological treatment is most commonly an activated sludge process-based aerobic treatment system. Anaerobic treatment has also been applied in many cases as a biological treatment where biogas is produced that can provide additional value-added by generating energy [17]. However, the effluent from such treatment can contain residual nutrients, which can cause eutrophication of the water bodies where these effluents are discharged. The algal treatment can be applied as a tertiary treatment to remove residual nutrients or can be used as the primary biological treatment.

In algal treatment, many previous studies have reported high treatment efficiency. Both photo-autotrophic and mixotrophic cultivation strategies have been implemented where nutrients are mostly assimilated in the algal biomass. For example, Wang *et al.* [7] examined the feasibility of *Chlorella* sp. to treat wastewater collected from various points of a municipal wastewater treatment plant and evaluated its growth using municipal wastewater as the sole substrate. The study found that the *Chlorella* sp. could effectively remove 74.7–82.4% of ammoniacal nitrogen, 62.5% of nitrate nitrogen, 83.2–90.6% of phosphate, and 83% of chemical oxygen demand (COD) from the four wastewater samples studied. The specific growth rate of the algal strain ranged between 0.343 and 0.948 per day. In addition, different metals that are present in the wastewater concentrate, such as aluminum, calcium, iron, magnesium, and manganese, were also removed effectively by the algal species.

Selvaratnam *et al.* [18] demonstrated algal treatment of municipal wastewater using *Galdieria sulphuraria* strain with a high biomass growth rate of 0.34–0.37 g/L.d. The study reported a total N and P removal rates of 4.7–5.0 mg/L.d and 1.5–1.7 mg/L.d, respectively. Furthermore, a feasibility study of mixotrophic culture mode was studied using the same algal strain in a pilot-scale setup [19]. The system performed exceptionally well with 16.5 (±3.6) mg/L.d of BOD removal rate, 6.09 (±0.92) mg/L.d ammoniacal nitrogen removal rate, and 1.40 (±0.57) mg/L.d phosphorus removal rate under the fed-batch mode of operation. The performance of the mixotrophic cultivation method was superior to the photoautotrophic system for almost all the wastewater parameters studied.

Simultaneous removal of nutrients from wastewater and CO_2 sequestration can also be achieved using algal treatment systems. Woertz *et al.* [20]

studied municipal wastewater treatment using green algae in a semi-continuous indoor culturing system with 2–4 days HRT. The systems attained around 99% ammonium and orthophosphate removal. During municipal sewage treatment, the highest algal lipid productivity was 24 mg/L.d under 3 d HRT. The results confirmed that CO_2 supplementation to the algal system improved nitrogen and phosphorus removal and increased the amount of lipid within the algal biomass. In a similar study, the addition of CO_2 improved the total nitrogen removal in high-rate algal ponds and improved the biomass concentration [21]. The overall nitrogen removal was nearly 60% with CO_2 addition. The proportion of algae in the biomass also improved from 55.6% to 80.5% due to CO_2 addition.

In another approach, using algal–bacterial consortium for municipal wastewater treatment resulted in a better settleable sludge at the end of 8 days of treatment using a stirred tank-type photobioreactor [22]. The COD, total Kjeldahl nitrogen and phosphate removal were 98.2 (± 1.3)%, 88.3 (±1.6)%, and 64.8 (±1.0)%, respectively, along with good biomass productivity of 10.9 (±1.1) $g/m^2{\cdot}d$. The total suspended solids were also reduced from 1.84–0.016 g/L. Similarly, Craggs *et al.* [8] investigated the performance of a large scale completely mixed open raceway algal-bacterial pond for nutrient removal from wastewater without any CO_2 addition. The study found nearly 50% BOD removal, ~87% filtered BOD removal, ~65% ammoniacal nitrogen removal, ~19% phosphorus removal. In addition, the biomass production was 8 g/m^2.day (as VSS), which has significant potential to be converted to value-added products.

13.2.2 Dairy Wastewater Treatment

The dairy industry is one of the most commonly found industries worldwide, processing and producing various products such as milk, butter, cheese, yogurt, cream, whey, casein, etc. [23]. Dairy industry wastewater commonly contains soluble organics, suspended solids, trace organics, oil, grease, high COD and BOD, pH generally in the alkaline range. However, it can swiftly change to the acidic range due to lactic acid production through the fermentation of sugars by native microorganisms [24]. In addition, large amounts of casein, inorganic salts, detergents, and sanitizers get added during the washing and cleaning operations, which could be in considerable amounts in the case of large-scale plants [24]. Due to such high nutrient content, their release into the environment without treatment can cause several problems. Although algal treatment does not fall under conventional treatment for dairy wastewater, it has been successfully implemented in many previous studies.

For example, Hena *et al.* [25] studied dairy wastewater treatment using two commercially available strains and ten native strains isolated from a dairy wastewater treatment plant. The native algal consortium successfully removed 98% of the nutrients from the wastewater. The biomass productivity was 153.54 t/ha.y, and the lipid constituent in the biomass was 16.89%. In another study, Valizadeh and Davarpanah [26] demonstrated a micro-photo bioreactor with *Chlorella vulgaris* algal strain to treat simulated dairy wastewater. The COD removal efficiency obtained in this study was 42.57% at 52 h of reaction time.

Chlorella vulgaris strain has successfully treated dairy industry wastewater [27]. The study obtained high removal for BOD (85.61%), COD (80.62%) and nitrogen (85.47%), but the total phosphorus and suspended solids removal values were less (65.9 and 29.1%, respectively). At the end of 10 days of cultivation, a very high biomass concentration of 1.23 g/L was obtained, later used to extract lipid and converted to biodiesel. According to the author, this could be a sustainable strategy for producing bioenergy with simultaneous treatment of dairy wastewater.

Woertz *et al.* [20] reported high treatment efficiency for dairy wastewater using green algae containing outdoors in a bench-scale batch system. The ammonium and orthophosphate removal gradually improved to reach the highest values of 96 and >99%, respectively, after 12 days of cultivation. The lipid content in the resulting algal biomass was 14–29% and mainly depended upon the wastewater composition. The CO_2 supplementation positively impacted the treatment efficiency and lipid accumulation.

A recent study evaluated a comparison between mixotrophic and heterotrophic cultivation modes for dairy wastewater treatment using two different strains of marine and freshwater microalgae, namely *Scenedesmus quadricauda* and *Tetraselmis suecica* [28]. The treatment efficiency was best for *S. quadricauda*, which could remove 92.15% of total nitrogen, 100% of phosphate, 100% of sulfate, and 76.77% of total organic carbon from dairy wastewater, at the end of two cultivation cycles. The biomass concentration was high for *T. suecica* in both cases (0.58 and 0.65 g/L), indicating the suitability of dairy wastewater treatment as a substrate for biomass production.

The previous studies show that dairy wastewater can be successfully treated using algal treatment systems. Moreover, dairy wastewater can be one of the most suitable substrates for algal biomass cultivation. However, in the case of very high-strength wastewater, complete removal of the pollutants may not be possible. In such cases, hybrid treatment technologies involving physicochemical treatment technologies or any other biological treatment systems can be included with the algal process to achieve effluent standards.

13.2.3 Food Industry Wastewater

The food industry or food processing industries are other commonly found industries throughout the world that generate a large amount of nutrient-rich wastewater. In addition, these industries consume a large amount of water, which could be recovered using wastewater treatment strategy, and it would be a much more sustainable approach [29]. Examples of such industries include meat processing and poultry products, dairy products, fruits and vegetable processing and preserving plants, grain products, sugar, related confectionaries, fats and oils, beverages and brewing, etc. The food industry wastewater is generally treatable using biological treatment systems as they do not contain much of any toxic or non-biodegradable compounds [30]. General characteristics of the food industry vary mainly depending upon the type of industry. As seen for other types of wastewater, wastewater generated from the food industry can also be treated using an algal system. Some selected studies on the algal treatment of wastewater from the different food industries are discussed below.

Yonezawa *et al.* [31] treated wastewater generated from a soybean curd manufacturing plant using *Botryococcus braunii*. In order to understand if the algae were capable of growing on this wastewater, AF-6 medium was supplemented with the wastewater at a range of 0-10% by volume. The results showed an excellent treatment efficiency with high biomass concentration at experiments with 1% and 2% (v/v) wastewater concentration. Another study evaluated the potential of *Chlorella pyrenoidosa* for treating soybean processing wastewater under batch and fed-batch mode of operation [32]. The system attained 77.8% soluble COD removal, 88.8% total nitrogen and 89.1% ammoniacal nitrogen removal, and 70.3% total phosphate removal. The maximum biomass production rate was 0.64 g/L.d. with 37% lipid content in the biomass, whereas the lipid productivity was 0.40 g/L.d.

Gupta *et al.* [33] studied raw food-processing industry wastewater using *Chlorella* sp. under the mixotrophic mode of cultivation. Over 90% TOC and COD removal were achieved for all the different dilutions of wastewater. The nitrate and phosphate removal was also high (60% and 40%, respectively) even for concentrated wastewater. The maximum biomass concentration was obtained with diluted wastewater.

Another study by Lutzu *et al.* [34] reported the capability of microalgae *Scenedesmus dimorphus* was able to successfully treat brewery wastewater containing 229 mg/L of nitrogen and 1.4–5.5 mg/L phosphorous. The removal was 99% of nitrogen and phosphorous from the wastewater within seven days of cultivation. Similarly, Travieso *et al.* [35] explored Chlorella vulgaris for its capability to treat distillery industry effluent and found a very high (98%) COD and BOD removal. Overall, the algal treatment could achieve a safe discharge limit for distillery wastewater in this study.

Optimum carbon to nitrogen ratio plays a critical role in brewery wastewater treatment and algal biomass growth. In a study, piggery wastewater was mixed with brewery wastewater to improve the treatment efficiency [36]. The optimum ratio was found to be 1:5 for piggery to brewery wastewater, at which the C/N ratio was 7.9 and resulted in the removal of 100% ammonia, 96% of total nitrogen, 90% of total phosphorus, and 93% of COD from the mixed wastewater. The biomass concentration was also good (2.85 g/L). A two-stage cultivation strategy was used to improve the algal treatment efficiency for brewery wastewater where photoautotrophic and mixotrophic modes were operated sequentially [37]. The two-stage system reported a 90% pollutant removal. The biomass concentration was also improved from 1.5 g/L in a single stage (photoautotrophic) system to 3.5 g/L in a two-stage system.

In a different strategy, raw and anaerobically treated food processing wastewater was treated using *Chlorella sorokiniana*, *Scenedesmus obliquus*, and *Scenedesmus abundans* [38]. In this study, mixed wastewater containing raw and anaerobically digested wastewater showed the best biomass growth for *S. obliquus* (211 mg/L.d). The anaerobic digest contains high amounts of nitrogen and phosphorous but lacks carbon, which was overcome by adding raw wastewater. The COD, total nitrogen, and phosphorus removal were 89%, 84%, and 70%, respectively.

13.2.4 Animal Farming Wastewater Treatment

Animal farming plays an essential role in the economic development of any country. Farm-generated wastewater is generally characterized by high COD, BOD, suspended solids, and other nutrients such as nitrogen and phosphorous [39]. In many places, such operations are at an industrial scale. However, they can also be carried out by unorganized sectors. In such cases, proper wastewater treatment is often not provided, and the unregulated release of such wastewater creates large-scale pollution in the surrounding areas [40]. Algal treatment to wastewater generated by livestock farming, including poultry, cattle, and piggery, has been reported in the literature. The algae can efficiently remove the nutrients present in such wastewater, and in many studies, biomass produced at the end of treatment was utilized for biodiesel production.

Zhu *et al.* [41] studied piggery wastewater treatment using *Chlorella zofingiensis* strain in a tubular bubble column photobioreactors. The effect of wastewater concentration on pollutant removal and biomass growth and lipid productivity was estimated. The best treatment efficiency was obtained at a diluted wastewater concentration of 1900 mg/L COD, with biomass productivity ranging from 106.28 to 296.16 mg/L.d. The specific growth rate of algal biomass was 0.287 to 0.492 per day at different wastewater concentrations. The pollutant removal was also high during the treatment. However, the values varied greatly depending on wastewater concentration (i.e., dilution). The COD, total nitrogen and total phosphate removal were 65.81–79.84%, 68.96–82.7% and 98.17–100%, respectively.

Hu *et al.* [42] demonstrated an algal-based semi-continuous system to treat acid digested swine manure wastewater using a native *Chlorella* sp. The optimum condition for algal cultivation was found to be eight times dilution rate and 2.26 days HRT. Further, up-gradation of the treatment system to a lab-scale photobioreactor could achieve high treatment efficiency and good biomass concentration. The lipid content in the biomass was also high (58.78%), indicating the potential for biodiesel production.

Liu *et al.* [43] explored *Botryococcus braunii* to treat swine wastewater using an aerated lagoon. The algae could efficiently remove 40.8% of nitrogen and 93.3% of phosphorous from the wastewater for 26 days of cultivation. The maximum biomass concentration obtained was 0.94 mg/L which was 1.73 times higher than that obtained using standard BG11 medium. Similarly, Luo *et al.* [44] reported a very high nitrogen (78.46%) and phosphorus (91.66%) removal efficiency by *Desmodesmus* sp. CHX1 in piggery wastewater.

Cattle farm wastewater was successfully treated using *Chlorella vulgaris* applying a two-stage process [45]. In this strategy, the biological treatment involving algae removed 62.3% COD, 81.1% ammoniacal nitrogen, and 85.2% total phosphorus during the first stage. In the second stage, activated carbon-based methods were applied, increasing the overall treatment efficiency to a maximum of 92.2% COD, 94.7% ammonia nitrogen, and 94.4% phosphorous.

Markou *et al.* [46] evaluated acid extract of poultry litter as growth media for growing two algal species viz. *Arthrospira platensis* and *Chlorella vulgaris*. The study reported very high nutrient removal from leachate with good biomass growth. Tan *et al.* [47] studied chicken compost as a growth medium for *Chlorella vulgaris* under batch and semi-continuous mode of operation in a photobioreactor. The biomass growth was higher in semi-continuous mode (0.0736 g/L.d) than in the batch mode of cultivation (0.0568 g/L.d) using chicken litter compost as substrate. The lipid extracted from the biomass obtained in this study was further converted to biodiesel.

However, as the pollutant load is generally high in animal farming wastewater, it is more suited for anaerobic digestion type treatment where bioenergy in the form of biogas can be recovered. Following the anaerobic digestion, the digest can also have a high nutrient concentration, which can be treated using algal systems. Following are some of the examples where this strategy is implemented. Anaerobically digested piggery wastewater was treated using microalgae under batch lab-scale experiment and large-scale open raceway pond [48]. The total nitrogen and phosphorous removal were more than 90%, whereas, in a large-scale setup, the removal efficiencies were 89.5 and 85.3% for nitrogen and phosphorous, respectively. Hence, a very high treatment efficiency was obtained using this strategy by removing nutrients from the anaerobic digest. In another study, anaerobically digested poultry litter is treated using three algal strains: *Chlorella minutissima, Chlorellasorokiniana*, and *Scenedesmus bijuga* individually and as a consortium [49]. The nitrogen and phosphorous removal efficiencies were 26–49% and 100%, respectively, with very high biomass growth (76 mg/L.d).

13.2.5 Petroleum Refinery Wastewater

Petroleum refineries and petrochemical plants are significant sources of environmental pollution worldwide. The general characteristics include high COD, high BOD, high total solids, oil and grease, high total dissolved salts, ammonia, nitrates, sulfides, waste catalyst, heavy metals, volatile organic compounds [50]. Wastewater generated by these industries is very toxic due to benzene, toluene, ethylbenzene, xylene, phenol, and other hydrocarbons that naturally occur in petroleum [51]. In addition, heavy metals such as lead, cadmium, nickel, zinc, arsenic, iron, etc., different sulfide compounds are also present in petroleum industry wastewater that contribute to its hazardous nature [52]. Although mainly Physico-chemical techniques are implemented for petroleum wastewater treatment, biological treatments have also been used. However, the biological treatment, including algal treatment, is quite challenging due to its hazardous nature. Following are some previous studies on the algal treatment of petroleum wastewater.

Abid *et al.* [53] reported simultaneous petroleum hydrocarbon degradation and CO_2 sequestration using an algal-bacterial consortium containing an airlift bioreactor. The microbes used in the study were native hydrocarbon-degrading strains along with *Spongiochloris* sp. of microalgae. The bioreactor obtained a high hydrocarbon removal efficiency (99.18%) in addition to a CO_2 fixation rate of 2.92 g/L.d. The algal biomass production rate was 1.5 ± 0.3 g/L.d.

Hodges *et al.* [54] demonstrated nutrient removal from petrochemical industry wastewater using a microalgal rotating biofilm-based bioreactor. The results showed a very high treatment efficiency for nitrogen (70.8–72.4%), phosphorus (50–55.6%), and total suspended solids (53.6–61.3%) in the bioreactor. The biomass production was also significantly higher at 4.11 g/M^2.day in comparison to that obtained using an open algal pond system (0.4 g/M^2.day). In another study, surfactant-mediated enhancement of petroleum wastewater treatment is demonstrated [55]. In this study algae, *Chlorella vulgaris* was utilized for biological treatment of petrochemical wastewater from an Iranian refinery. The study observed complete total nitrogen and phosphate removal along with 27% of total petroleum hydrocarbon removal. The improved hydrocarbon removal efficiency in this work was due to the addition of surfactant Tween 60.

Chavan and Mukherji [56] studied the effect of phenol addition on treating diesel oil-containing wastewater in a lab-scale rotating biological contactor reactor with algal-bacterial biofilm consisting predominantly of *Burkholderia cepacia* and a freshwater algal culture. The phenol loading rate ranged from 0.11–0.69 g phenol/m^2.day with 21 h of HRT. The study observed complete phenol removal even when phenol loading was increased. However, with an increase in phenol loading, total petroleum hydrocarbon removal reduced from 99% to 94%, along with a reduction in total COD removal. According to the authors, algal culture in the biofilm made it feasible to operate the RBC at high organic loading. Other additional benefits of algal presence were superior immobilization of bacterial biomass, production of O_2, and alkalinity.

Kalhor *et al.* [57] examined the capability of *Chlorella vulgaris* for the degradation of petroleum industry hydrocarbons. Experiments were performed with 10 and 20 g/L of crude oil in water with 7 and 14 days of experimental periods. *C. vulgaris* was extremely good, with 94% light and 88% heavy hydrocarbons removed from the simulated wastewater. Apart from that, the biomass growth of the algae was not affected due to the presence of crude oil, and a high biomass concentration was achieved at the end of the experiments.

13.2.6 Polycyclic Aromatic Hydrocarbons Containing Wastewater

Polycyclic aromatic hydrocarbons (PAHs) contain two or more benzene rings in different arrangements, categorized as a priority pollutant by the United States Environmental Protection Agency (US EPA). Due to their hydrophobic nature and poor water solubility, they are less biodegradable and adsorb on organic matter and accumulate in living beings [58]. As these compounds are toxic, carcinogenic, mutagenic, their accumulation or even exposure to the living being can cause huge damage to their general health [59]. Some of the most commonly available and environmentally important PAHs are naphthalene, anthracene, phenanthrene, pyrene, chrysene, benzo[a]pyrene, etc. The sources of PAHs in the environment mainly include coal and petroleum industry where during fossil fuels (coal and petroleum) or wood-burning (pyrolytic), crude oil extraction and petro-refining process, oil spills, etc. can cause the release of PAHs in the environment [60]. As already mentioned that these compounds are less biodegradable; treatment of PAHs using an algal system involves mainly adsorption or photodegradation, and sometimes a combination of both these methods.

Luo *et al.* [10] performed a comparative study with live and dead cells of microalgae *Selenastrum capricornutum* for removal of seven different high molecular weight PAHs, namely benz[a]anthracene, benzo[b]fluoranthene, benzo[k]fluoranthene, benzo[a]pyrene, dibenzo[a,h]anthracene, indeno[1,2,3-c,d]pyrene and benzo[g,h,i]perylene in the presence of gold and white light irradiation. For live algal cells, there was not much difference in the removal efficiencies of PHAs. However, for dead cells, white light was better suited for treatment. A significant difference between live and dead cells for PAH removal was also reported, but the difference depended on the compound's molecular structure. For example, benz[a] anthracene and benzo[a]pyrene were better converted by dead cells. The mechanism by which dead algal biomass converts PAHs was found to be mainly photodegradation, where dead cells act like photosensitizers.

Process parameters greatly impact the adsorption process, and since algal mediated removal mechanism involves adsorption, Zhang *et al.* [61] explored this aspect in their study. The removal of PAHs by adsorption using marine algae *Ulva prolifera* was examined. The study reported equilibrium adsorptive capacities of 1.27, 1.97, and 2.49 mg/kg of biomass for naphthalene, phenanthrene, and benzo[a] pyrene, respectively, on the *U. prolifera* biomass. Critical parameters such as PAH concentration, pH, and salinity showed great influence on the PAH adsorption process using the algal biomass.

Often heavy metals are also present along with PAHs as co-pollutants, and their presence can affect the biological treatment due to an increase in toxicity. Ke *et al.* [62] studied the impact of heavy metal addition on the biological removal of five polycyclic aromatic hydrocarbons (PAHs), namely fluorene (FLU), phenanthrene (PHE), fluoranthene (FLA), pyrene (PYR), and benzo-pyrene (BAP) by *Selenastrum capricornutum*. The metal stress caused higher removal of PAHs, mainly for two lower molecular weight PAHs, i.e., FLU (99%) and PHE (89%). However, biodegradation of the higher molecular weight compounds was unaffected by heavy metal presence.

13.2.7 Heavy Metal-Containing Wastewater Treatment

Heavy metal pollution is primarily caused by mining and metallurgical operations [63]. Other than that, agriculture, coal burning, petroleum refining, power plants, chemical and electronics industry, solid waste disposal, and landfills also contribute to heavy metal pollution to various extents [64]. Due to the severe hazardous effects of heavy metals on animals, human health, and the overall environment, there is very strict regulation on their release to the environment [65]. Conventional treatment of heavy metals involves both physicochemical and biological methods. In the case of algal treatment, the heavy metal is generally removed by adsorption on the algal biomass [66]. However, bioaccumulation of the heavy metal to a lesser extent also plays a critical role in the removal mechanisms [67].

Pena-Castro *et al.* [66] examined heavy metal (Cr, Cd, and Cu) removal using *Scenedesmus incrassatulus* under the continuous mode of operation with single metal and multi-metal conditions. The heavy metal removal efficiencies varied greatly (24.1–52.7%) with an order of Cd>Cu>Cr removal. However, during the multi-metal condition, Cd and Cr's removal efficiencies increased compared to their individual metal experiments. In another study, cadmium removal using *Scenedesmus obliquus* strain was evaluated [68]. The mechanism was found to be the heavy metal adsorption on the algal surface with a maximum adsorption capacity of 6.04 mg of Cd/g of biomass during the experiments with heat-inactivated algal biomass. The pH was found to be best suited for the adsorption of Cd on algal biomass.

Arora *et al.* [9] examined four different microalgal species for their arsenic (As) treatment capability and found that two strains, namely *Chorella minutissima* and *Scenedesmus* sp., have very high tolerance (500 mg/L) to As. The high bioaccumulation factor of these strains indicated that the governing mechanism for As removal is bioaccumulation. The As removal efficiency was nearly 70–73% and 60–66% for *Scenedesmus* sp. and *C. minutissima* species, respectively.

Chan *et al.* [69] studied copper and zinc removal using two pure strains of *Chlorella vulgaris* and *Spirulina maxima* and a mixed algal culture from wastewater treatment plants. The maximum removal efficiency for copper and zinc were 81.7% and 94.1%, respectively, over a period of 10 days.

However, the adsorption was fast, and equilibrium concentration reached within 6 h of incubation, indicating a very high potential of heavy metal removal by the algal system.

Algal treatment had been applied to remove heavy metals from high saline landfill leachate samples [70], where four halophilic marine algae, namely *Nanochloropsis*, *Pavlova lutheri*, *Tetraselmis chuii*, and *Chaetoceros muelleri*, were studied for their capability to remove heavy metal as well as produce lipid. The study reported 95% removal of heavy metals from the wastewater under high saline conditions within ten days of operation in a photobioreactor. Among the four strains, *N. gaditana* and *C. muelleri* thrived in the heavy metal-containing saline wastewater and accumulated the highest amount of lipid.

In a recent study, two commonly available algal strains viz. *Chlorella vulgaris* and *Scenedesmus spinosus* were examined to remove copper (Cu) and Molybdenum (Mo) from mining wastewater in northern Chile [67]. From the results, it was found that *C. vulgaris* were more tolerant to heavy metals. The metal removal values were also very high, with 64.7 and 99.9 % for Cu and Mo, respectively, in real mining wastewater. In contrast, the Cu and Mo removal efficiencies were slightly lower at 55 and 80.3 % in synthetic wastewater. The protein content in the algal biomass showed an increased profile in the presence of heavy metals in comparison to the control experiment. However, lipid content did not change significantly.

Algal treatment of heavy metal-containing wastewater showed promising results from the previously published literature. However, there are a few challenges such as processing or disposal of heavy metal adsorbed algal biomass, desorption and recovery of heavy metal from the biomass, etc. Developing a continuous treatment system can also be challenging as long-term exposure of heavy metal to the algal biomass can exhaust its adsorption capacities without performing simultaneous metal desorption. Furthermore, biological activities can get inhibited irreversibly due to long-term metal toxicity to the biomass.

13.2.8 Textile Industry Wastewater

The textile industry is one of the significant industries that generate vast amounts of wastewater containing high recalcitrant dyes [71]. In addition, it contains high COD, dissolved solids, nitrogen, phosphorous, inorganic compounds, and minute quantities of different heavy metals such as chromium, arsenic, iron, copper, zinc, etc. [72]. Hence, releasing them without treatment can possess a significant threat to the environment. Along with other conventional Physico-chemical and biological treatment methods, algal treatment has also been successfully used for treating textile industry wastewater.

For example, Wu *et al.* [73] examined bio removal of textile wastewater containing supranol red dye using *Chlorella* sp. with CO_2 sparging. The total nitrogen and COD removal was very high (>75%). The algal biomass cultivated using textile wastewater contained around 20% fatty acid methyl ester, which was useful for the production of biodiesel. In another study, a different strain of *Chlorella vulgaris* was used for remediation of textile wastewater in a high-rate algae pond system [74]. The study reported 44.4-45.1% ammonium nitrogen, 33% phosphate, 38.3–62.3% of COD along with 41.8–50% of color removal. The dye removal mechanism by the algae was mainly by adsorption with both monolayer and multilayer adsorption on the algal cell surface.

Pathak *et al.* [71] reported dye (Methylene blue) removal from textile wastewater using *Chlorella pyrenoidosa*. The maximum phosphate, nitrate, and BOD removal during the study was 87%, 82%, and 63%, respectively. The dye removal was mainly by adsorption. However, dry biomass was found to be more effective for dye adsorption. The biomass growth of the algal strain was in the range of 8.1–14 µg/ml.d using textile wastewater at a different concentration ranging from 25% to 100%.

Immobilized form of algal cells has also been found to be suited for textile wastewater treatment. Immobilization improves resistance towards pollutants present in the wastewater and helps in biomass separation from the wastewater after the treatment. In a study by Chu *et al.* [75], immobilized *Chlorella vulgaris* was used for textile dye (supranol red, lanaset red, and levafix navy blue) removal from wastewater. Immobilized cells in alginate (2% by volume) could remove 44% of Lanaset Red and 48.9% color from textile wastewater which was higher than results obtained using suspended algal cells. Aeration was counterproductive as it increased color intensity in the effluent rather than reducing it. In a similar study, Wu *et al.* [76] immobilized two species of *Chlorella* species in alginate (0.5–5% by volume) for textile wastewater treatment. Microalgae immobilized beads of 3% alginate showed the best results with more than 50% COD, ammoniacal nitrogen, and color removal from wastewater.

Mixed algal cultures have been used for textile wastewater treatment. In such cases, without isolating an individual strain, a consortium is often enriched using the target wastewater. The algal species that survive and grow using the selected wastewater get enriched in the consortium. Aragaw and Asmare [77] reported textile industry wastewater treatment using a mixed culture of freshwater algae. The maximum COD, BOD, and TDS removal observed were 91.5%, 91.9%, and 89.1%, respectively, along with

82.6% decolorization, after 20 days of incubation. The dominant strains present in the mixed culture were identified as *Scenedesmus* sp., *Chlorella* sp., *Synedra* sp., and *Achnanthidium* sp. Nirmalakhandan *et al.* [19] examined the potential of six microalgal species belonging *Chlorellaceae* family to grow on textile wastewater. The microalgal species could be able to efficiently remove chromogenic substances from wastewater (47.1–70%) along with different heavy metals (Al, Cu, V, Pb, and Se) present in textile wastewater. Another study examined mixed algal consortia to treat textile industry wastewater by cultivating them under fed-batch mode [78]. For adapting the algal strains to the textile wastewater, the cultivation time was gradually decreased from an initial 30 d to 10 d during later stages. The study found 68–72% of color removal along with 95% of phosphorous and 70% nitrogen removal from the wastewater.

Other than conventional photobioreactor and open pond systems, microbial fuel cell containing microalgal biofilm (*Chlorella vulgaris*) as cathode has also been used for textile wastewater treatment [79]. This system produced a maximum volumetric power density of 123.2 ± 27.5 mW/m^3. The algal MFC obtained very high removal efficiencies for COD (92–98%), zinc (98%), and chromium (54–80%).

13.2.9 Emerging Micropollutant Containing Wastewater Treatment

Micropollutants have been categorized as contaminants of emerging concern (CECs) and one of the major concerns for global pollution prevention authorities [80]. The sources of such emerging pollutants are household products, personal care products, pharmaceuticals such as hormones, antibiotics, pesticides, etc. [81]. As their name suggests, they are found in ng/L to mg/L concentrations. In most cases, they cannot be treated by conventional wastewater treatment systems such as the activated sludge process and need special polishing steps involving specialized treatment [81]. In this context, algae can be used to treat such pollutants to prevent their release into the environment. The literature observed that the removal mechanism for such pollutants could be degradation or adsorption onto the algal cell surface or a combination of both.

For example, de Godos *et al.* [82] studied tetracycline degradation using a high rate algal pond. The tetracycline removal efficiency in the high rate algal pond was 69% during the steady-state condition. The degradation mechanism was examined under the batch mode of operation, and photodegradation and biosorption were found to be governing mechanisms for tetracycline removal from wastewater. The presence of tetracycline did not affect biomass growth or COD removal. The only negative impact was deflocculation of the algal biomass that led to poor settling ability and affected the biomass separation from the treated water.

Guo *et al.* [11] studied the degradation of an important antibiotic compound, 7-amino cephalosporanic acid (7-ACA), using three microalgal isolates belonging to *Chlorella* sp., *Chlamydomonas* sp., and *Mychonastes* sp. During the study, it was found that tough 7-ACA has a slightly negative effect on biomass growth, the lipid accumulation was not affected. The algal 7-ACA removal mechanism involved adsorption, photolysis, and hydrolysis. Guo and Chen [83] examined the capability of *Chlorella pyrenoidosa* species to degrade four antibiotic compounds, cefradine, cefalexin, ceftazidime, and cefixime, reported a removal efficiency of 89.9%, 94.9%, 89.7%, and 100%, respectively. Furthermore, an algal-activated sludge system was evaluated to treat these antibiotics, which showed 97.91% of overall removal efficiency [83]. These removal efficiencies were much higher (53-74%) than those obtained using only conventional activated sludge systems, indicating the superior treatment potential of algal systems.

Peng *et al.* [84] examined the degradation of two steroid hormones, namely progesterone and norgestrel, using two microalgae, *Scenedesmus obliquus*, and *Chlorella pyrenoidosa*. A very high removal efficiency (>95%) for progesterone was obtained with both strains. In the case of norgestrel, *S. obliquus* could be able to degrade it completely, whereas only a 40% degradation could be achieved with *C. pyrenoidosa* strain. The major and minor degradation products from the degradation were identified, and according to the authors, hydroxylation, reduction, and oxidation played a prominent role in the transformation of these compounds.

Hom-Diaz *et al.* [85] tested two algal strains, namely *Selenastrumcapricornutum* and *Chlamydomonas reinhardtii*, to study bio removal of the hormones β-estradiol (E2) and 17α-ethinylestradiol (EE2) using concentrated anaerobic digest as the nutrient. The study obtained 88% and 100% removal efficiencies for E2 and EE2 compounds by *S. capricornutum*. In the case of *C. reinhardtii*, both the compounds are completely removed when cultured without anaerobic digest. However, digest experiments resulted in lowering removal efficiency to 76%. Both adsorption and biodegradation played a role in the removal of these compounds. According to the author, the presence of these compounds did not affect biomass growth. Bai and Acharya [86] evaluated the potential of *Nannochloris* sp. to degrade three pharmaceutical compounds viz. trimethoprim (TMP), sulfamethoxazole (SMX), and triclosan (TCS). The TCS was completely removed by *Nannochloris* sp. However, the SMX and TMP are not significantly degraded in this study.

13.2.10 Saline Wastewater Treatment

Algae has been used for treating high salt-containing wastewater. In many such cases, due to salt-induced stress, it could be advantageous for further bioenergy production from harvested biomass. A halophilic marine microalgae strain of *Chlorella vulgaris* was tested to treat saline wastewater containing 0-50 g/L salt concentration under the batch mode of operation [87]. The microalgae successfully removed nitrogen (92.2–96.6%) and phosphorus (99%) within an experimental period of 20 days, even at high salt concentrations. The lipid accumulation started after the nitrogen level was reduced below 5 mg/L in the wastewater, with dry cell weight having the highest lipid content of about 40%.

Meng *et al.* [88] explored aerobic algal-bacterial granules for a saline wastewater (1-4% salt concentration) treatment. High salt concentration was found to positively affect nitrogen and phosphorus removal. The granule stability also showed improved features at the elevated salt concentration for the algal-bacterial consortium in comparison to only sludge granules. *Halomonas* and *Nitzschia* species were predominantly present in the algal-bacterial granules at high salt concentrations.

In a different strategy, integration of desalination, petroleum industry wastewater treatment, and bioelectricity production was achieved in a microbial desalination cell [89]. The microalgae *Scenedesmus abundans* was used as bio-cathode in the system. The system reported a high COD removal efficiency (61.3–68%) with maximum voltage and desalination efficiency of 654 mv and 55.3%, respectively.

13.3 Resource Recovery from Algal Biomass

As already mentioned that the significant advantage of algal wastewater treatment is resource recovery from the algal biomass produced utilizing wastewater as substrate. Different physical, chemical, and biological methods are used to harvest algal biomass after complete wastewater treatment (Figure 13.2). The main components in algal biomass, i.e., lipid (bio-oil), carbohydrate, protein, and pigments, are valuable products. For example, lipids can be converted to biodiesel or can be used in skincare products, national oils, or any other specialty chemical and personal care products [90]. Carbohydrates and proteins have nutritional value and can be used for food-feed applications. Pigment and other bioactive compounds have

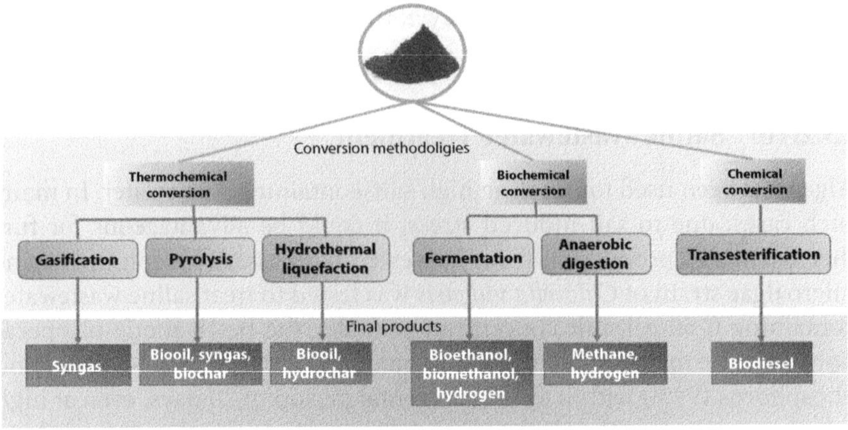

Figure 13.3 Resource recovery from algal biomass using different methods.

applications in human and animal health products. Furthermore, following the biomass recovery, algal biomass can be used for the production of biofuel or other products for which different conversion routes (Figure 13.3) have been explored in the literature. The following section discusses the different methods available for algal biomass conversion to value-added products.

13.3.1 Thermochemical Conversion

Thermochemical conversion can be either gasification, pyrolysis, or hydrothermal liquefaction. Depending upon the type of biomass used and the desired end product, either of these processes can be used for processing algal biomass. This differentiation be route is used for the conversion of biomass to bio-crude oil. Based on the type of biomass used and of process condition, the oil yield on biomass and its quality may vary. The criteria for selecting an efficient methodology lie in the purity of end-product, process cost, and eco-friendliness [91].

Gasification operates at a temperature ranging from 600°C to 700°C, and under this process, biomass gets converted to a gaseous mixture known as syngas containing hydrogen, carbon monoxide, carbon dioxide [92]. The produced gas composition can be altered by changing the experimental conditions such as applied pressure and oxygen presence. For algal biomass, this process has been applied primarily to produce hydrogen-rich gases [93].

Pyrolysis is a thermal decomposition process that produces liquid bio-oil, gases, and solid products from biomass, applied under anaerobic conditions and at a high-temperature range (350-550°C). Formerly pyrolysis was extensively used to produce energy from lignocellulosic biomass. However, in recent times, this technology has been used to convert algal biomass due to the process efficiency and superior quality of products [94]. Depending upon the process condition, pyrolysis can be of different types, namely slow pyrolysis, fast pyrolysis, flash pyrolysis, and catalytic pyrolysis [94].

Hydrothermal liquefaction is used for converting biomass into bio-oil. This process is operated at temperatures ranging from 230–350°C and high pressure (5–20 MPa) [95]. The main advantage of hydrothermal treatment over other thermochemical methods is the suitability of wet biomass, which makes the process easier and reduces the cost of biomass drying [96]. Hydrothermal treatment of algal biomass includes bio-oil, hydrochar and gaseous products.

13.3.2 Biochemical Conversion

Algal biomass can be used for biogas production using anaerobic digestion in which different groups of microorganisms work in tandem through a series of successive and simultaneous biochemical reactions to produce the final product (methane) [97]. The chemical composition of algal biomass and the experimental conditions (temperature, pH, etc.) primarily decide the biogas yield. However, certain bottlenecks need to be addressed to implement this technology successfully. One of the major problems with algal biomass as the substrate for biogas production is the sessional variation in the algal biomass composition due to the variable nature of the wastewater that is used to grow the algae [98]. Also, a critical factor carbon-nitrogen ratio of algal biomass is often not precisely suitable for anaerobic digestion, which can lead to unwanted product formation and inhibition of methane production in such systems. Other common problems such as the accumulation of volatile fatty acids, ammonia or hydrogen sulfide can cause a system failure. In order to improve biogas production and its yield from algal biomass, pretreatment technologies have also been used. However, unlike lignocellulose compounds, only mild treatment methods are sufficient enough in this case [99].

Algal biomass can also be used for biohydrogen production using dark fermentation. Whole algal biomass or lipid extracted biomass both had been reported as a substrate for efficient biohydrogen production [100]. However, in order to overcome the inhibition in the hydrolysis step due to complex algal biomass structure, physical, chemical, and biological pretreatment have been used to improve biohydrogen production from algal biomass. Another aspect of dark fermentative biohydrogen production from algal biomass is the negative energy balance, i.e., less net energy generation from the process can be difficult for practical implementation. To overcome such a problem, recently, biorefinery concepts, particularly those involving cascading biorefinery with multiple products using zero waste strategy, have gained a lot of research interest [101].

Bioethanol, another attractive biofuel, can also be produced using algal biomass as a renewable substrate. In this process, carbohydrates present in algal biomass can be converted to ethanol by fermentation [101]. Commonly used algal species for wastewater treatment such as *Chlorella*, *Chlamydomonas*, *Scenedesmus* contain high amounts of starch and have very little lignin content, which makes them ideal for ethanol production. However, pretreatment technologies, including Physico-chemical and biological treatment, have been reported in the literature to enhance ethanol production from algal biomass. Similar to ethanol production, algal biomass can be utilized to produce biobutanol. For this, common strains such as *Clostridium acetobutylicum* can be applied. From the economic point of view, biobutanol is more advantageous in comparison to bioethanol as butanol is immiscible in water, has a higher energy content, and has low Reid vapor pressure [102]. The main drawback of biobutanol production is the solvent toxicity to the microbial strain used, which could be avoided by fast recovery of products from the fermenter as soon as they are produced in order to avoid their accumulation.

13.3.3 Chemical Conversion

Chemical conversion of algal biomass is mainly for the production of biodiesel. Algal lipids have been well recognized as the alternative and green substrate for biodiesel production [103]. Microalgal biomass is rich in lipids that can reach up to 75% of its cell dry weight, making their use as a renewable substrate for biodiesel production extremely beneficial. In this process, lipids are extracted from algal biomass and reacted with methanol, converted to methyl esters, i.e., biodiesel. Glycerol is produced as a waste byproduct of the process. Different techniques are used for extraction of bio-oil from algal biomass, including mechanical pressing, solvent (n-hexane, chloroform, benzene, diethyl ether, ethanol, etc.) based extraction, and advanced methods such as supercritical fluid extraction [104]. However, solvent extraction using hexane is the most commonly used method as per the literature survey [105]. The transesterification reaction can be catalyzed by either chemical catalysts (acids and alkalis) or biocatalysts (lipase enzyme). In another approach, whole algal biomass can be subjected to transesterification reaction in the presence of methanol and potassium hydroxide. This would eliminate the solvent extraction step, which reduces the process's time and cost, making it more economical [106].

13.4 Concluding Remarks

Algal wastewater treatment has excellent potential for checking pollution and providing alternative sustainable options for resource recovery. This system can simultaneously treat industrial waste gas streams along with wastewater remediation. The bottlenecks of algal treatment systems are mainly related to their cultivation systems and the effect of pollutant toxicity on waste treatment efficiency and biomass growth. Many research works focusing on solving these problems have been undertaken recently, as indicated by the high number of publications in this area. Genetic engineering and metabolic engineering-based approaches have also been embarked upon to construct more toxic resistant strains and improve biomass yield and lipid content in the biomass. Algal-derived biofuel is the only renewable energy source that can completely replace fossil fuel-derived energy. However, to achieve that economics of the algal process need to be improved significantly. One promising approach is implementing the waste-fed algal biorefinery concept, which can integrate various biomass processing and conversion methods and produce a wide range of products in the form of bioenergy food-feed and chemicals all derived from algal biomass cultivated using wastewater.

References

1. Sinharoy, A., Pakshirajan, K., Lens, P.N.L., Biological sulfate reduction using gaseous substrates to treat acid mine drainage. *Curr. Pollut. Rep.*, 6, 328–344, 2020. https://doi.org/10.1007/s40726-020-00160-6.
2. Sinharoy, A. and Lens, P.N.L., Biological removal of selenate and selenite from wastewater: Options for selenium recovery as nanoparticles. *Curr. Pollut. Rep.*, 6, 230–249, 2020. https://doi.org/10.1007/s40726-020-00146-4.
3. Sinharoy, A. and Pakshirajan, K., Methane free biohydrogen production from carbon monoxide using a continuously operated moving bed biofilm

reactor. *Int. J. Hydrog. Energy*, 46, 306–313, 2021. https://doi.org/10.1016/j.ijhydene.2020.09.250.

4. Umamaheswari, J. and Shanthakumar, S., Efficacy of microalgae for industrial wastewater treatment: A review on operating conditions, treatment efficiency and biomass productivity. *Rev. Environ. Sci. Biotechnol.*, 15, 2, 265–284, 2016. https://doi.org/10.1007/s11157-016-9397-7.
5. Arun, S., Sinharoy, A., Pakshirajan, K., Lens, P.N.L., Algae-based microbial fuel cells for wastewater treatment and recovery of value-added products. *Renew. Sust. Energ. Rev.*, 132, 110041, 2020. https://doi.org/10.1016/j.rser.2020.110041.
6. Ramasamy, S., Arun, S., Pakshirajan, K., An overview of algal photobioreactors for resource recovery from waste, in: *Bioreactors*, L. Singh, A. Yousuf, D.M. Mahapatra (Eds.), pp. 215–248, Elsevier, The Netherlands, 2020.
7. Wang, L., Min, M., Li, Y., Chen, P., Chen, Y., Liu, Y., Wang, Y., Ruan, R., Cultivation of green algae Chlorella sp. in different wastewaters from municipal wastewater treatment plant. *Appl. Biochem. Biotechnol.*, 162, 4, 1174–1186, 2010. https://doi.org/10.1007/s12010-009-8866-7.
8. Craggs, R., Sutherland, D., Campbell, H., Hectare-scale demonstration of high rate algal ponds for enhanced wastewater treatment and biofuel production. *J. Appl. Phycol.*, 24, 3, 329–337, 2012. https://doi.org/10.1007/s10811-012-9810-8.
9. Arora, N., Gulati, K., Patel, A., Pruthi, P.A., Poluri, K.M., Pruthi, V., A hybrid approach integrating arsenic detoxification with biodiesel production using oleaginous microalgae. *Algal Res.*, 24, 29–39, 2017. https://doi.org/10.1016/j.algal.2017.03.012.
10. Luo, L., Wang, P., Lin, L., Luan, T., Ke, L., Tam, N.F.Y., Removal and transformation of high molecular weight polycyclic aromatic hydrocarbons in water by live and dead microalgae. *Process Biochem.*, 49, 10, 1723–1732, 2014. https://doi.org/10.1016/j.procbio.2014.06.026.
11. Guo, W.Q., Zheng, H.S., Li, S., Du, J.S., Feng, X.C., Yin, R.L., Wu, Q.L., Ren, N.Q., Chang, J.S., Removal of cephalosporin antibiotics 7-ACA from wastewater during the cultivation of lipid-accumulating microalgae. *Bioresour. Technol.*, 221, 284–290, 2016. https://doi.org/10.1016/j.biortech.2016.09.036.
12. Wang, M., Keeley, R., Zalivina, N., Halfhide, T., Scott, K., Zhang, Q., van der Steen, P., Ergas, S.J., Advances in algal-prokaryotic wastewater treatment: a review of nitrogen transformations, reactor configurations and molecular tools. *J. Environ. Manage.*, 217, 845–857, 2018. https://doi.org/10.1016/j.jenvman.2018.04.021.
13. Milledge, J.J. and Heaven, S., A review of the harvesting of micro-algae for biofuel production. *Rev. Environ. Sci. Biotechnol.*, 12, 2, 165–178, 2013. https://doi.org/10.1007/s11157-012-9301-z.
14. Show, K.Y. and Lee, D.J., Algal biomass harvesting, in: *Biofuels From Algae*, A. Pandey, D.J. Lee, Y. Chisti, C.R. Soccol (Eds.), pp. 85–110, Elsevier, The Netherlands, 2014.

15. Prajapati, S.K., Kaushik, P., Malik, A., Vijay, V.K., Phycoremediation coupled production of algal biomass, harvesting and anaerobic digestion: Possibilities and challenges. *Biotechnol. Adv.*, 31, 8, 1408–1425, 2013. https://doi.org/10.1016/j.biotechadv.2013.06.005.
16. Jetten, M.S., Horn, S.J., van Loosdrecht, M.C., Towards a more sustainable municipal wastewater treatment system. *Water Sci. Technol.*, 35, 9, 171–180, 1997. https://doi.org/10.1016/S0273-1223(97)00195-9.
17. Chan, Y.J., Chong, M.F., Law, C.L., Hassell, D.G., A review on anaerobic-aerobic treatment of industrial and municipal wastewater. *Chem. Eng. J.*, 155, 1-2, 1–18, 2009. https://doi.org/10.1016/j.cej.2009.06.041.
18. Selvaratnam, T., Pegallapati, A., Montelya, F., Rodriguez, G., Nirmalakhandan, N., Lammers, P.J., Van Voorhies, W., Feasibility of algal systems for sustainable wastewater treatment. *Renew. Energy*, 82, 71–76, 2015. https://doi.org/10.1016/j.renene.2014.07.061.
19. Nirmalakhandan, N., Selvaratnam, T., Henkanatte-Gedera, S.M., Tchinda, D., Abeysiriwardana-Arachchige, I.S.A., Delanka-Pedige, H.M.K., Munasinghe-Arachchige, S.P., Oyebamiji, O.O., Boeing, W.J., Holguin, F.O., Ilori, O., Amund, O., Green microalgae cultured in textile wastewater for biomass generation and biodetoxification of heavy metals and chromogenic substances. *Bioresour. Technol. Rep.*, 7, 100247, 2019. https://doi.org/10.1016/j.biteb.2019.100247.
20. Woertz, I., Feffer, A., Lundquist, T., Nelson, Y., Algae grown on dairy and municipal wastewater for simultaneous nutrient removal and lipid production for biofuel feedstock. *J. Environ. Eng.*, 135, 11, 1115–1122, 2009. https://doi.org/10.1061/(ASCE)EE.1943-7870.0000129.
21. Park, J.B.K. and Craggs, R.J., Nutrient removal in wastewater treatment high rate algal ponds with carbon dioxide addition. *Water Sci. Technol.*, 63, 8, 1758–1764, 2011. https://doi.org/10.2166/wst.2011.114.
22. Su, Y., Mennerich, A., Urban, B., Municipal wastewater treatment and biomass accumulation with a wastewater-born and settleable algal-bacterial culture. *Water Res.*, 45, 11, 3351–3358, 2011. https://doi.org/10.1016/j.watres.2011.03.046.
23. Kolev Slavov, A., General characteristics and treatment possibilities of dairy wastewater–a review. *Food Technol. Biotechnol.*, 55, 1, 14–28, 2017. https://doi.org/10.17113/ft b.55.01.17.4520.
24. Shete, B.S. and Shinkar, N.P., Dairy industry wastewater sources, characteristics & its effects on environment. *Int. J. Curr. Eng. Technol.*, 3, 5, 1611–1615, 2013.
25. Hena, S., Fatimah, S., Tabassum, S., Cultivation of algae consortium in a dairy farm wastewater for biodiesel production. *Water Resour. Ind.*, 10, 1–14, 2015. https://doi.org/10.1016/j.wri.2015.02.002.
26. Valizadeh, K. and Davarpanah, A., Design and construction of a micro-photo bioreactor in order to dairy wastewater treatment by micro-algae:

Parametric study. *Energy Sources A: Recovery Util. Environ. Eff.*, 42, 5, 611–624, 2020. https://doi.org/10.1080/15567036.2019.1588425.

27. Choi, H.J., Dairy wastewater treatment using microalgae for potential biodiesel application. *Environ. Eng. Res.*, 21, 4, 393–400, 2016. https://doi.org/10.4491/eer.2015.151.
28. Daneshvar, E., Zarrinmehr, M.J., Koutra, E., Kornaros, M., Farhadian, O., Bhatnagar, A., Sequential cultivation of microalgae in raw and recycled dairy wastewater: Microalgal growth, wastewater treatment and biochemical composition. *Bioresour. Technol.*, 273, 556–564, 2019. https://doi.org/10.1016/j.biortech.2018.11.059.
29. Barbera, M. and Gurnari, G. (Eds.), *Wastewater Treatment And Reuse In Ihe Food Industry*, Springer Nature, Switzerland, 2018.
30. Tekerlekopoulou, A.G., Economou, C.N., Tatoulis, T.I., Akratos, C.S., Vayenas, D.V., Wastewater treatment and water reuse in the food industry, in: *The Interaction of Food Industry and Environment*, C. Galanakis (Ed.), pp. 245–280, Academic Press, Cambridge, Massachusetts, USA, 2020.
31. Yonezawa, N., Matsuura, H., Shiho, M., Kaya, K., Watanabe, M.M., Effects of soybean curd wastewater on the growth and hydrocarbon production of *Botryococcus braunii* strain BOT-22. *Bioresour. Technol.*, 109, 304–307, 2012. https://doi.org/10.1016/j.biortech.2011.07.090.
32. Hongyang, S., Yalei, Z., Chunmin, Z., Xuefei, Z., Jinpeng, L., Cultivation of Chlorella pyrenoidosa in soybean processing wastewater. *Bioresour. Technol.*, 102, 21, 9884–9890, 2011. https://doi.org/10.1016/j.biortech.2011.08.016.
33. Gupta, S., Pandey, R.A., Pawar, S.B., Microalgal bioremediation of food-processing industrial wastewater under mixotrophic conditions: Kinetics and scale-up approach. *Front. Chem. Sci. Eng.*, 10, 4, 499–508, 2016. https://doi.org/10.1007/s11705-016-1602-2.
34. Lutzu, G.A., Zhang, W., Liu, T., Feasibility of using brewery wastewater for biodiesel production and nutrient removal by *Scenedesmus dimorphus*. *Environ. Technol.*, 37, 1568–1581, 2016. https://doi.org/10.1080/09593330.2015.1121292.
35. Travieso, L., Benítez, F., Sánchez, E., Borja, R., León, M., Raposo, F., Rincón, B., Assessment of a microalgae pond for post-treatment of the effluent from an anaerobic fixed bed reactor treating distillery wastewater. *Environ. Technol.*, 29, 985–992, 2008. https://doi.org/10.1080/09593330802166228.
36. Zheng, H., Liu, M., Lu, Q., Wu, X., Ma, Y., Cheng, Y., Addy, M., Liu, Y., Ruan, R., Balancing carbon/nitrogen ratio to improve nutrients removal and algal biomass production in piggery and brewery wastewaters. *Bioresour. Technol.*, 249, 479–486, 2018. https://doi.org/10.1016/j.biortech.2017.10.057.
37. Farooq, W., Lee, Y.C., Ryu, B.G., Kim, B.H., Kim, H.S., Choi, Y.E., Yang, J.W., Two-stage cultivation of two Chlorella sp. strains by simultaneous treatment of brewery wastewater and maximizing lipid productivity. *Bioresour. Technol.*, 132, 230–238, 2013. https://doi.org/10.1016/j.biortech.2013.01.034.

38. Gupta, S. and Pawar, S.B., An integrated approach for microalgae cultivation using raw and anaerobic digested wastewaters from food processing industry. *Bioresour. Technol.*, 269, 571–576, 2018. https://doi.org/10.1016/j.biortech.2018.08.113.
39. Hooda, P.S., Edwards, A.C., Anderson, H.A., Miller, A., A review of water quality concerns in livestock farming areas. *Sci. Total Environ.*, 250, 1–3, 143–167, 2000. https://doi.org/10.1016/S0048-9697(00)00373-9.
40. Wang, H., Magesan, G.N., Bolan, N.S., An overview of the environmental effects of land application of farm effluents. *New Z. J. Agric. Res.*, 47, 4, 389–403, 2004. https://doi.org/10.1080/00288233.2004.9513608.
41. Zhu, L., Wang, Z., Shu, Q., Takala, J., Hiltunen, E., Feng, P., Yuan, Z., Nutrient removal and biodiesel production by integration of freshwater algae cultivation with piggery wastewater treatment. *Water Res.*, 47, 13, 4294–4302, 2013. https://doi.org/10.1016/j.watres.2013.05.004.
42. Hu, B., Zhou, W., Min, M., Du, Z., Chen, P., Ma, X., Liu, Y., Lei, H., Shi, J., Ruan, R., Development of an effective acidogenically digested swine manure-based algal system for improved wastewater treatment and biofuel and feed production. *Appl. Energy*, 107, 255–263, 2013. https://doi.org/10.1016/j.apenergy.2013.02.033.
43. Liu, J., Ge, Y., Cheng, H., Wu, L., Tian, G., Aerated swine lagoon wastewater: A promising alternative medium for *Botryococcus braunii* cultivation in open system. *Bioresour. Technol.*, 139, 190–194, 2013. https://doi.org/10.1016/j.biortech.2013.04.036.
44. Luo, L., Shao, Y., Luo, S., Zeng, F.J., Tian, G.M., Nutrient removal from piggery wastewater by *Desmodesmus* sp.CHX1 and its cultivation conditions optimization. *Environ. Technol.*, 3330, 1–8, 2018. https://doi.org/10.1080/09593330.2018.1449903.
45. Lv, J., Liu, Y., Feng, J., Liu, Q., Nan, F., Xie, S., Nutrients removal from undiluted cattle farm wastewater by the two-stage process of microalgae-based wastewater treatment. *Bioresour. Technol.*, 264, 311–318, 2018. https://doi.org/10.1016/j.biortech.2018.05.085.
46. Markou, G., Iconomou, D., Muylaert, K., Applying raw poultry litter leachate for the cultivation of Arthrospira platensis and Chlorella vulgaris. *Algal Res.*, 13, 79–84, 2016.
47. Tan, X.B., Lam, M.K., Uemura, Y., Lim, J.W., Wong, C.Y., Ramli, A., Kiew, P.L., Lee, K.T., Semi-continuous cultivation of Chlorella vulgaris using chicken compost as nutrients source: Growth optimization study and fatty acid composition analysis. *Energy Convers. Manage.*, 164, 363–373, 2018. https://doi.org/10.1016/j.enconman.2018.03.020.
48. Wang, M., Yang, Y., Chen, Z., Chen, Y., Wen, Y., Chen, B., Removal of nutrients from undiluted anaerobically treated piggery wastewater by improved microalgae. *Bioresour. Technol.*, 222, 130–138, 2016. https://doi.org/10.1016/j.biortech.2016.09.128.

49. Singh, M., Reynolds, D.L., Das, K.C., Microalgal system for treatment of effluent from poultry litter anaerobic digestion. *Bioresour. Technol.*, 102, 23, 10841–10848, 2011. https://doi.org/10.1016/j.biortech.2011.09.037.
50. Jasmine, J. and Mukherji, S., Characterization of oily sludge from a refinery and biodegradability assessment using various hydrocarbon-degrading strains and reconstituted consortia. *J. Environ. Manage.*, 149, 118–125, 2015. https://doi.org/10.1016/j.jenvman.2014.10.007.
51. Varjani, S., Joshi, R., Srivastava, V.K., Ngo, H.H., Guo, W., Treatment of wastewater from petroleum industry: Current practices and perspectives. *Environ. Sci. Pollut. Res.*, 27, 22, 27172–27180, 2020. https://doi.org/10.1007/s11356-019-04725-x.
52. Namdari, A., Roayaei, E., Jaafarzadeh, N., Shariat, M., Measurement the significant heavy metals of Petroleum Desalination Influent in an Iranian on-shore desalination plant. *Pet. Sci. Technol.*, 35, 7, 681–686, 2017. https://doi.org/10.1080/10916466.2016.1270302.
53. Abid, A., Saidane, F., Hamdi, M., Feasibility of carbon dioxide sequestration by Spongiochlorissp microalgae during petroleum wastewater treatment in airlift bioreactor. *Bioresour. Technol.*, 234, 297–302, 2017. https://doi.org/10.1016/j.biortech.2017.03.041.
54. Hodges, A., Fica, Z., Wanlass, J., VanDarlin, J., Sims, R., Nutrient and suspended solids removal from petrochemical wastewater via microalgal biofilm cultivation. *Chemosphere*, 174, 46–48, 2017. https://doi.org/10.1016/j.chemosphere.2017.01.107.
55. Madadi, R., Pourbabaee, A.A., Tabatabaei, M., Zahed, M.A., Naghavi, M.R., Treatment of petrochemical wastewater by the green algae *Chlorella vulgaris*. *Int. J. Environ. Res.*, 10, 4, 555–560, 2016. https://doi.org/10.22059/IJER.2016.59684.
56. Chavan, A. and Mukherji, S., Effect of co-contaminant phenol on performance of a laboratory-scale RBC with algal-bacterial biofilm treating petroleum hydrocarbon-rich wastewater. *J. Chem. Technol. Biotechnol.*, 85, 6, 851–859, 2010. https://doi.org/10.1002/jctb.2378.
57. Kalhor, A.X., Movafeghi, A., Mohammadi-Nassab, A.D., Abedi, E., Bahrami, A., Potential of the green alga Chlorella vulgaris for biodegradation of crude oil hydrocarbons. *Mar. Pollut. Bull.*, 123, 1–2, 286–290, 2017. https://doi.org/10.1016/j.marpolbul.2017.08.045.
58. Mahanty, B., Pakshirajan, K., Dasu, V.V., Understanding the complexity and strategic evolution in PAH remediation research. *Crit. Rev. Environ. Sci. Technol.*, 41, 19, 1697–1746, 2011. https://doi.org/10.1080/10643389.2010.481586.
59. Kim, K.H., Jahan, S.A., Kabir, E., Brown, R.J., A review of airborne polycyclic aromatic hydrocarbons (PAHs) and their human health effects. *Environ. Int.*, 60, 71–80, 2013. https://doi.org/10.1016/j.envint.2013.07.019.
60. Kannan, K., Johnson-Restrepo, B., Yohn, S.S., Giesy, J.P., Long, D.T., Spatial and temporal distribution of polycyclic aromatic hydrocarbons in sediments

from Michigan inland lakes. *Environ. Sci. Technol.*, 39, 13, 4700–4706, 2005. https://doi.org/10.1021/es050064f.
61. Zhang, C., Lu, J., Wu, J., Adsorptive removal of polycyclic aromatic hydrocarbons by detritus of green tide algae deposited in coastal sediment. *Sci. Total Environ.*, 670, 320–327, 2019. https://doi.org/10.1016/j.scitotenv.2019.03.296.
62. Ke, L., Luo, L., Wang, P., Luan, T., Tam, N.F.Y., Effects of metals on biosorption and biodegradation of mixed polycyclic aromatic hydrocarbons by a freshwater green alga *Selenastrumcapricornutum. Bioresour. Technol.*, 101, 18, 6950–6961, 2010. https://doi.org/10.1016/j.biortech.2010.04.011.
63. Hazarika, J., Pakshirajan, K., Sinharoy, A., Syiem, M.B., Bioremoval of Cu (II), Zn (II), Pb (II) and Cd (II) by *Nostoc muscorum* isolated from a coal mining site. *J. Appl. Phycol.*, 27, 4, 1525–1534, 2015. https://doi.org/10.1007/s10811-014-0475-3.
64. Sinharoy, A. and Pakshirajan, K., Heavy metal sequestration by sulfate reduction using carbon monoxide as the sole carbon and energy source. *Process Biochem.*, 82, 135–143, 2019. https://doi.org/10.1016/j.procbio.2019.04.002.
65. Roy, A.S., Hazarika, J., Manikandan, N.A., Pakshirajan, K., Syiem, M.B., Heavy metal removal from multicomponent system by the cyanobacterium Nostoc muscorum: Kinetics and interaction study. *Appl. Biochem. Biotechnol.*, 175, 8, 3863–3874, 2015. https://doi.org/10.1007/s12010-015-1553-y.
66. Pena-Castro, J.M., Martınez-Jerónimo, F., Esparza-Garcıa, F., Canizares-Villanueva, R.O., Heavy metals removal by the microalga *Scenedesmus incrassatulus* in continuous cultures. *Bioresour. Technol.*, 94, 2, 219–222, 2004. https://doi.org/10.1016/j.biortech.2003.12.005.
67. Urrutia, C., Yañez-Mansilla, E., Jeison, D., Bioremoval of heavy metals from metal mine tailings water using microalgae biomass. *Algal Res.*, 43, 101659, 2019. https://doi.org/10.1016/j.algal.2019.101659.
68. Monteiro, C.M., Castro, P.M., Malcata, F.X., Use of the microalga Scenedesmus obliquus to remove cadmium cations from aqueous solutions. *World J. Microbiol. Biotechnol.*, 25, 9, 1573–1578, 2009. https://doi.org/10.1007/s11274-009-0046-y.
69. Chan, A., Salsali, H., McBean, E., Heavy metal removal (copper and zinc) in secondary effluent from wastewater treatment plants by microalgae. *ACS. Sustain. Chem. Eng.*, 2, 2, 130–137, 2014. https://doi.org/10.1021/sc400289z.
70. Richards, R.G. and Mullins, B.J., Using microalgae for combined lipid production and heavy metal removal from leachate. *Ecol. Modell.*, 249, 59–67, 2013. https://doi.org/10.1016/j.ecolmodel.2012.07.004.
71. Pathak, V.V., Kothari, R., Chopra, A.K., Singh, D.P., Experimental and kinetic studies for phytoremediation and dye removal by *Chlorella pyrenoidosa* from textile wastewater. *J. Environ. Manage.*, 163, 270–277, 2015. https://doi.org/10.1016/j.jenvman.2015.08.041.
72. Fazal, T., Mushtaq, A., Rehman, F., Khan, A.U., Rashid, N., Farooq, W., Rehman, M.S.U., Xu, J., Bioremediation of textile wastewater and successive

biodiesel production using microalgae. *Renew. Sust. Energ. Rev.*, 82, 3107–3126, 2018. https://doi.org/10.1016/j.rser.2017.10.029.

73. Wu, J.Y., Lay, C.H., Chen, C.C., Wu, S.Y., Lipid accumulating microalgae cultivation in textile wastewater: Environmental parameters optimization. *J. Taiwan Inst. Chem. Eng.*, 79, 1–6, 2017. https://doi.org/10.1016/j.jtice.2017.02.017.
74. Lim, S.L., Chu, W.L., Phang, S.M., Use of chlorella vulgaris for bioremediation of textile wastewater. *Bioresour. Technol.*, 101, 19, 7314–7322, 2010. https://doi.org/10.1016/j.biortech.2010.04.092.
75. Chu, W.L., See, Y.C., Phang, S.M., Use of immobilised *chlorella vulgaris* for the removal of colour from textile dyes. *J. Appl. Phycol.*, 21, 6, 641, 2009. https://doi.org/10.1007/s10811-008-9396-3.
76. Wu, J.Y., Lay, C.H., Chiong, M.C., Chew, K.W., Chen, C.C., Wu, S.Y., Zhou, D., Kumar, G., Show, P.L., Immobilized chlorella species mixotrophic cultivation at various textile wastewater concentrations. *J. Water Process. Eng.*, 38, 101609, 2020. https://doi.org/10.1016/j.jwpe.2020.101609.
77. Aragaw, T.A. and Asmare, A.M., Phycoremediation of textile wastewater using indigenous microalgae. *Water Pract. Technol.*, 13, 2, 274–284, 2018. https://doi.org/10.2166/wpt.2018.037.
78. Kumar, G., Huy, M., Bakonyi, P., Belafi-Bako, K., Kim, S.H., Evaluation of gradual adaptation of mixed microalgae consortia cultivation using textile wastewater via fed-batch operation. *Biotechnol. Rep.*, 20, e00289, 2018. https://doi.org/10.1016/j.btre.2018.e00289.
79. Logroño, W., Pérez, M., Urquizo, G., Kadier, A., Echeverría, M., Recalde, C., Rákhely, G., Single chamber microbial fuel cell (SCMFC) with a cathodic microalgal biofilm: A preliminary assessment of the generation of bioelectricity and biodegradation of real dye textile wastewater. *Chemosphere*, 176, 378–388, 2017. https://doi.org/10.1016/j.chemosphere.2017.02.099.
80. Barbosa, M.O., Moreira, N.F., Ribeiro, A.R., Pereira, M.F., Silva, A.M., Occurrence and removal of organic micropollutants: An overview of the watch list of EU decision 2015/495. *Water Res.*, 94, 257–279, 2016. https://doi.org/10.1016/j.watres.2016.02.047.
81. Kanaujiya, D.K., Paul, T., Sinharoy, A., Pakshirajan, K., Biological treatment processes for the removal of organic micropollutants from wastewater: A review. *Curr. Pollut. Rep.*, 5, 3, 112–128, 2019. https://doi.org/10.1007/s40726-019-00110-x.
82. de Godos, I., Muñoz, R., Guieysse, B., Tetracycline removal during wastewater treatment in high-rate algal ponds. *J. Hazard. Mater.*, 229, 446–449, 2012. https://doi.org/10.1016/j.jhazmat.2012.05.106.
83. Guo, R. and Chen, J., Application of alga-activated sludge combined system (AASCS) as a novel treatment to remove cephalosporins. *Chem. Eng. J.*, 260, 550–556, 2015. https://doi.org/10.1016/j.cej.2014.09.053.
84. Peng, F.Q., Ying, G.G., Yang, B., Liu, S., Lai, H.J., Liu, Y.S., Chen, Z.F., Zhou, G.J., Biotransformation of progesterone and norgestrel by two

freshwater microalgae (*Scenedesmus* obliquus and *Chlorella pyrenoidosa*): Transformation kinetics and products identification. *Chemosphere*, 95, 581–588, 2014. https://doi.org/10.1016/j.chemosphere.2013.10.013.
85. Hom-Diaz, A., Llorca, M., Rodríguez-Mozaz, S., Vicent, T., Barcelo, D., Blánquez, P., Microalgae cultivation on wastewater digestate: β-estradiol and 17α-ethynylestradiol degradation and transformation products identification. *J. Environ. Manage.*, 155, 106–113, 2015. https://doi.org/10.1016/j.jenvman.2015.03.003.
86. Bai, X. and Acharya, K., Removal of trimethoprim, sulfamethoxazole, and triclosan by the green alga *Nannochloris* sp. *J. Hazard. Mater*, 315, 70–75, 2016. https://doi.org/10.1016/j.jhazmat.2016.04.067.
87. Shen, Q.H., Gong, Y.P., Fang, W.Z., Bi, Z.C., Cheng, L.H., Xu, X.H., Chen, H.L., Saline wastewater treatment by *Chlorella vulgaris* with simultaneous algal lipid accumulation triggered by nitrate deficiency. *Bioresour. Technol.*, 193, 68–75, 2015. https://doi.org/10.1016/j.biortech.2015.06.050.
88. Meng, F., Huang, W., Liu, D., Zhao, Y., Huang, W., Lei, Z., Zhang, Z., Application of aerobic granules-continuous flow reactor for saline wastewater treatment: Granular stability, lipid production and symbiotic relationship between bacteria and algae. *Bioresour. Technol.*, 295, 122291, 2020. https://doi.org/10.1016/j.biortech.2019.122291.
89. Ashwaniy, V.R.V. and Perumalsamy, M., Reduction of organic compounds in petro-chemical industry effluent and desalination using *Scenedesmus abundant* algal microbial desalination cell. *J. Environ. Chem. Eng.*, 5, 6, 5961–5967, 2017. https://doi.org/10.1016/j.jece.2017.11.017.
90. Mohan, S.V., Hemalatha, M., Chakraborty, D., Chatterjee, S., Ranadheer, P., Kona, R., Algal biorefinery models with self-sustainable closed-loop approach: Trends and prospective for blue-bioeconomy. *Bioresour. Technol.*, 295, 122128, 2020. https://doi.org/10.1016/j.biortech.2019.122128.
91. Nigam, P.S. and Singh, A., Production of liquid biofuels from renewable resources. *Prog. Energy Combust. Sci.*, 37, 1, 52–68, 2011. https://doi.org/10.1016/j.pecs.2010.01.003.
92. Paul, T., Sinharoy, A., Baskaran, D., Pakshirajan, K., Pugazhenthi, G., Lens, P.N.L., Bio-oil production from oleaginous microorganisms using hydrothermal liquefaction: A biorefinery approach. *Crit. Rev. Environ. Sci. Technol.*, 52, 3, 356–394, 2022. https://doi.org/10.1080/10643389.2020.1820803.
93. Raheem, A., Ji, G., Memon, A., Sivasangar, S., Wang, W., Zhao, M., Taufiq-Yap, Y.H., Catalytic gasification of algal biomass for hydrogen-rich gas production: Parametric optimization via central composite design. *Energy Convers. Manage.*, 158, 235–245, 2018. https://doi.org/10.1016/j.enconman.2017.12.041.
94. Das, P., Chandramohan, V.P., Mathimani, T., Pugazhendhi, A., A comprehensive review on the factors affecting thermochemical conversion efficiency of algal biomass to energy. *Sci. Total Environ.*, 766, 144213, 2020. https://doi.org/10.1016/j.scitotenv.2020.144213.

95. Gollakota, A.R.K., Kishore, N., Gu, S., A review on hydrothermal liquefaction of biomass. *Renew. Sust. Energ. Rev.*, 81, 1378–1392, 2018. https://doi.org/10.1016/j.rser.2017.05.178.
96. Paul, T., Sinharoy, A., Pakshirajan, K., Pugazhenthi, G., Lipid-rich bacterial biomass production using refinery wastewater in a bubble column bioreactor for bio-oil conversion by hydrothermal liquefaction. *J. Water Process. Eng.*, 37, 101462, 2020. https://doi.org/10.1016/j.jwpe.2020.101462.
97. Montingelli, M.E., Tedesco, S., Olabi, A.G., Biogas production from algal biomass: A review. *Renew. Sust. Energ. Rev.*, 43, 961–972, 2015. https://doi.org/10.1016/j.rser.2014.11.052.
98. Sialve, B., Bernet, N., Bernard, O., Anaerobic digestion of microalgae as a necessary step to make microalgal biodiesel sustainable. *Biotechnol. Adv.*, 27, 4, 409–416, 2009. https://doi.org/10.1016/j.biotechadv.2009.03.001.
99. Gonzalez-Fernandez, C., Salve, B., Bernet, N., Steyer, J.P., Comparison of ultrasound and thermal pretreatment of Scenedesmus biomass on methane production. *Bioresour. Technol.*, 110, 610–616, 2012. https://doi.org/10.1016/j.biortech.2012.01.043.
100. Sinharoy, A., Kumar, M., Pakshirajan, K., An overview of bioreactor configurations and operational strategies for dark fermentative biohydrogen production, in: *Bioreactors*, L. Singh, A. Yousuf, D.M. Mahapatra (Eds.), pp. 249–288, Elsevier, The Netherlands, 2020.
101. Özçimen, D., İnan, B., Biernat, K., An overview of bioethanol production from algae, in: *Biofuels-Status and Perspective*, K. Biernat (Ed.), pp. 141–162, IntechOpen Limited, London, 2015.
102. Yeong, T.K., Jiao, K., Zeng, X., Lin, L., Pan, S., Danquah, M.K., Microalgae for biobutanol production–Technology evaluation and value proposition. *Algal Res.*, 31, 367–376, 2018. https://doi.org/10.1016/j.algal.2018.02.029.
103. Chisti, Y., Biodiesel from microalgae. *Biotechnol. Adv.*, 25, 3, 294–306, 2007. https://doi.org/10.1016/j.biotechadv.2007.02.001.
104. Roy, A.S., Chingkheihunba, A., Pakshirajan, K., An overview of production, properties, and uses of biodiesel from vegetable oil, in: *Green Fuels Technology*, C. Soccol, S. Brar, C. Faulds, L. Ramos (Eds.), pp. 83–105, Springer, Cham, 2016.
105. Rawat, I., Kumar, R.R., Mutanda, T., Bux, F., Biodiesel from microalgae: A critical evaluation from laboratory to large scale production. *Appl. Energy*, 103, 444–467, 2013. https://doi.org/10.1016/j.apenergy.2012.10.004.
106. Haas, M.J., Scott, K.M., Foglia, T.A., Marmer, W.N., The general applicability of *in situ* transesterification for the production of fatty acid esters from a variety of feedstocks. *J. Am. Oil Chem. Soc.*, 84, 10, 963–970, 2007. https://doi.org/10.1007/s11746-007-1119-4.

14

Sustainable Treatment of Pharmaceutical Effluents Using Microorganisms and Enzymes

Arpita Vats[1*] **and Sanjay Kumar**[2]

[1]*Department of Chemical and Environmental Engineering, University of Nottingham, United Kingdom*
[2]*Department of Biochemical Engineering and Biotechnology, Indian Institute of Technology, Delhi, India*

Abstract
Pharmaceutically derived micro-pollutants are emerging contaminants of great concern due to their spiked levels in an aquatic environment, hazardous to human health. The water treatment technologies in practice are inadequate and ineffective to eliminate pharmaceuticals, and therefore developing an efficient and economical treatment method becomes imperative. The disadvantages associated with several physicochemical treatment approaches such as incomplete removal of pollutants, formation of toxic by-products, and high operational costs necessitate the search for better alternatives. This chapter reviews the possibility of alleviating pharmaceutically derived pollutants with the help of a low-cost, environment-friendly biological approach using microorganisms or the enzymes derived from these. The scalability of these treatments is also discussed, along with the challenges and mitigation.

Keywords: Pharmaceuticals, hazardous, water treatment, microorganisms, enzymes

**Corresponding author*: arpita.vats@nottingham.ac.uk; arpitavats.iitd@gmail.com

Sanjay Kumar, Narendra Kumar and Shahid-ul-Islam (eds.) Role of Microbes in Industrial Products and Processes, (363–396) © 2023 Scrivener Publishing LLC

14.1 Introduction

Clean and safe drinking water is a necessity for all living beings. An increase in the human population and advancement in medical and pharmaceutical sciences has resulted in a steep rise in the production of pharmaceutical compounds. Many of these pharmaceutical (biologically active) compounds (PhACs) are not fully metabolized by the body and excreted out in wastewater. A massive spike of pharmaceuticals has been witnessed in the world's water bodies, drawing everyone's attention [1–3]. These pharmaceutical compounds are categorized as persistent pollutants and are of extreme concern due to their recalcitrance towards breakdown/removal from the environment, attributed to their complex chemical structures [4]. They are often regarded as micropollutants due to their micro- to nanogram per liter concentrations in the aquatic environment. These can still interfere with living beings' endocrine system, causing cancerous tumors and other birth defects in new born babies [1]. It is of particular concern for antibiotics, whose ill effects manifest as pathogenic resistance and challenge global health authorities [5, 6]. The current treatment technology in wastewater treatment plants was not designed to remove such PhACs [7, 8], which leads to their ultimate discharge/disposal in the surface and groundwater (Figure 14.1). Therefore, there is an urgent need to develop an effective treatment approach to remove the PhACs at the source (pharmaceutical industries/hospitals discharge) or in municipal wastewater treatment plants.

Globally, several organizations like the United States Environmental Protection Agency, the European Union, the International Programme

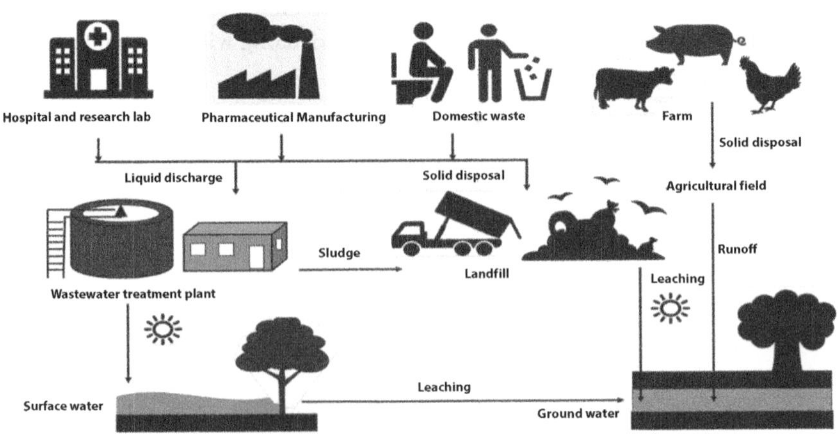

Figure 14.1 Primary sources of PhACs in the environment are domestic discharge (feces/urine or expired medicine in bins), pharmaceutical industries, hospitals, animal waste.

on Chemical Safety, and the World Health Organization have published legal directives to curb the PhACs in the environment. Additionally, many countries have signed agreements with the global environmental authorities for modifying existing treatment technologies. Based on their watch lists of pollutants, some of the PhACs categorized into hormones, non-steroidal anti-inflammatory drugs, and antibiotics are discussed in this chapter (Table 14.1). Hormones like estrone and estradiol form a part of contraceptive pills and other medications ingested by a large human population. Likewise, most anti-inflammatory drugs (ibuprofen) and ointments (diclofenac) are available globally without prescriptions to relieve pain and other inflammations. These compounds are either taken orally or applied to the skin, which may be released as it is or in partially metabolized form into the domestic wastewater through urine/feces or skin (runoff). Their toxicological impacts have been widely studied; for instance, Green *et al.* [9] found diclofenac poisoning to be the primary cause of the sharp decline of the vulture's population in the Indian subcontinent. Unregulated sale and consumption of antibiotics, particularly in developing countries, results in their bioaccumulation and antimicrobial resistance [10], as seen in the emergence of highly resistant superbugs in India [11, 12].

Physico-chemical techniques like ozonation, ion exchange, adsorption, electrochemical, membrane separation, and other advanced oxidation processes have shown promising results in removing these PhACs from wastewater [4], but the limitations like incomplete removal, increased toxicity of by-products, and huge energy costs associated with the process (operational, maintenance, and equipment perspectives) make these uneconomical and less preferred. On the other hand, the biological treatment approach is seen as favorable. It can effectively degrade PhACs, is environment friendly, low-cost, sustainable compared to Physico-chemical techniques, and is highly suitable for low-income group countries in meeting the global environmental standards. At present, most of the wastewater treatment plants in developing countries only include primary (physical separation of waste from water influent) and secondary (microbial degradation of organics) treatment of wastewater which is insufficient to remove PhACs. Installation of a more efficient and advanced tertiary set up demands for high installation and operational costs, which are unaffordable for most low-income group countries. Therefore, developing an efficient and low-cost biological treatment approach is much needed. This chapter focuses on the recent developments in this area and would discuss the feasibility and limitations of applying microorganisms and their enzymes in removing PhACs from the environment.

Table 14.1 List of PhACs, their chemical structure, use, and toxic effects.

Category	Name	Structure	Use	Ecotoxicity
Hormones	Estrone	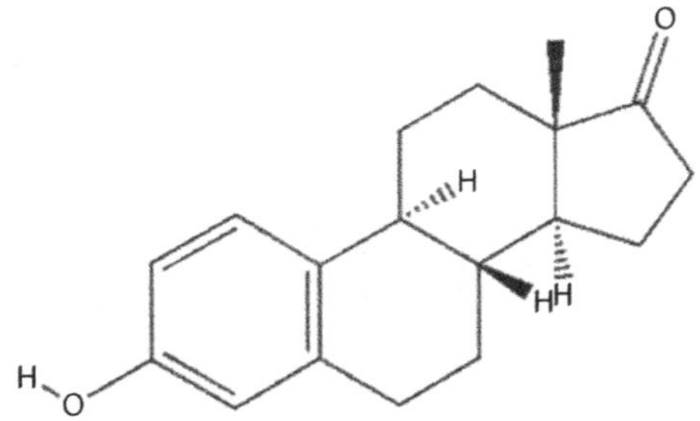	Contraceptive pills	Endocrine disruption; Impair fertility and reproductive functions [13]
	Estradiol	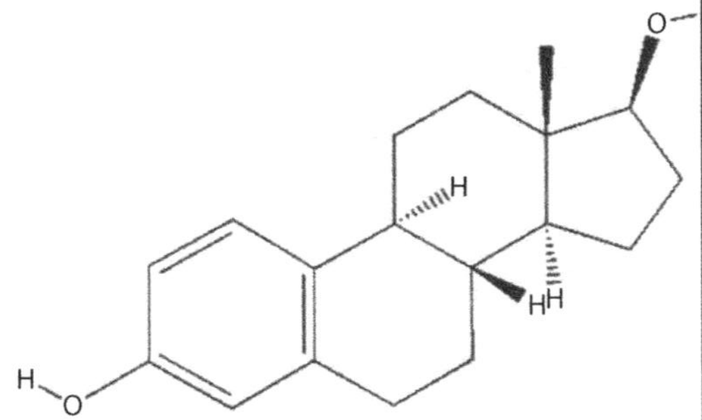		

(*Continued*)

Table 14.1 List of PhACs, their chemical structure, use, and toxic effects. (*Continued*)

Category	Name	Structure	Use	Ecotoxicity
Antibiotics	Sulfamethoxazole	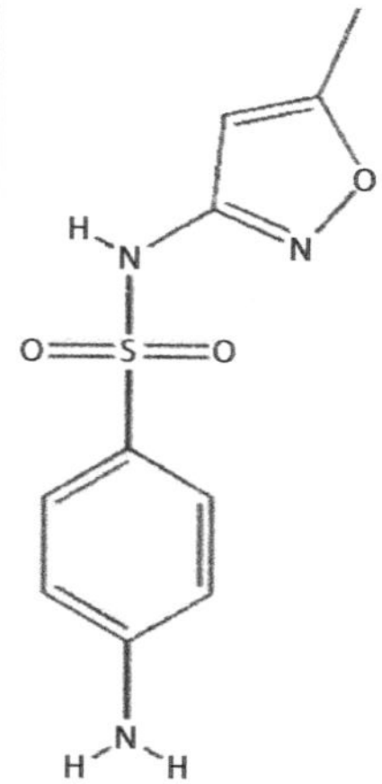	Treat bacterial infections/Kill pathogens	Pathogenic resistance; Toxic to the aquatic community [14]
	Ciprofloxacin	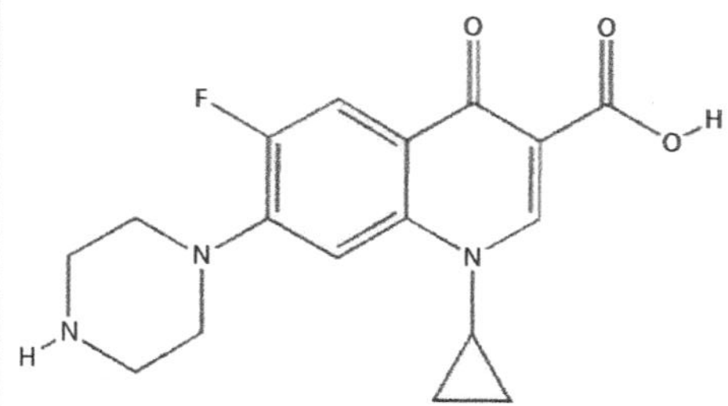		

(*Continued*)

Table 14.1 List of PhACs, their chemical structure, use, and toxic effects. (*Continued*)

Category	Name	Structure	Use	Ecotoxicity
	Azithromycin	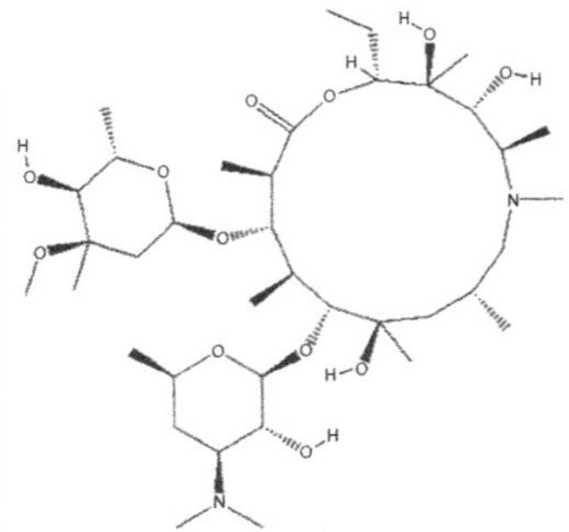		
	Erythromycin	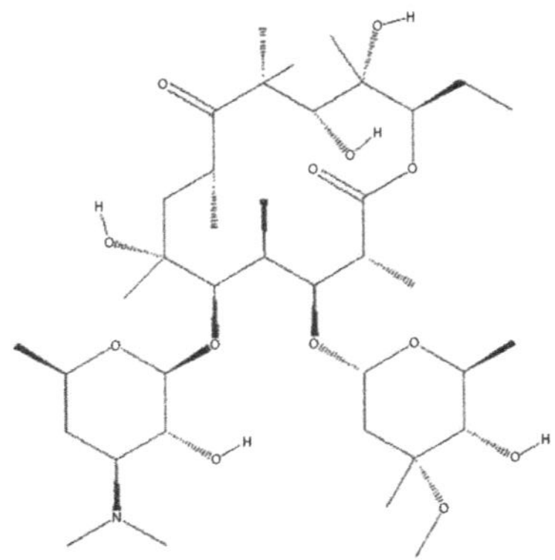		

(*Continued*)

Table 14.1 List of PhACs, their chemical structure, use, and toxic effects. (*Continued*)

Category	Name	Structure	Use	Ecotoxicity
NSAIDs*	Diclofenac	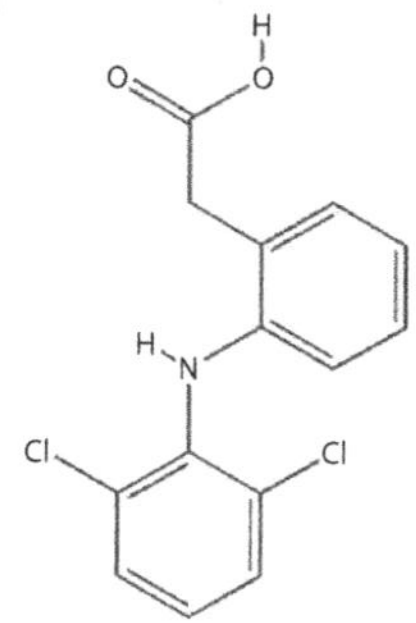	Treat pain and inflammations	Toxic to aquatic animals like crustaceans, fishes, mussels [15]
	Ibuprofen	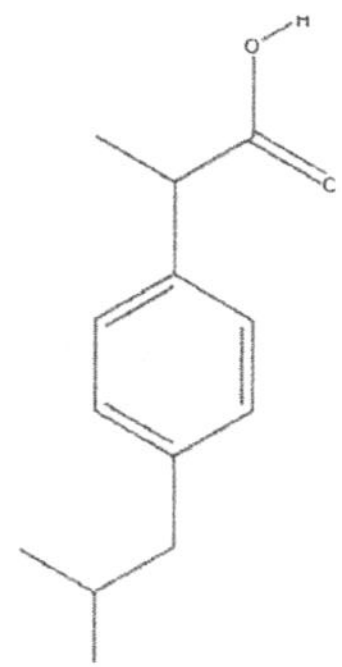	Treat pain, fever, and inflammation	Toxic to aquatic animals, crustaceans, and fishes [16]

(*Continued*)

Table 14.1 List of PhACs, their chemical structure, use, and toxic effects. (*Continued*)

Category	Name	Structure	Use	Ecotoxicity
Analgesic and antipyretic	Acetaminophen	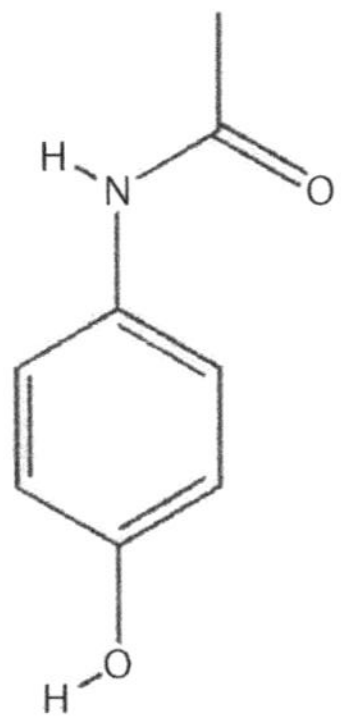	Treat pain and fever	Neurotoxic to aquatic animals like crustaceans, rotifers, and fishes [17]

*NSAIDs, Nonsteroidal anti-inflammatory drugs.

14.2 Microorganisms for Removal of Pharmaceuticals

For a long time, microorganisms have been used for remediation purposes in the secondary level of the water treatment process. They remove pollutants either via biodegradation or bio-sorption (Figure 14.2). While biodegradation occurs by two principle mechanisms, i.e., either by co-metabolism, in which pharmaceutical pollutant is degraded by microbial enzymes or by sole substrate degradation, in which targeted compounds become sole carbon and energy source for microbes [18]; Bio-sorption includes both absorption (entry of pollutants inside the biomass) and adsorption (adhesion of pollutants to the bio-mass surface) [19]. Few biodegradation processes result in complete mineralization of pollutants to CO_2, whereas others end up in the formation of small molecular weight transformants. Recent studies have shown the exceptional ability of a few microorganisms, includ ing bacteria, fungi, and algae, to degrade PhACs (Table 14.2). In some of the instances, enhanced removal of a PhAC by the whole organism compared to extracted enzymes is attributed to the additional/synergistic role played by intracellular enzymes as well as sorption or mycelium bound enzymes in addition to the extracellular enzymes [20].

Bacteria are preferred over other microbial candidates for bioremediation due to their fast growth rate and ease of handling cultures while optimizing process parameters. These are widely utilized during secondary treatment of wastewater under both aerobic as well as anaerobic conditions when dissolved organic concentrations in wastewater are low and high, respectively. Both metabolic as well as co-metabolic removal of PhACs, is seen in bacteria. Wang *et al.* [23] showed a linear increase in growth of *Acinetobacter* with sulfamethoxazole concentration, which stopped after the complete removal of sulfamethoxazole, indicating metabolic removal

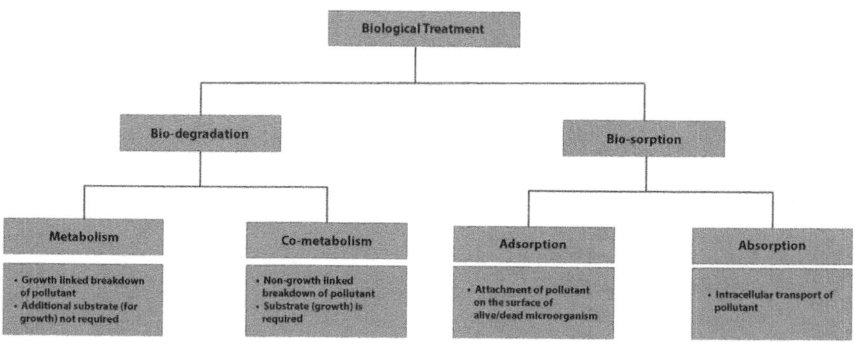

Figure 14.2 Various approaches of removing PhACs during biological treatment.

Table 14.2 List of microorganisms that can degrade PhACs.

Organisms	Pollutant (concentration)	Removal	Reference
BACTERIA			
Comamonas *Arcobacter* *Dysgonomonas* *Macellibacteroides* *Actinomyces* *Comamonas* *Castellaniella* *Desulfovibrio* *Enterococcus* *Peptostreeptococcus*	Ciprofloxacin (1 mg L^{-1}) Estradiol (1 mg L^{-1})	85% (8 Days) 84% (8 Days)	[21]
Klebsiella sp.	Diclofenac (70 mg L^{-1})	100% (72h)	[22]
Acinetobacter sp.	Sulfamethoxazole (240 mg L^{-1})	100% (7h)	[23]
Ochrobactrum sp.	Erythromycin A (100 mg L^{-1})	97% (72h)	[24]
Patulibacter sp.	Ibuprofen (50 μg L^{-1})	92% (90h)	[25]
FUNGI			
Pleurotus djamor	Diclofenac (10 mg L^{-1})	99% (6h)	[26]
Penicillium oxalicum	Diclofenac (100 μM)	100% (24h)	[27]
Trametes versicolor	Estrone (5 mg L^{-1}) Estradiol (5 mg L^{-1})	98% (12h)	[28]
Irpex lacteus *Trametes versicolor*	Ciprofloxacin (10 μg L^{-1})	100% (10 days) 100% (14 days)	[29]
Phanerochaete chrysosporium	Sulfamethoxazole (10 mg L^{-1})	53% (24h)	[30]

(Continued)

Table 14.2 List of microorganisms that can degrade PhACs. (*Continued*)

Organisms	Pollutant (concentration)	Removal	Reference
ALGAE			
Picocystis sp. *Graesiellla sp.*	Diclofenac (25 mg L^{-1})	73% (5 days) 52% (5 days)	[31]
Chlamydomonas mexicana	Ciprofloxacin (2 mg L^{-1})	13% (11 days)	[32]
Chlorella sorokiniana	Diclofenac (147 μg L^{-1}) Ibuprofen (317 μg L^{-1}) Acetaminophen (337 μg L^{-1})	100% (30 days) 100% (10 days) 100% (10 days)	[33]
Chlorella sorokiniana	Acetaminophen (25 mg L^{-1})	67% (4 days)	[34]

of SMX by *Acinetobacter*. Similarly, Jiang *et al.* [35] and Stylianou *et al.* [22] demonstrated utilization of sulfamethoxazole and diclofenac as sole carbon sources, respectively. *Labrys portucalensis* could degrade diclofenac both metabolically (70% in 30 days) as well as co-metabolically (~100% in 6 days) with acetate supplied as an additional carbon source in the latter [36]. Clearly, an additional growth substrate, in this case, improved the removal efficiency of diclofenac along with reducing the time period for degradation.

Fungi, mainly basidiomycetes have been vastly explored for their exceptional performance in removing PhACs compared to ascomycetes and zygomycetes [37]. The removal of PhACs by fungi is a combination of biodegradation and sorption as shown by the removal of diclofenac by 6 different fungal strains [38]. According to the authors, sorption of the pharmaceutical depends on their hydrophobicity and the fungal surface. Biodegradation of pollutants by fungi is mainly contributed by intracellular enzymes, i.e., CYP450 and extracellular ligninolytic enzymes, laccase, lignin peroxidase, and manganese peroxidase (discussed in detail in section 14.3). Fungal species are often used as fungal pellets for treatment purposes [39]. *Irpex lacteus* and *Trametes versicolor* could completely degrade ciprofloxacin after 10-14 days of incubation [29]. *Trametes versicolor* also showed significant removal (>98% in 12 h) of estrone and estradiol after initially grown for 8 days [28]. Enhanced removal of PhACs under co-cultivation of fungi was shown by Bankole *et al.* [40], who obtained an effective removal (~99%) of diclofenac and ibuprofen in 72h by co-culturing *Ganoderma applanatum* and *Laetiporus sulphureus.* According to the authors, the improved removal efficiencies in co-cultures compared to the individual fungal cultures were attributed to increased induction of oxidoreductase enzymes in the co-cultures.

Recently, Algae-mediated removal of PhACs has gained scientific attention as the treatment is eco-friendly, driven by solar energy, cutting down operational costs, and hence sustainable. The removal of PhACs can occur via biodegradation, bio-adsorption, and bioaccumulation in the algal cells [41]. The algae can be acclimatized in the presence of PhACs to improve the tolerance as well as biodegradation capacity [42, 43]. Mixotrophic algae can metabolize pollutants both metabolically and co-metabolically depending on the availability of carbon sources and nutrients in the environment [44], thus providing great flexibility during the treatment process and overcoming the limitation of supplying carbon and other nutrients for growth like in bacteria and fungi. Effective removal of PhACs like acetaminophen [32, 45], diclofenac [46], ibuprofen, and sulfamethoxazole [47] by various algal species is shown in the literature. However, the only limitation of using algae is the long growth periods up to several days, which slows down the treatment process. Using algae-bacterial consortiums can combine the effectivity of both organisms to remove PhACs and may reduce the treatment time as transformants formed from bacterial action can serve as algal growth promoters [48].

14.2.1 Scaling Up Microbial Treatment

To make biological treatment commercially viable and industrially scalable, it is essential to develop efficient bioreactor systems for the application of microorganisms. Both conventional and advanced bioreactors (Figure 14.3) that can be used to remove PhACs from wastewater are discussed here.

14.2.1.1 Activated Sludge Process

The activated sludge process is a low-cost biological treatment approach involving microorganisms to remove organics from the wastewater. Conventionally used in wastewater treatment plants, the raw, untreated

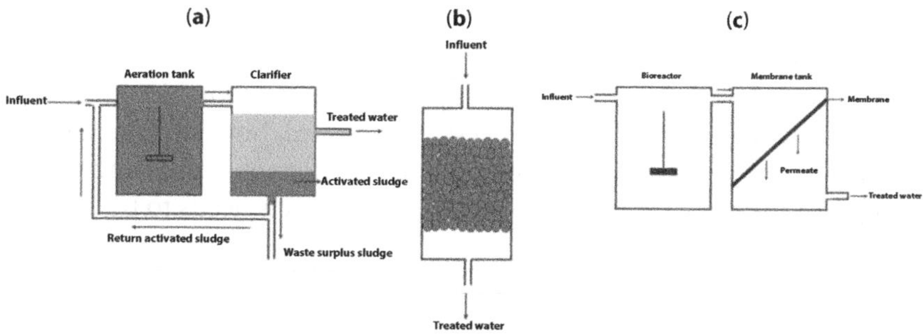

Figure 14.3 Types of bioreactors for removal of PhACs: (a) Activated sludge reactor (b) Fixed bed biofilm reactor, and (c) Membrane reactor.

water influent is mixed with an appropriately selected microbial consortium in an aerated reactor followed by recycling of the sludge prior to the subsequent cycles (Figure 14.3a). Predominantly, removal by this process is either via biodegradation or adsorption onto the microbial flocs as shown by the removal of sulfamethoxazole (~57%), ciprofloxacin (~95%), ibuprofen (~98%) [49], estrone (>98%) and estradiol (>98%) [50]. The selection of microbial consortium is extremely crucial in this process, as seen in the inefficiency of many treatment plants in completely removing PhACs by activated sludge process [51, 52]. Some of these treatment plants cannot remove the PhACs at all and hence require modifications in their current treatment method. Incomplete removal of PhACs and not meeting the expected permissible limits is the biggest disadvantage associated with activated sludge processes which has led to developing more advanced reactors.

14.2.1.2 *Biofilm Reactor*

In a biofilm reactor, immobilized microbial mass degrades PhACs dissolved in wastewater that gradually percolates through the support material (Figure 14.3b). Immobilization of microbial cells offers many advantages over suspended organisms, such as it prevents cell washout, increases the sludge retention time, and offers protection against shear damage. Abtahi *et al.* [53] showed higher co-metabolic removal of diclofenac and estradiol from municipal wastewater by biofilm reactor compared to suspended biomass. Similarly, in a study by Falas *et al.* [54], biofilm carriers from 7 wastewater treatment plants demonstrated considerably higher removal rates (per unit biomass) of diclofenac and ibuprofen compared to the activation sludges. Significant removal of sulfamethoxazole (>90%) and ciprofloxacin

(>80%) was also achieved in an anaerobic fixed bed biofilm reactor utilizing a consortium of methanogenic and fermentative bacteria [55]. A careful consideration for low-cost support materials for biofilm attachment can make the treatment process economical, however, the concern regarding the diffusion limitations (in biofilms) for substrates, and oxygen along with the separation of transformation products needs to be carefully understood and addressed.

14.2.1.3 Membrane-Based Bioreactors

Recently, membrane-based reactors have drawn everyone's attention due to their high removal capacities for PhACs. The microorganisms can either be immobilized or retained on the membrane in these reactors (Figure 14.3c). The PhAC removal in membrane bioreactors (MBR) is due to both physical retentions as well as biodegradation [56]. MBR provides relatively high sludge retention time and biomass concentration than the conventional activated sludge process, thus demonstrating higher efficiency in biodegradation. Tiwari *et al.* [57] demonstrated an effective removal (~90%) of ibuprofen and estradiol from synthetic hospital wastewater using a submerged membrane bioreactor. A study by Kim *et al.* [58] showed sorption to be the primary removal mechanism for ciprofloxacin while azithromycin and sulfamethoxazole were removed by degradation in a membrane bioreactor. A membrane reactor inoculated with aerobic granular sludge was used to treat water-containing pharmaceuticals and exhibited a removal rate of 79% for sulfamethoxazole and 63% for ibuprofen [59]. Though MBR showcases higher removal efficiencies than the activated sludge process in meeting the expected levels of discharges, fouling of membrane poses challenges [60] and increases the treatment costs.

14.2.1.4 Hybrid Approach

A combination of treatments increases the removal efficiency of pollutants and in some cases, completely eliminates both the pollutant as well as toxicity associated with its transformants. A combination of attached biofilm and activated sludge process as hybrid approach effectively (>50%) removed ibuprofen, sulfamethoxazole, azithromycin from municipal wastewater [61]. MBR coupled with activated carbon improved the efficiency of sulfamethoxazole (95%) and diclofenac (82%) removal as activated carbon, being porous, and having a large surface area provides support for the bacterial growth along with the absorption of low molecular weight contaminants [62]. A 3-step biological-ozone-biological process enhanced the removal (~100%) of diclofenac, ibuprofen, and sulfamethoxazole from wastewater effluent along with the elimination of toxicity [63]. In an attempt to combine the Physico-chemical approach with the biological, ozonation and ultrasound were applied as a pretreatment to a stream of PhACs containing wastewater prior to the membrane reactor, and the results showed >80% removal of sulfamethoxazole and diclofenac with lesser fouling compared to using MBR alone [64].

Though microorganisms have formed an integral part of wastewater treatment for decades, complete and effective removal of PhACs has always been a challenge. The slow growth of microorganisms like fungi, algae, and most importantly, adsorption of PhACs onto the surface of microorganisms defeats the purpose as it only concentrates the PhACs into the sludge that ends up in the landfills. The use of microbial enzyme extracts can overcome these limitations and can further speed up the treatment process.

14.3 Microbial Enzymes for Removal of Pharmaceuticals

As microbial enzymes play a significant role in biodegradation and are mainly responsible for the breakdown of pollutants, these can be applied in the free form as purified or crude enzyme extracts. Unlike microorganisms, extracted enzymes do not require continuous addition of nutrients for growth and can reach high reaction kinetics sooner, thus making the treatment process economical to use on a large scale. Moreover, specificity and selectivity of enzymes towards their substrates minimize the unfavorable side reactions, thus making enzymatic treatment a benign approach that requires lower energy, moderate working conditions, and produces less or non-toxic by-products. Oxidoreductases are attractive candidates for an effective breakdown of recalcitrant pharmaceuticals, as seen by several studies published in the last decade [65]. The ligninolytic enzymes from fungus, namely laccases and peroxidases, originally secreted to break down the lignin component of the biomass, are relatively less specific towards organic compounds as they use a free radical mechanism to catalyze the breakdown of a wide range of pollutants. Along with laccase and peroxidase, enzymes like tyrosinase and esterase are shown to degrade pharmaceuticals effectively (Figure 14.4). Both crude extracts and purified enzymes have demonstrated exceptional performance in removing PhACs (Table 14.3). Though the use of crude enzyme mixture is economical and may yield better removal due to the synergy between enzymes,

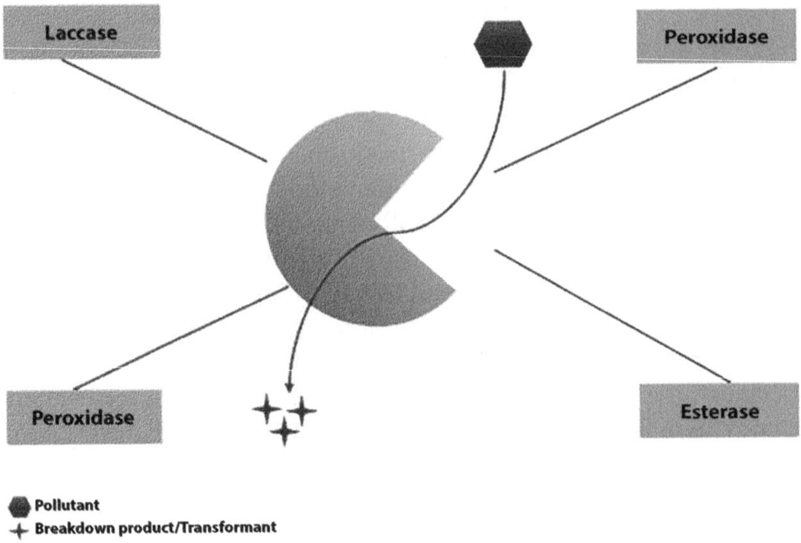

Figure 14.4 Enzymatic degradation of PhACs.

the difficulty of optimizing operational parameters and the presence of unnecessary culture components in crude extracts favors the use of purified enzymes. Nonetheless, the effectiveness of enzyme treatment is well established and discussed in detail here.

Laccase (EC 1.10.3.2) is a metalloprotein consisting of a trinuclear copper center that oxidizes its substrates via a free radical mechanism by extracting electrons from the substrate (pollutant) by T1Cu, which is then passed on to the T2-T3 copper center where O_2 is reduced to water [75]. It is often referred to as a green enzyme since it only requires molecular oxygen as a cofactor for its activity. The catalytic ability of laccase to degrade pollutants depends on the presence of electron-donating/withdrawing functional groups in the pollutant's structure, which also determines its redox potential. Laccase can only catalyze hydrogen/electron-donating pollutants, like phenolics (-OH) and aromatic amines (-NH2), and is unable to act upon many high redox pollutants containing electron-withdrawing functional groups [76]. Laccases are highly effective over phenolic contaminants like hormones [77], as shown by the complete removal of estrone and estradiol from the municipal wastewater within 24h [78]. A comparison between bacterial (*Streptomyces*) and fungal (*Trametes versicolor*) laccase was made by Margot *et al.* [79] for degradation of diclofenac and found fungal laccase to be a better candidate in terms of specific activity and enzyme stability.

Table 14.3 List of microbial enzymes that can degrade PhACs.

Predominant enzyme	Organism	Pollutant (concentration)	(%) Removal	Immobilization technique	Reference
CRUDE EXTRACT					
Laccase	*Trametes versicolor*	Diclofenac (500 μg L^{-1})	100% (5h)	Biochar	[66]
Laccase	*Trametes versicolor*	Diclofenac (1 mg L^{-1})	100% (24h)	-	[67]
Laccase	*Phanerochaete chrysosporium*	Sulfamethoxazole (10 mg L^{-1})	42% (24h)	Carbon nanotubes	[30]
Versatile peroxidase	*Bierkandera adusta*	Sulfamethoxazole (2.5 mg L^{-1})	80% (7h)	-	[68]
PURIFIED ENZYME					
Manganese peroxidase	*Trametes polyzona*	Ciprofloxacin (25 mg L^{-1})	73% (24h)	-	[69]
Laccase	*Aspergillus oryzae*	Estrone (20 μg L^{-1}) Estradiol (20 μg L^{-1}) Diclofenac (20 μg L^{-1}) Ibuprofen (20 μg L^{-1})	100% (60h) 100% (60h) 98% (60h) 70% (60h)	-	[70]

(Continued)

Table 14.3 List of microbial enzymes that can degrade PhACs. (*Continued*)

Predominant enzyme	Organism	Pollutant (concentration)	(%) Removal	Immobilization technique	Reference
Tyrosinase Laccase	Mushroom *Trametes versicolor*	Acetaminophen Ibuprofen (10–35 μg L^{-1}) Ciprofloxacin	85% (8h) 90% (24h) 60% (24h)	Enzyme cross-linked aggregates	[71]
Laccase	*Trametes versicolor*	Acetaminophen (1 mg L^{-1}) Diclofenac (1 mg L^{-1})	99% (18h) 85% (18h)	Fe_3O_4/SiO_2-Hybrid nanocomposites	[72]
Tyrosinase	Mushroom	Acetaminophen (100 μg L^{-1})	1005 (8h)	Enzyme cross-linked aggregates	[73]
Esterase	*Escherichia coli*	Erythromycin (100 mg L^{-1})	52% (16h)	-	[74]

A similar observation of complete removal of diclofenac by *Trametes* laccase was made later by Lonappan *et al.* [67] and Alharbi *et al.* [80].

Though laccase is very efficient in removing phenolic contaminants, pharmaceuticals, including antibiotics that are non-phenolic, are difficult to degrade. The poor degradation of non-phenolic PhACs is generally attributed to the presence of strong electron-withdrawing groups, such as amide (-CONR2), carboxylic (-COOH), halogen (-X), and nitro (-NO2) in the molecular structure, which corresponds to the high redox potential compared to laccase [81]. For such compounds, low molecular weight chemical molecules (acting as electron shuttle) are required that mediates the transfer of electrons as these small molecules can swiftly enter the active site of the enzyme and gets oxidized first (to form a cationic radical), in turn, extracts electrons from the bulkier pollutants (that cannot enter the enzyme's catalytic pocket) to return back to their native state. This was shown by Guardado *et al.* [82] for the degradation of sulfamethoxazole and ciprofloxacin. Three laccases from *Trametes versicolor, Myceliophtora thermophila*, and *Pycnoporus sanguineus* were compared for the degradation of antibiotics and could only do so in the presence of mediators, namely 2,2′-azino-bis(3-ethylbenzothiazoline-6-sulfonic acid, coumaric acid, and syringaldehyde. Similar observations were made by Ding *et al.* [83] for degradation of sulfonamide antibiotics where > 97% degradation was achieved on the addition of syringaldehyde and 1-hydroxy benzotriazole as mediators. Though the addition of mediators speeds up the enzyme kinetics, it may lead to the formation of toxic metabolites and increase the chemical oxygen demand of the wastewater and, therefore, highly discouraged [84]. Using crude mixtures can be a green alternative over mediator supplementation, as shown by Tran *et al.* [76] incomplete removal of ibuprofen and diclofenac by crude laccase extract of *T. versicolor* and *Aspergillus oryzae* compared to the purified laccases (20-50% removal) from these organisms. According to the authors, the degradation enhancement in crude extract could be due to natural mediators in the culture medium.

The catalytic mechanism of lignin peroxidase (LiP, EC 1.11.1.14) and manganese peroxidase (MnP, EC 1.11.1.13) involves H_2O_2 induced oxidation of LiP/MnP to form electron deficient cation radical form of the enzyme which extracts electrons from the pollutants to return back to its native form [85]. Peroxidases can catalyze both phenolic as well as non-phenolic pollutants which have lower redox potential than peroxidases. MnP from *Trametes polyzona* could successfully degrade ciprofloxacin (~73%) without the assistance of any external mediator [69]. Versatile peroxidase, on the other hand, combined the substrate specificity of MnP and LiP and was able to completely remove estrogens (2.5 mg/L) in less than

5 min, whereas other peroxidases and laccase take up to a few hours for complete removal of estrogens [86].

Apart from laccase and peroxidases, Tyrosinase (EC 1.14.18.1), a bi-cupric oxidoreductase, is also effective over phenolic pollutants. Mechanistically, tyrosinase catalyzes the hydroxylation of mono-phenolic substrates (to its corresponding o-diphenol) using O_2 and subsequent formation of ortho-quinone (that polymerizes spontaneously) by releasing H_2O molecules [87]. Like laccases, it only requires O_2 as a cofactor but differs in the oxidation mechanism as it generates quinones instead of free radicals (by laccase). Tyrosinase appears as a promising candidate for PhACs removal as these have been found very effective over phenolic contaminants such as acetaminophen [88, 89]. Along with the laccase, it was recently used to degrade (60–100%) a mixture of 14 pharmaceuticals spiked into municipal wastewater including ciprofloxacin, acetaminophen, and ibuprofen [90].

Llorca *et al.* [91] reported 52% degradation of erythromycin, a beta-lactam antibiotic by a bacterial esterase in 16h. The removal percentage of erythromycin increased to >90% when the enzyme was applied in the immobilized form [92].

For enzyme applications at a large scale, it is necessary to ensure that enzymes are not washed out, can be recovered (and reused), and are stable over long periods, all of these can be achieved by immobilizing enzymes. The storage and operational stability of the enzyme are increased by immobilizing as it decreases the enzyme's flexibility, reducing its propensity to unfold and denature under the influence of harsh conditions like heat, organic solvents, or by autolysis [93]. Also, immobilizing enzymes for continuous processes becomes even more crucial as it prevents them from washing out, increases their retention time and substrate availability, enhances temperature and pH stability, and protects inhibitory molecules [94, 95]. Due to numerous benefits of immobilization, recent development in finding new support materials and techniques of immobilizing oxidoreductases have been on the rise in the past decade [96, 97]. Carrier-based (adsorption, covalent binding), carrier-free (cross-linking and encapsulation) immobilization approaches for oxidoreductases have been extensively documented in the literature. Degradation of diclofenac and acetaminophen by laccase and peroxidase was enhanced on (covalent-) immobilization over carbon nanofibers with an increase in stability and performance of enzyme [98, 99]. Also, application of 3 co-immobilized enzymes, laccase, versatile peroxidase, and glucose oxidase as cross-linked enzyme aggregates over a cocktail of pharmaceuticals including acetaminophen, diclofenac, and ciprofloxacin showed over 80% removal of pollutants without the assistance of any mediator, combining the effectivity of laccase and peroxidase [100].

14.3.1 Scaling Up Enzymatic Treatment

For large-scale operations, the research based on using immobilized enzymes in bioreactors is gaining momentum. Mainly, two configurations of reactors are most popular for the application of enzymes, enzymatic bioreactors (EBRs) and enzymatic membrane reactors (EMRs). Though both mostly consist of immobilized enzymes, EBRs require an additional step of separating (enzyme) reaction products from the mixture whereas, in EMR, biocatalysis is simultaneous with membrane separation of the products from the reaction mixture (Figure 14.5). In EMR, the enzyme can either be immobilized over a membrane that also acts as a selective barrier for transformants (Figure 14.5b) or it can be added as a free enzyme (to water stream containing pollutants) in a tank attached to another tank in a series (Figure 14.5c) containing membrane as a selective barrier to retain enzyme [101]. A packed bed reactor (EBR) containing laccase immobilized on granular activated carbon efficiently removed (~96%) sulfamethoxazole and (~98%) diclofenac [102]. The removal of diclofenac was contributed solely by biodegradation, whereas sulfamethoxazole was removed by both adsorption (~60%) and degradation (~36%). The authors also reported high residual laccase activities over a broad range of pH and temperature for immobilized laccase compared to its free form. A hybrid membrane bioreactor containing combined cross-linked aggregates of tyrosinase and laccase showed improved removal efficiency for antibiotics (~100%) compared to the batch experiments (~60%) after 24h [71]. In a recent study by

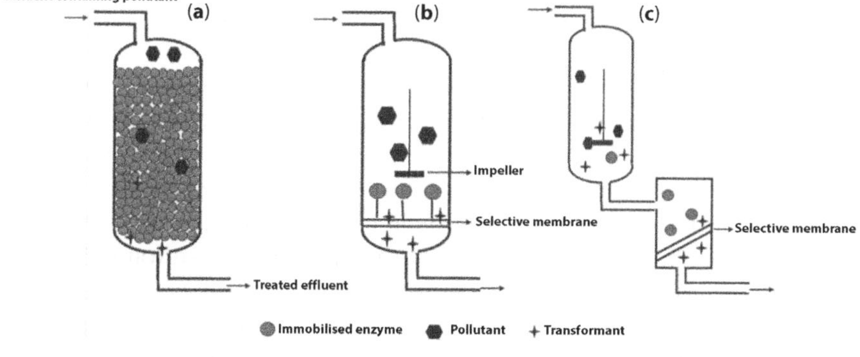

Figure 14.5 Types of enzymatic reactors: (a) Packed bed enzyme reactor (EBR); Enzymatic membrane reactor (EMR) with membrane-separation in the (b) Same compartment and (c) Separate compartment.

Asif *et al.* [70], laccase from *Aspergillus oryzae* was used in EMR to degrade 30 diverse micro-pollutants and observed >99% degradation for phenolic contaminants like estrone and estradiol, whereas non-phenolic pollutants like ibuprofen and diclofenac showed a moderate removal (60-70%) which was enhanced in the presence of mediators, syringaldehyde and violuric acid.

14.4 Factors Affecting Removal of Pharmaceuticals

Removal efficiency depends on the physical and chemical properties (solubility, biodegradability) of PhACs, operational parameters (sludge and hydraulic retention time, pH, and temperature), microbial community as well as enzyme specificities. The structure (stereo-chemistry) and functional group of the pollutant and its redox potential, can influence biodegradation efficiency [103, 104]. The complexity and stability of pollutants also govern biodegradability; for instance, the unsaturated pollutants and those with short side chains are easily biodegraded than the aromatic or highly branched, long side-chain pollutants. In bioreactors, high sludge retention time and low hydraulic retention time favor biodegradation by increasing the contact time between the microbes (in the sludge) and the pollutants (in the effluent stream). The acidity, metal content, organic/inorganic content of wastewater can also affect the growth of microbes and enzymatic efficiency to degrade pollutants [105] as seen in higher degradation of ionizable pollutants, ibuprofen, and sulfamethoxazole at acidic pH [103].

14.5 Challenges and Future Perspectives

Despite of the infinite advantages associated with the biological treatment approach, its large-scale application to remove pharmaceuticals is limited by several factors. The conventional activated sludge treatment cannot completely remove pharmaceuticals and therefore more efficient organisms need to be explored that can be applied as consortiums to improve the removal efficiency. The complexity in nature of wastewater in treatment plants often inactivate microorganisms or enzymes and make the treatment process inefficient, however, it may be easier to treat PhACs at source, like before its discharge from hospitals or pharmaceutical manufacturing industries, where the composition of effluent stream is known, and the challenges of unknowns can be bypassed. Also, an additional tertiary step designed to remove PhACs can be included in treatment plants as the effluent after the secondary treatment is relatively clean. To identify and measure the PhACs in wastewater, it is necessary to have an efficient monitoring system as the trace amounts of PhACs often go un-measured or detected. Advanced technology and highly sensitive instruments like LC-MS enable overcoming this challenge. The design of highly sensitive low-cost biosensors [106] may further reduce the cost and make the analysis affordable. Replacing buffers with wastewater for optimizations may help develop a more realistic and practical approach. High catalytic efficiencies and selectivity may be achieved using enzyme mixtures/or combinations instead of individual users.

Apart from degrading pharmaceuticals, it is essential to assess the potential toxicity and fate of the metabolites resulting from biodegradation to ascertain the success of the potential application of such processes. As degradation of PhACs by enzymes take place through different pathways resulting in the formation of several intermediates and end-products, it is not sufficient to only focus on the disappearance of the parent compounds, but also essential to look into the degradation pathways and toxicity of the transformation products [84], which can be done using extremely advanced analytical techniques like mass spectrometry (GC-MS/LC-MS).

14.6 Conclusion

Water, being the most valuable resource and essential for the existence of life should be free from all types of hazardous contaminants for the wellbeing of both humans as well as animals. Ubiquitously present and unseen, pharmaceutical pollutants have emerged as the most worrisome contaminants in recent years. Developing an effective biological treatment approach is preferred over other Physico-chemical methods due to its cost-effectiveness and mild operational conditions without the use of advanced instrument/equipment and toxic solvents compared to conventional technologies. Broad substrate specificities of oxidoreductase enzymes make these attractive candidates for remediation of extremely recalcitrant pollutants. Modification of conventional treatment processes by adopting a hybrid approach that includes a combination of one or more conventional/advanced biological treatments may be very effective in the complete eradication of pharmaceutical contaminants, provided both technical as well as economic feasibility are kept in mind.

Acknowledgements

Arpita Vats would like to acknowledge the Royal Society and Science and Engineering Research Board, India for providing research funding. She also acknowledges the University of Nottingham's support in accessing literature facilities.

References

1. Ebele, A.J., Abdallah, M.A.E., Harrad, S., Pharmaceuticals and personal care products (PPCPs) in the freshwater aquatic environment. *Emerg. Contam.*, 3, 1–16, 2017. https://doi.org/10.1016/j.emcon.2016.12.004.
2. de Oliveira, M., Frihling, B.E.F., Velasques, J., Filho, F.J.C.M., Cavalheri, P.S., Migliolo, L., Pharmaceuticals residues and xenobiotics contaminants: Occurrence, analytical techniques and sustainable alternatives for wastewater treatment. *Sci. Total Environ.*, 705, 135568, 2020. https://doi.org/10.1016/j.scitotenv.2019.135568.
3. Majumder, A., Gupta, B., Gupta, A.K., Pharmaceutically active compounds in aqueous environment: A status, toxicity and insights of remediation. *Environ. Res.*, 176, 108542, 2019. https://doi.org/10.1016/j.envres.2019.108542.
4. Patel, M., Kumar, R., Kishor, K., Mlsna, T., Pittman, C.U., Mohan, D., PharPatel, M., Kumar, R., Kishor, K. *et al.*, Pharmaceuticals of emerging concern in aquatic systems: Chemistry, occurrence, effects, and removal methods. *Chem. Rev.*, 119, 3510–3673, 2019. https://doi.org/10.1021/acs.chemrev.8b00299.
5. Danner, M.C., Robertson, A., Behrends, V., Reiss, J., Antibiotic pollution in surface fresh waters: Occurrence and effects. *Sci. Total Environ.*, 664, 793–804, 2019. https://doi.org/10.1016/j.scitotenv.2019.01.406.
6. Klein, E.Y., Van Boeckel, T.P., Martinez, E.M., Pant, S., Gandra, S., Levin, S.A. *et al.*, Global increase and geographic convergence in antibiotic consumption between 2000 and 2015. *Proc. Natl. Acad. Sci.*, 115, E3463–70, 2018. https://doi.org/10.1073/pnas.1717295115.
7. Tran, N.H., Reinhard, M., Gin, K.Y.H., Occurrence and fate of emerging contaminants in municipal wastewater treatment plants from different geographical regions-A review. *Water Res.*, 133, 182–207, 2018. https://doi.org/10.1016/j.watres.2017.12.029.
8. Meyer, M.F., Powers, S.M., Hampton, S.E., An evidence synthesis of pharmaceuticals and personal care products (PPCPs) in the environment: Imbalances among compounds, sewage treatment techniques, and ecosystem types. *Environ. Sci. Technol.*, 53, 12961–73, 2019. https://doi.org/10.1021/acs.est.9b02966.

9. Green, R.E., Newton, I., Shultz, S., Cunningham, A.A., Gilbert, M., Pain, D.J. *et al.*, Diclofenac poisoning as a cause of vulture population declines across the Indian subcontinent. *J. Appl. Ecol.*, 41, 793–800, 2004. https://doi.org/10.1111/j.0021-8901.2004.00954.x.
10. Ayukekbong, J.A., Ntemgwa, M., Atabe, A.N., The threat of antimicrobial resistance in developing countries: Causes and control strategies. *Antimicrob. Resist. Infect. Control*, 6, 47, 2017. https://doi.org/10.1186/s13756-017-0208-x.
11. Srivastava, R.K., Ichhpujani, R.I., Khare, S., Rai, A., Chauhan, L.S., Superbug-the so-called NDM-1. *Indian J. Med. Res.*, 133, 458–60, 2011.
12. Oberoi, L., Singh, N., Sharma, P., Aggarwal, A., ESBL, MBL and Ampc β lactamases producing superbugs - Havoc in the intensive care units of Punjab India. *J. Clin. Diagn. Res.*, 7, 70–3, 2013. https://doi.org/10.7860/JCDR/2012/5016.2673.
13. Adeel, M., Song, X., Wang, Y., Francis, D., Yang, Y., Environmental impact of estrogens on human, animal and plant life: A critical review. *Environ. Int.*, 99, 107–19, 2017. https://doi.org/10.1016/j.envint.2016.12.010.
14. Felis, E., Kalka, J., Sochacki, A., Kowalska, K., Bajkacz, S., Harnisz, M. *et al.*, Antimicrobial pharmaceuticals in the aquatic environment-occurrence and environmental implications. *Eur. J. Pharmacol.*, 866, 172813, 2020. https://doi.org/10.1016/j.ejphar.2019.172813.
15. Lonappan, L., Brar, S.K., Das, R.K., Verma, M., Surampalli, R.Y., Diclofenac and its transformation products: Environmental occurrence and toxicity-A review. *Environ. Int.*, 96, 127–38, 2016. https://doi.org/10.1016/j.envint.2016.09.014.
16. Marchlewicz, A., Guzik, U., Hupert-Kocurek, K., Nowak, A., Wilczyńska, S., Wojcieszyńska, D., Toxicity and biodegradation of ibuprofen by *Bacillus thuringiensis* B1(2015b). *Environ. Sci. Pollut. Res.*, 24, 7572–84, 2017. https://doi.org/10.1007/s11356-017-8372-3.
17. Zur, J., Pinski, A., Marchlewicz, A., Hupert-Kocurek, K., Wojcieszynska, D., Guzik, U., Organic micropollutants paracetamol and ibuprofen—Toxicity, biodegradation, and genetic background of their utilization by bacteria. *Environ. Sci. Pollut. Res.*, 25, 21498–524, 2018. https://doi.org/10.1007/s11356-018-2517-x.
18. Tran, N.H., Urase, T., Ngo, H.H., Hu, J., Ong, S.L., Insight into metabolic and co-metabolic activities of autotrophic and heterotrophic microorganisms in the biodegradation of emerging trace organic contaminants. *Bioresour. Technol.*, 146, 721–31, 2013. https://doi.org/10.1016/j.biortech.2013.07.083.
19. Costa, F., Lago, A., Rocha, V., Barros, Ó, Costa, L., Vipotnik, Z. *et al.*, A review on biological processes for pharmaceuticals wastes abatement - A growing threat to modern society. *Environ. Sci. Technol.*, 53, 7185–202, 2019. https://doi.org/10.1021/acs.est.8b06977.
20. Nguyen, L.N., Hai, F.I., Yang, S., Kang, J., Leusch, F.D.L., Roddick, F. *et al.*, Removal of pharmaceuticals, steroid hormones, phytoestrogens, UV-filters, industrial chemicals and pesticides by *Trametes versicolor*: Role of

biosorption and biodegradation. *Int. Biodeterior. Biodegradation*, 88, 169–75, 2014. https://doi.org/10.1016/j.ibiod.2013.12.017.

21. Martins, M., Sanches, S., Pereira, I.A.C., Anaerobic biodegradation of pharmaceutical compounds: New insights into the pharmaceutical-degrading bacteria. *J. Hazard. Mater.*, 357, 289–97, 2018. https://doi.org/10.1016/j.jhazmat.2018.06.001.
22. Stylianou, K., Hapeshi, E., Vasquez, M.I., Fatta-Kassinos, D., Vyrides, I., Diclofenac biodegradation by newly isolated. *Klebsiella sp. KSC: Microbial intermediates ecotoxicological assessment. J. Environ. Chem. Eng.*, 6, 3242–8, 2018. https://doi.org/10.1016/j.jece.2018.04.052.
23. Wang, S. and Wang, J., Biodegradation and metabolic pathway of sulfamethoxazole by a novel strain *Acinetobacter* sp. *Appl. Microbiol. Biotechnol.*, 102, 425–32, 2018. https://doi.org/10.1007/s00253-017-8562-4.
24. Zhang, W., Qiu, L., Gong, A., Yuan, X., Isolation and characterization of a high-efficiency erythromycin A-degrading *Ochrobactrum* sp. strain. *Mar. Pollut. Bull.*, 114, 896–902, 2017. https://doi.org/10.1016/j.marpolbul.2016.10.076.
25. Almeida, B., Kjeldal, H., Lolas, I., Knudsen, A.D., Carvalho, G., Nielsen, K.L. *et al.*, Quantitative proteomic analysis of ibuprofen-degrading. *Patulibacter sp. strain I11. Biodegradation*, 24, 615–30, 2013.
26. Cruz-Ornelas, R., Sánchez-Vázquez, J.E., Amaya-Delgado, L., Guillén-Navarro, K., Calixto-Romo, A., Biodegradation of NSAIDs and their effect on the activity of ligninolytic enzymes from *Pleurotus djamor. 3 Biotech.*, 9, 373, 2019. https://doi.org/10.1007/s13205-019-1904-4.
27. Olicón-Hernández, D.R., Camacho-Morales, R.L., Pozo, C., González-López, J., Aranda, E., Evaluation of diclofenac biodegradation by the ascomycete fungus *Penicillium oxalicum* at flask and bench bioreactor scales. *Sci. Total Environ.*, 662, 607–14, 2019. https://doi.org/10.1016/j.scitotenv.2019.01.248.
28. Shreve, M.J., Brockman, A., Hartleb, M., Prebihalo, S., Dorman, F.L., Brennan, R.A., The white-rot fungus *Trametes* versicolor reduces the estrogenic activity of a mixture of emerging contaminants in wastewater treatment plant effluent. *Int. Biodeterior. Biodegradation*, 109, 132–40, 2016. https://doi.org/10.1016/j.ibiod.2016.01.018.
29. CvanCarova, M., Moeder, M., Filipova, A., Cajthaml, T., Biotransformation of fluoroquinolone antibiotics by ligninolytic fungi-metabolites, enzymes and residual antibacterial activity. *Chemosphere*, 136, 311–20, 2014. https://doi.org/10.1016/j.chemosphere.2014.12.012.
30. Guo, X.L., Zhu, Z.W., Li, H.L., Biodegradation of sulfamethoxazole by *Phanerochaete chrysosporium. J. Mol. Liq.*, 198, 169–72, 2014. https://doi.org/10.1016/j.molliq.2014.06.017.
31. Ben Ouada, S., Ben Ali, R., Cimetiere, N., Leboulanger, C., Ben Ouada, H., Sayadi, S., Biodegradation of diclofenac by two green microalgae: *Picocystis* sp. and *Graesiella* sp. *Ecotoxicol. Environ. Saf.*, 186, 109769, 2019. https://doi.org/10.1016/j.ecoenv.2019.109769.

32. Xiong, J.Q., Kurade, M.B., Kim, J.R., Roh, H.S., Jeon, B.H., Ciprofloxacin toxicity and its co-metabolic removal by a freshwater microalga *Chlamydomonas mexicana*. *J. Hazard. Mater.*, 323, 212–9, 2017. https://doi.org/10.1016/j.jhazmat.2016.04.073.
33. de Wilt, A., Butkovskyi, A., Tuantet, K., Leal, L.H., Fernandes, T.V., Langenhoff, A. *et al.*, Micropollutant removal in an algal treatment system fed with source separated wastewater streams. *J. Hazard. Mater.*, 304, 84–92, 2016. https://doi.org/10.1016/j.jhazmat.2015.10.033.
34. Escapa, C., Coimbra, R.N., Paniagua, S., García, A.I., Otero, M., Nutrients and pharmaceuticals removal from wastewater by culture and harvesting of *Chlorella sorokiniana*. *Bioresour. Technol.*, 185, 276–84, 2015. https://doi.org/10.1016/j.biortech.2015.03.004.
35. Jiang, B., Li, A., Cui, D., Cai, R., Ma, F., Wang, Y., Biodegradation and metabolic pathway of sulfamethoxazole by *Pseudomonas psychrophila* HA-4, a newly isolated cold-adapted sulfamethoxazole-degrading bacterium. *Appl. Microbiol. Biotechnol.*, 98, 4671–81, 2014. https://doi.org/10.1007/s00253-013-5488-3.
36. Moreira, I.S., Bessa, V.S., Murgolo, S., Piccirillo, C., Mascolo, G., Castro, P.M.L., Biodegradation of diclofenac by the bacterial strain *Labrys portucalensis* F11. *Ecotoxicol. Environ. Saf.*, 152, 104–13, 2018. https://doi.org/10.1016/j.ecoenv.2018.01.040.
37. Olicón-Hernández, D.R., González-López, J., Aranda, E., Overview on the biochemical potential of filamentous fungi to degrade pharmaceutical compounds. *Front. Microbiol.*, 8, 1792, 2017. https://doi.org/10.3389/fmicb.2017.01792.
38. Lucas, D., Castellet-Rovira, F., Villagrasa, M., Badia-Fabregat, M., Barceló, D., Vicent, T. *et al.*, The role of sorption processes in the removal of pharmaceuticals by fungal treatment of wastewater. *Sci. Total Environ.*, 610–611, 1147–53, 2018. https://doi.org/10.1016/j.scitotenv.2017.08.118.
39. Marco-Urrea, E., Pérez-Trujillo, M., Cruz-Morató, C., Caminal, G., Vicent, T., Degradation of the drug sodium diclofenac by *Trametes versicolor* pellets and identification of some intermediates by NMR. *J. Hazard. Mater.*, 176, 836–42, 2010. https://doi.org/10.1016/j.jhazmat.2009.11.112.
40. Bankole, P.O., Adekunle, A.A., Jeon, B.H., Govindwar, S.P., Novel cobiomass degradation of NSAIDs by two wood rot fungi, *Ganoderma applanatum* and *Laetiporus sulphureus*: Ligninolytic enzymes induction, isotherm and kinetic studies. *Ecotoxicol. Environ. Saf.*, 203, 110997, 2020. https://doi.org/10.1016/j.ecoenv.2020.110997.
41. Xiong, J.Q., Kurade, M.B., Jeon, B.H., Can microalgae remove pharmaceutical contaminants from water? *Trends Biotechnol.*, 36, 30–44, 2018. https://doi.org/10.1016/j.tibtech.2017.09.003.
42. Xiong, J.Q., Kurade, M.B., Jeon, B.H., Biodegradation of levofloxacin by an acclimated freshwater microalga, *Chlorella vulgaris*. *Chem. Eng. J.*, 313, 1251–7, 2017. https://doi.org/10.1016/j.cej.2016.11.017.

43. Osundeko, O., Dean, A.P., Davies, H., Pittman, J.K., Acclimation of microalgae to wastewater environments involves increased oxidative stress tolerance activity. *Plant Cell Physiol.*, 55, 1848–57, 2014. https://doi.org/10.1093/pcp/pcu113.
44. Subashchandrabose, S.R., Ramakrishnan, B., Megharaj, M., Venkateswarlu, K., Naidu, R., Mixotrophic cyanobacteria and microalgae as distinctive biological agents for organic pollutant degradation. *Environ. Int.*, 51, 59–72, 2013. https://doi.org/10.1016/j.envint.2012.10.007.
45. Escapa, C., Coimbra, R.N., Neuparth, T., Torres, T., Santos, M.M., Otero, M., Acetaminophen removal from water by microalgae and effluent toxicity assessment by the zebrafish embryo bioassay. *Water*, 11, 1929, 2019. https://doi.org/10.3390/w11091929.
46. Escapa, C., Coimbra, R.N., Paniagua, S., García, A.I., Otero, M., Comparative assessment of diclofenac removal from water by different microalgae strains. *Algal Res.*, 18, 127–34, 2016. https://doi.org/10.1016/j.algal.2016.06.008.
47. Xiong, J.Q., Kim, S.J., Kurade, M.B., Govindwar, S., Abou-Shanab, R.A.I., Kim, J.R. *et al.*, Combined effects of sulfamethazine and sulfamethoxazole on a freshwater microalga, *Scenedesmus obliquus*: Toxicity, biodegradation, and metabolic fate. *J. Hazard. Mater.*, 370, 138–46, 2019. https://doi.org/10.1016/j.jhazmat.2018.07.049.
48. Mamta, Bhushan, S., Rana, M.S., Raychaudhuri, S., Simsek, H., Prajapati, S.K., Algae-and bacteria-driven technologies for pharmaceutical remediation in wastewater, in: *Removal of Toxic Pollutants Through Microbiological and Tertiary Treatment*, pp. 373–408, Elsevier, New Perspectives, 2020, https://doi.org/10.1016/B978-0-12-821014-7.00015-0.
49. Min, X., Li, W., Wei, Z., Spinney, R., Dionysiou, D.D., Seo, Y. *et al.*, Sorption and biodegradation of pharmaceuticals in aerobic activated sludge system: A combined experimental and theoretical mechanistic study. *Chem. Eng. J.*, 342, 211–9, 2018. https://doi.org/10.1016/j.cej.2018.01.012.
50. Bernardelli, J.K.B., José Belli, T., Emanuelly da Costa, R., Angeles Lobo-Recio, M., Damasceno Silveira, D., Dutra de Armas, R. *et al.*, Bacterial community structure applied to hormone degradation. *J. Environ. Eng.*, 145, 04019086, 2019. https://doi.org/10.1061/(asce)ee.1943-7870.0001607.
51. Jelic, A., Gros, M., Ginebreda, A., Cespedes-Sánchez, R., Ventura, F., Petrovic, M. *et al.*, Occurrence, partition and removal of pharmaceuticals in sewage water and sludge during wastewater treatment. *Water Res.*, 45, 1165–76, 2011. https://doi.org/10.1016/j.watres.2010.11.010.
52. Salgado, R., Marques, R., Noronha, J.P., Carvalho, G., Oehmen, A., Reis, M.A.M., Assessing the removal of pharmaceuticals and personal care products in a full-scale activated sludge plant. *Environ. Sci. Pollut. Res.*, 19, 1818–1827, 2012. https://doi.org/10.1007/s11356-011-0693-z.
53. Abtahi, S.M., Petermann, M., Juppeau Flambard, A., Beaufort, S., Terrisse, F., Trotouin, T. *et al.*, Micropollutants removal in tertiary moving bed biofilm reactors (MBBRs): Contribution of the biofilm and suspended

biomass. *Sci. Total Environ.*, 643, 1464–80, 2018. https://doi.org/10.1016/j.scitotenv.2018.06.303.

54. Falås, P., Baillon-Dhumez, A., Andersen, H.R., Ledin, A., La Cour Jansen, J., Suspended biofilm carrier and activated sludge removal of acidic pharmaceuticals. *Water Res.*, 46, 1167–75, 2012. https://doi.org/10.1016/j.watres.2011.12.003.
55. Carneiro, R.B., Mukaeda, C.M., Sabatini, C.A., Santos-Neto, Á.J., Zaiat, M., Influence of organic loading rate on ciprofloxacin and sulfamethoxazole biodegradation in anaerobic fixed bed biofilm reactors. *J. Environ. Manage.*, 273, 111170, 2020. https://doi.org/10.1016/j.jenvman.2020.111170.
56. Asif, M.B., Hou, J., Price, W.E., Chen, V., Hai, F.I., Removal of trace organic contaminants by enzymatic membrane bioreactors: Role of membrane retention and biodegradation. *J. Membr. Sci.*, 611, 118345, 2020. https://doi.org/10.1016/j.memsci.2020.118345.
57. Tiwari, B., Sellamuthu, B., Piché-Choquette, S., Drogui, P., Tyagi, R.D., Vaudreuil, M.A. *et al.*, The bacterial community structure of submerged membrane bioreactor treating synthetic hospital wastewater. *Bioresour. Technol.*, 286, 121362, 2019. https://doi.org/10.1016/j.biortech.2019.121362.
58. Kim, M., Guerra, P., Shah, A., Parsa, M., Alaee, M., Smyth, S.A., Removal of pharmaceuticals and personal care products in a membrane bioreactor wastewater treatment plant. *Water Sci. Technol.*, 69, 2221–9, 2014. https://doi.org/10.2166/wst.2014.145.
59. Wang, X.C., Shen, J.M., Chen, Z.L., Zhao, X., Xu, H., Removal of pharmaceuticals from synthetic wastewater in an aerobic granular sludge membrane bioreactor and determination of the bioreactor microbial diversity. *Appl. Microbiol. Biotechnol.*, 100, 8213–23, 2016. https://doi.org/10.1007/s00253-016-7577-6.
60. Le-Clech, P., Chen, V., Fane, T.A.G., Fouling in membrane bioreactors used in wastewater treatment. *J. Membr. Sci.*, 284, 17–53, 2006. https://doi.org/10.1016/j.memsci.2006.08.019.
61. Tang, K., Ooi, G.T.H., Torresi, E., Kaarsholm, K.M.S., Hambly, A., Sundmark, K. *et al.*, Municipal wastewater treatment targeting pharmaceuticals by a pilot-scale hybrid attached biofilm and activated sludge system (Hybas™). *Chemosphere*, 259, 127397, 2020. https://doi.org/10.1016/j.chemosphere.2020.127397.
62. Xiao, Y., Yaohari, H., De Araujo, C., Sze, C.C., Stuckey, D.C., Removal of selected pharmaceuticals in an anaerobic membrane bioreactor (AnMBR) with/without powdered activated carbon (PAC). *Chem. Eng. J.*, 321, 335–45, 2017. https://doi.org/10.1016/j.cej.2017.03.118.
63. de Wilt, A., van Gijn, K., Verhoek, T., Vergnes, A., Hoek, M., Rijnaarts, H. *et al.*, Enhanced pharmaceutical removal from water in a three-step bio-ozone-bio process. *Water Res.*, 138, 97–105, 2018. https://doi.org/10.1016/j.watres.2018.03.028.

64. Prado, M., Borea, L., Cesaro, A., Liu, H., Naddeo, V., Belgiorno, V. *et al.*, Removal of emerging contaminant and fouling control in membrane bioreactors by combined ozonation and sonolysis. *Int. Biodeterior. Biodegradation*, 119, 577–86, 2017. https://doi.org/10.1016/j.ibiod.2016.10.044.
65. Naghdi, M., Taheran, M., Brar, S.K., Kermanshahi-Pour, B.A., Verma, M., Surampalli, R.Y., Removal of pharmaceutical compounds in water and wastewater using fungal oxidoreductase enzymes. *Environ. Pollut.*, 234, 190–213, 2018. https://doi.org/10.1016/j.envpol.2017.11.060.
66. Lonappan, L., Liu, Y., Rouissi, T., Brar, S.K., Verma, M., Surampalli, R.Y., Adsorptive immobilization of agro-industrially produced crude laccase on various micro-biochars and degradation of diclofenac. *Sci. Total Environ.*, 640–641, 1251–8, 2018. https://doi.org/10.1016/j.scitotenv.2018.06.005.
67. Lonappan, L., Rouissi, T., Laadila, M.A., Brar, S.K., Hernandez Galan, L., Verma, M. *et al.*, Agro-industrial-produced laccase for degradation of diclofenac and identification of transformation products. *ACS Sustain. Chem. Eng.*, 5, 5772–81, 2017. https://doi.org/10.1021/acssuschemeng.7b00390.
68. Eibes, G., Debernardi, G., Feijoo, G., Moreira, M.T., Lema, J.M., Oxidation of pharmaceutically active compounds by a ligninolytic fungal peroxidase. *Biodegradation*, 22, 539–550, 2011. https://doi.org/10.1007/s10532-010-9426-0.
69. Lueangjaroenkit, P., Teerapatsakul, C., Sakka, K., Sakka, M., Kimura, T., Kunitake, E. *et al.*, Two manganese peroxidases and a laccase of *Trametes polyzona* KU-RNW027 with novel properties for dye and pharmaceutical product degradation in redox mediator-free system. *Mycobiology*, 47, 217–29, 2019. https://doi.org/10.1080/12298093.2019.1589900.
70. Asif, M.B., Hai, F.I., Kang, J., van de Merwe, J.P., Leusch, F.D.L., Price, W.E. *et al.*, Biocatalytic degradation of pharmaceuticals, personal care products, industrial chemicals, steroid hormones and pesticides in a membrane distillation-enzymatic bioreactor. *Bioresour. Technol.*, 247, 528–36, 2018. https://doi.org/10.1016/j.biortech.2017.09.129.
71. Ba, S., Haroune, L., Soumano, L., Bellenger, J.P., Jones, J.P., Cabana, H., A hybrid bioreactor based on insolubilized tyrosinase and laccase catalysis and microfiltration membrane remove pharmaceuticals from wastewater. *Chemosphere*, 201, 749–55, 2018. https://doi.org/10.1016/j.chemosphere.2018.03.022.
72. Pylypchuk, I.V., Kessler, V.G., Seisenbaeva, G.A., Simultaneous removal of acetaminophen, diclofenac, and Cd(II) by *Trametes versicolor* laccase immobilized on Fe3O4/SiO2-DTPA Hybrid nanocomposites. *ACS Sustain. Chem. Eng.*, 6, 9979–89, 2018. https://doi.org/10.1021/acssuschemeng.8b01207.
73. Ba, S., Haroune, L., Cruz-Morató, C., Jacquet, C., Touahar, I.E., Bellenger, J.P. *et al.*, Synthesis and characterization of combined cross-linked laccase and tyrosinase aggregates transforming acetaminophen as a model phenolic compound in wastewaters. *Sci. Total Environ.*, 487, 748–55, 2014. https://doi.org/10.1016/j.scitotenv.2013.10.004.

74. Llorca, M., Rodríguez-Mozaz, S., Couillerot, O., Panigoni, K., de Gunzburg, J., Bayer, S. *et al.*, Identification of new transformation products during enzymatic treatment of tetracycline and erythromycin antibiotics at laboratory scale by an on-line turbulent flow liquid-chromatography coupled to a high-resolution mass spectrometer LTQ-Orbitrap. *Chemosphere*, 119, 90–8, 2015. https://doi.org/10.1016/j.chemosphere.2014.05.072.
75. Riva, S., Laccases: Blue enzymes for green chemistry. *Trends Biotechnol.*, 24, 219–26, 2006. https://doi.org/10.1016/j.tibtech.2006.03.006.
76. Tran, N.H., Urase, T., Kusakabe, O., Biodegradation characteristics of pharmaceutical substances by whole fungal culture *Trametes versicolor* and its Laccase. *J. Water Environ. Technol.*, 8, 125–40, 2010. https://doi.org/10.2965/jwet.2010.125.
77. Bilal, M. and Iqbal, H.M.N., Persistence and impact of steroidal estrogens on the environment and their laccase-assisted removal. *Sci. Total Environ.*, 690, 447–59, 2019. https://doi.org/10.1016/j.scitotenv.2019.07.025.
78. Spina, F., Cordero, C., Schilirò, T., Sgorbini, B., Pignata, C., Gilli, G. *et al.*, Removal of micropollutants by fungal laccases in model solution and municipal wastewater: Evaluation of estrogenic activity and ecotoxicity. *J. Clean. Prod.*, 100, 185–94, 2015. https://doi.org/10.1016/j.jclepro.2015.03.047.
79. Margot, J., Bennati-Granier, C., Maillard, J., Blánquez, P., Barry, D.A., Holliger, C., Bacterial versus fungal laccase: Potential for micropollutant degradation. *AMB Express*, 3, 63, 2013. https://doi.org/10.1186/2191-0855-3-63.
80. Alharbi, S.K., Nghiem, L.D., van de Merwe, J.P., Leusch, F.D.L., Asif, M.B., Hai, F.I. *et al.*, Degradation of diclofenac, trimethoprim, carbamazepine, and sulfamethoxazole by laccase from *Trametes versicolor*: Transformation products and toxicity of treated effluent. *Biocatal. Biotransformation*, 37, 399–408, 2019. https://doi.org/10.1080/10242422.2019.1580268.
81. D'Acunzo, F., Galli, C., Gentili, P., Sergi, F., Mechanistic and steric issues in the oxidation of phenolic and non-phenolic compounds by laccase or laccase-mediator systems. The case of bifunctional substrates. *New J. Chem.*, 30, 583–91, 2006. https://doi.org/10.1039/b516719a.
82. Parra Guardado, A.L., Belleville, M.P., Rostro Alanis M de, J., Parra Saldivar, R., Sanchez-Marcano, J., Effect of redox mediators in pharmaceuticals degradation by laccase: A comparative study. *Process Biochem.*, 78, 123–31, 2019. https://doi.org/10.1016/j.procbio.2018.12.032.
83. Ding, H., Wu, Y., Zou, B., Lou, Q., Zhang, W., Zhong, J. *et al.*, Simultaneous removal and degradation characteristics of sulfonamide, tetracycline, and quinolone antibiotics by laccase-mediated oxidation coupled with soil adsorption. *J. Hazard. Mater.*, 307, 350–8, 2016. https://doi.org/10.1016/j.jhazmat.2015.12.062.
84. Becker, D., Giustina, S.V.D., Rodriguez-Mozaz, S., Schoevaart, R., Barcelo., D., de Cazes, M. *et al.*, Removal of antibiotics in wastewater by enzymatic treatment with fungal laccase–degradation of compounds does not

always eliminate toxicity. *Bioresour. Technol.*, 219, 500–9, 2016. https://doi.org/10.1016/j.biortech.2016.08.004.
85. Battistuzzi, G., Bellei, M., Bortolotti, C.A., Sola, M., Redox properties of heme peroxidases. *Arch. Biochem. Biophys.*, 500, 21–36, 2010. https://doi.org/10.1016/j.abb.2010.03.002.
86. Taboada-Puig, R., Eibes, G., Lloret, L., Lú-Chau, T.A., Feijoo, G., Moreira, M.T. *et al.*, Fostering the action of versatile peroxidase as a highly efficient biocatalyst for the removal of endocrine disrupting compounds. *N. Biotechnol.*, 33, 187–95, 2016. https://doi.org/10.1016/j.nbt.2015.05.003.
87. Durán, N., Rosa, M.A., D'Annibale, A., Gianfreda, L., Applications of laccases and tyrosinases (phenoloxidases) immobilized on different supports: A review. *Enzyme Microb. Technol.*, 31, 907–31, 2002. https://doi.org/10.1016/S0141-0229(02)00214-4.
88. Ba, S., Haroune, L., Cruz-Morató, C., Jacquet, C., Touahar, I.E., Bellenger, J.P. *et al.*, Synthesis and characterization of combined cross-linked laccase and tyrosinase aggregates transforming acetaminophen as a model phenolic compound in wastewaters. *Sci. Total Environ.*, 487, 748–55, 2014. https://doi.org/10.1016/j.scitotenv.2013.10.004.
89. Ba, S. and Kumar, V.V., Recent developments in the use of tyrosinase and laccase in environmental applications. *Crit. Rev. Biotechnol.*, 37, 819–32, 2017. https://doi.org/10.1080/07388551.2016.1261081.
90. Ba, S., Haroune, L., Soumano, L., Bellenger, J.P., Jones, J.P., Cabana, H., A hybrid bioreactor based on insolubilized tyrosinase and laccase catalysis and microfiltration membrane remove pharmaceuticals from wastewater. *Chemosphere*, 201, 749–55, 2018. https://doi.org/10.1016/j.chemosphere.2018.03.022.
91. Llorca, M., Rodríguez-Mozaz, S., Couillerot, O., Panigoni, K., de Gunzburg, J., Bayer, S. *et al.*, Identification of new transformation products during enzymatic treatment of tetracycline and erythromycin antibiotics at laboratory scale by an on-line turbulent flow liquid-chromatography coupled to a high-resolution mass spectrometer LTQ-Orbitrap. *Chemosphere*, 119, 90–8, 2015. https://doi.org/10.1016/j.chemosphere.2014.05.072.
92. de Cazes, M., Belleville, M.P., Petit, E., Salomo, M., Bayer, S., Czaja, R. *et al.*, Erythromycin degradation by esterase (EreB) in enzymatic membrane reactors. *Biochem. Eng. J.*, 114, 70–8, 2016. https://doi.org/10.1016/j.bej.2016.06.029.
93. Thompson, M.P., Peñafiel, I., Cosgrove, S.C., Turner, N.J., Biocatalysis using immobilized enzymes in continuous flow for the synthesis of fine chemicals. *Org. Process Res. Dev.*, 23, 9–18, 2019. https://doi.org/10.1021/acs.oprd.8b00305.
94. Basso, A. and Serban, S., Industrial applications of immobilized enzymes—A review. *Mol. Catal.*, 479, 110607, 2019. https://doi.org/10.1016/j.mcat.2019.110607.

95. Mateo, C., Palomo, J.M., Fernandez-Lorente, G., Guisan, J.M., Fernandez-Lafuente, R., Improvement of enzyme activity, stability and selectivity via immobilization techniques. *Enzyme Microb. Technol.*, 40, 1451–63, 2007. https://doi.org/10.1016/j.enzmictec.2007.01.018.
96. Zdarta, J., Meyer, A.S., Jesionowski, T., Pinelo, M., Multi-faceted strategy based on enzyme immobilization with reactant adsorption and membrane technology for biocatalytic removal of pollutants: A critical review. *Biotechnol. Adv.*, 37, 107401, 2019. https://doi.org/10.1016/j.biotechadv.2019.05.007.
97. Zdarta, J., Meyer, A.S., Jesionowski, T., Pinelo, M., Developments in support materials for immobilization of oxidoreductases: A comprehensive review. *Adv. Colloid Interface Sci.*, 258, 1–20, 2018. https://doi.org/10.1016/j.cis.2018.07.004.
98. Xu, R., Si, Y., Li, F., Zhang, B., Enzymatic removal of paracetamol from aqueous phase: horseradish peroxidase immobilized on nanofibrous membranes. *Environ. Sci. Pollut. Res.*, 22, 3838–46, 2015. https://doi.org/10.1007/s11356-014-3658-1.
99. Xu, R., Tang, R., Zhou, Q., Li, F., Zhang, B., Enhancement of catalytic activity of immobilized laccase for diclofenac biodegradation by carbon nanotubes. *Chem. Eng. J.*, 262, 88–95, 2015. https://doi.org/10.1016/j.cej.2014.09.072.
100. Touahar, I.E., Haroune, L., Ba, S., Bellenger, J.P., Cabana, H., Characterization of combined cross-linked enzyme aggregates from laccase, versatile peroxidase and glucose oxidase, and their utilization for the elimination of pharmaceuticals. *Sci. Total Environ.*, 481, 90–9, 2014. https://doi.org/10.1016/j.scitotenv.2014.01.132.
101. de Cazes, M., Belleville, M.P., Mougel, M., Kellner, H., Sanchez-Marcano, J., Characterization of laccase-grafted ceramic membranes for pharmaceuticals degradation. *J. Membr. Sci.*, 476, 384–93, 2015. https://doi.org/10.1016/j.memsci.2014.11.044.
102. Nguyen, L.N., Hai, F.I., Dosseto, A., Richardson, C., Price, W.E., Nghiem, L.D., Continuous adsorption and biotransformation of micropollutants by granular activated carbon-bound laccase in a packed-bed enzyme reactor. *Bioresour. Technol.*, 210, 108–16, 2016. https://doi.org/10.1016/j.biortech.2016.01.014.
103. Tadkaew, N., Sivakumar, M., Khan, S.J., McDonald, J.A., Nghiem, L.D., Effect of mixed liquor pH on the removal of trace organic contaminants in a membrane bioreactor. *Bioresour. Technol.*, 101, 1494–500, 2010. https://doi.org/10.1016/j.biortech.2009.09.082.
104. Wijekoon, K.C., Hai, F.I., Kang, J., Price, W.E., Guo, W., Ngo, H.H. *et al.*, The fate of pharmaceuticals, steroid hormones, phytoestrogens, UV-filters and pesticides during MBR treatment. *Bioresour. Technol.*, 144, 247–54, 2013. https://doi.org/10.1016/j.biortech.2013.06.097.
105. Yang, S., Hai, F.I., Nghiem, L.D., Price, W.E., Roddick, F., Moreira, M.T. *et al.*, Understanding the factors controlling the removal of trace organic contaminants by white-rot fungi and their lignin modifying enzymes: A critical

review. *Bioresour. Technol.*, 141, 97–108, 2013. https://doi.org/10.1016/j.biortech.2013.01.173.

106. Rozi, N., Zaid, M.H.M., Tahrim, N.A., Ikeda, M., Hanifah, S.A., Polymer-based biosensor for estrogenic endocrine-disrupting chemicals in water. *Int. J. Environ. Anal. Chem.*, 2020. https://doi.org/10.1080/03067319.2020.1746776.

Index

Printed and bound by CPI Group (UK) Ltd, Croydon, CR0 4YY

17/09/2026

14964275-0006